Modern Algebra
An Introduction

Modern Algebra
An Introduction

Fourth Edition

John R. Durbin

John Wiley & Sons, Inc.
New York / Chichester / Weinheim / Brisbane / Singapore / Toronto

Acquisition Editor	*Barbara Holland*
Associate Editor	*Sharon Prendergast*
Production Editor	*Ken Santor*
Cover Design	*David Levy*
Cover Illustration	*Norm Christiansen*
Illustration Coordinator	*Gene Aiello*

This book was set in Times Roman by Eigentype Compositors, and printed and bound by Hamilton Printing Company. The cover was printed by Phoenix Color Corporation.

This book is printed on acid-free paper.

The paper in this book was manufactured by a mill whose forest management programs include sustained yield harvesting of its timberlands. Sustained yield harvesting principles ensure that the numbers of trees cut each year does not exceed the amount of new growth.

Library of Congress Cataloging in Publication Data:
Durbin, John R.
    Modern algebra : an introduction / John R. Durbin – 4th ed.
        p.   cm.
    Includes index.
    ISBN 0-471-32147-8 (alk. paper)
    1. Algebra, Abstract.   I. Title.
    QA162.D87   2000
    512'.12–dc21   99-13830
                                                             CIP

Printed in the United States of America

10 9 8 7 6 5 4 3 2 1

To Jane

PREFACE

This book is an introduction to modern (abstract) algebra for undergraduates. The first six chapters present the core of the subject, the basic ideas of groups, rings, and fields. The remainder is designed to be as flexible as possible. A diagram of chapter dependencies, preceding the Contents, suggests a number of options for introducing variety, applications, and depth.

A first course in modern algebra often has the additional goal of introducing the axiomatic method and the construction of proofs. I have tried to keep that constantly in mind. For example, the first chapter treats ideas that are important but neither abstract nor complicated, and provides practice in handling mathematical statements—their meaning, quantification, negation, and proof. I believe this chapter should be covered carefully, except for students who are already comfortable with the ideas and with proofs.

With the fourth edition the attempt to help students has been carried further in several ways:

- The Advice for Students section, following this preface, gives suggestions on how to approach the subject and this book, both of which may be different from what they have seen before.
- The appendix on proofs has been expanded to include discussions of deductive reasoning and methods of proofs.
- Advice on dealing with troublesome ideas (such as how to start a proof and how to handle quantifiers) has been added at several key places in the text and in the solutions appendix.

Major changes for the fourth edition, in addition to those making the book more student-friendly, include a rearrangement to move some traditional topics closer to the front, the addition of a section to give more substance to the introduction to Galois theory, and updated lists of references and discussions of topics such as Fermat's Last Theorem and the finite simple groups. As in earlier editions, the book has an optional Introduction giving an informal orientation to the subject.

Each problem set is arranged with the more straightforward choices at the beginning. These problems are grouped in pairs, with solutions to those with odd numbers in the solutions appendix; this will help students know if they are on the right track. Each problem is meant to require at least some thought; space is too valuable to do otherwise. The solutions manual indicates problems that may offer the most challenge. Some of the problems, especially later in the book, contain fairly substantial extensions of the material in the text.

There is a consistent attempt to arrange the material so that each section can be covered in one lecture of reasonable length, although, by the nature of the subject, students may occasionally need time to review and consolidate what they have learned.

The fourth edition preserves the style of the earlier editions, which reflects the author's philosophy that the best textbooks concentrate on presenting core ideas clearly, concisely, and with few distractions.

John R. Durbin

ADVICE FOR STUDENTS

What is modern algebra? Why is it important? What does it take to learn it? How can this book help? These are entirely reasonable questions. This section and the Introduction, which precedes Chapter I, will help provide answers.

Modern algebra is sometimes called algebraic structures or abstract algebra, or merely, in the context of advanced mathematics, algebra. Although the name may suggest just a new way to present the algebra that precedes calculus, it is in fact much broader and deeper than that.

The Introduction discusses some of the history and applications of modern algebra, to give a glimpse of what the subject is about. Please spend at least some time with the Introduction, even though it is optional. Other references to history occur throughout the book, and each of the last six chapters covers at least one application.

Applications are important. But the case for modern algebra rests on more than the applications that can be presented in this book. The ideas and ways of thought of the subject permeate nearly every part of modern mathematics. Moreover, no subject is better suited to cultivate the ability to handle abstract ideas—to understand and deal with the essential elements of a problem or a subject. This includes the ability to read mathematics, to ask the right questions, to solve problems, to use deductive reasoning, and to write mathematics so that it is correct, to the point, and clear. All these things make learning modern algebra worthwhile—for future teachers and graduate students, for computer science students and many others who will use mathematics, and for some who will simply appreciate its intellectual appeal. This book has been written with all such readers in mind.

ON READING MATHEMATICS

First, remember that mathematics must be read slowly and thoughtfully. That is true of this book. The book is intended to be as clear as possible. But that does not mean as easy as possible. Part of the reward of learning the subject comes from the exercise of questioning and concentration. It would be a disservice if the book could be read without thought. Sometimes, if you don't understand something, it's better to go on; but in general you should work to understand each step before going to the next. That may require referring back to something covered earlier; if so, the review is probably useful. It may require writing out details that have been omitted; if so, the thought will help reinforce what you have learned. Learning this subject is no different from learning music, a sport, or anything else. It takes effort and patience and perseverance to do it well.

ON ASKING THE RIGHT QUESTIONS

What does it mean to ask the right questions? As a start, continually ask yourself if you understand what you have just read. If it's a definition, make sure you know all the terms that you've previously seen. Try to think of examples that satisfy the definition, as well as some that don't. If you can't think of any, then from trying you'll at least appreciate examples when they are given.

When you read a theorem, ask yourself what the hypothesis and conclusion are. (These terms are reviewed in Appendix B.) Be sure you know what all the words mean, looking back if necessary. Then, before looking at the proof that's given, think about how you would try to do it yourself. Some proofs can be hard. But you'll have an advantage if you at least think about a general strategy before reading the proof in the book. (Again, see Appendix B.)

When you have finished an example or a proof or a section, ask yourself what you have learned. There are no lists to help you review in this book. Putting together such lists is part of the learning process.

Finally, there is one question that is always worthwhile: Could you explain it to someone else? This applies to everything you read, to every solution you find, and to every proof that you write. If you can't explain it to someone with a reasonable background, then you may not fully understand it. Most teachers will admit that they did not fully understand a subject until they taught it; and even then they may learn something new every time around. You can get some of the advantage of this simply by imagining how you would explain the subject yourself.

ON SOLVING PROBLEMS

The best way to learn mathematics is to solve problems. Or at least to make serious attempts to solve them. Even if you don't succeed, you will be forced to think about ideas that might help with a solution, and about how those ideas are related. Since mathematics is about ideas and how they are related, that is extremely valuable.

In each section of the book the problems preceding the double line usually occur in pairs, with solutions for the odd-numbered problems in Appendix E. The problems preceding the double line tend to be more straightforward than the others, and provide way to become familiar with the basic ideas in the section.

A good source for general advice on problem solving is the book *How to Solve It* by G. Pólya, which is listed at the end of Appendix B.

ON PROVING THEOREMS

In this book solving problems includes proving theorems. Suggestions about proofs appear at a number of places in the book. In particular, please read Appendix B for the help it can provide and since the text makes frequent use of its ideas. Included in Appendix B are some of the "nuts and bolts" of proofs and deductive reasoning. This includes important terms, basic facts about logic, and advice on strategy.

Do not become discouraged if constructing proofs seems difficult. It is difficult for nearly everyone. Don't be deceived by the form in which mathematicians generally display their finished products. Behind a polished proof of any significance there has often been a great deal of struggle and frustration and ruined paper. There is no other way to discover good mathematics, and for most of us there is no other way to learn it.

ON WRITING MATHEMATICS

After you have constructed a proof or solved a problem, it is good to remember that no one can be expected to know what is in your mind or what you have discarded. They have only what you write. For communicating proofs it would be hard to find sounder advice than that given by Quintilian, 1900 years ago:

One should not aim at being possible to understand, but at being impossible to misunderstand.

The Greek Alphabet		
A	α	alpha
B	β	beta
Γ	γ	gamma
Δ	δ	delta
E	ϵ	epsilon
Z	ζ	zeta
H	η	eta
Θ	θ	theta
I	ι	iota
K	κ	kappa
Λ	λ	lambda
M	μ	mu
N	ν	nu
Ξ	ξ	xi
O	o	omicron
Π	π	pi
P	ρ	rho
Σ	σ	sigma
T	τ	tau
Y	υ	upsilon
Φ	ϕ	phi
X	χ	chi
Ψ	ψ	psi
Ω	ω	omega

CHAPTER DEPENDENCIES

CONTENTS

Arabic (13th century).

San Francesco in
Assisi (13th century).

San Francesco in
Assisi (13th century).

Figure 1

INTRODUCTION

Modern algebra—like any other branch of mathematics—can be mastered only by working up carefully from the most basic ideas and examples. But that takes time, and some of the goals will not be clear until you reach them. This section is meant to help sustain you along the way. You may read this all at first, or begin with Chapter 1 and then return here at your leisure. The purpose is simply to convey a feeling for how modern algebra developed and for the kinds of problems it can help solve.

Please note that this section is not a survey of all of modern algebra. There is no discussion of the applications of algebra in modern theoretical physics, for instance, or of some of the deeper applications of algebra in mathematics itself; in fact, such applications are not covered even in the text. The examples here have been chosen because they can be understood without special background. Even at that, by its very purpose this section must occasionally be vague; we can worry about details and proofs when we get to the text itself.

SYMMETRY

Symmetrical designs like those in Figure 1 have been used for decoration throughout history. Each of these designs is built up from a basic irreducible component; such components for the designs in Figure 1 are shown in Figure 2. In each case the plane can be filled without overlap if the basic component is repeated by appropriate combinations of rotation (twisting), translation (sliding), and reflection (such as interchanging left and right).

Although the artistic possibilities for the components are unlimited, there is another sense in which the possibilities are *not* unlimited. Notice that the lower two examples in Figure 1 will look the same if the page is turned upside down, but the top example will not. Because of this we could say that the top example has a different symmetry type from the

Figure 2

1

other two examples. Continuing, notice that the middle example lacks the strict left-right symmetry of the top and bottom examples. (For instance, in the irreducible component for the middle example the ring passes under the white band on the left, but over the white band on the right.) We can conclude, then, that the three examples somehow represent three different symmetry types—different combinations of rotations, translations, and reflections are needed to build up the three different designs from their basic components. If we were to look at more designs, we would find examples of still other symmetry types, and we would also find many examples that could be distinguished from one another, but not on the basis of symmetry type alone. At some stage we would feel the need for a more precise definition of "symmetry type." There is such a definition, and it turns out that in terms of that definition there are exactly 17 different symmetry types of plane-filling designs. Figure 55.9 shows one example of each type. Although each of the 17 types occurs in decorations from ancient civilizations, it was only in the nineteenth century that these possibilities were fully understood. The key to making "symmetry type" precise, and also to determining the number of different symmetry types, is the idea of a *group*.

GROUPS

The idea of a group is one of the focal points of modern algebra. Like all significant mathematical ideas, this idea is general and abstract, and it is interesting and important because of the cumulative interest and importance of its many special cases. Roughly, a group is a set of elements that can be combined through some operation such as addition or multiplication, subject to some definite rules like those that govern ordinary addition of numbers. The elements may be something other than numbers, however, and the operation something other than the usual operations of arithmetic. For instance, the elements of the groups used to study symmetry are things like rotations, translations, and reflections. The precise definition of *group* is given in Section 5.

CRYSTALLOGRAPHY

Think again about symmetry type, but now move from two dimensions up to three. One easy example here is given by moving a cube repeatedly along the directions perpendicular to its faces. Other examples can be very complicated, and the general problem of finding all symmetry types was not easy. Like the problem in two dimensions, however, it was also settled in the nineteenth century: in contrast to the 17 different symmetry types of plane-filling designs, there are 230 different types of symmetry for figures that fill three-dimensional space. Again, groups provide the key. And with this use of groups we have arrived at an application to science, for ideas used to solve this three-dimensional problem are just what are needed to classify crystals according to symmetry type. The symmetry type of a crystal is a measure of the regular pattern in which the component atoms or molecules arrange themselves. Although this pattern is an internal property of the crystal, which may require x-ray techniques for analysis, symmetry is often evident from the external or surface features of the crystal. Figure 3 illustrates this with a picture of galena crystals: the repeated occurrence of the shape on the left, in the picture on the right, is a consequence of the internal symmetry of galena. (Galena is the chief ore mineral of

Figure 3

lead, and has properties that make it useful in electronics.) Classification by symmetry type is at the heart of crystallography, and is used in parts of physics, chemistry, and mineralogy.

COMBINATORICS

Although one of the preceding connections of groups is with design, and the other is with science, they both have to do with symmetry. Here is another example. The numbers 1 through 6 can be placed on the faces of a die (cube) in 720 different ways. But only 30 of these ways are distinguishable—if the numbers are put on more than 30 dice, then at least two of the dice can be made to look the same through rotation. In Figure 4, for example, the middle arrangement differs from the left-hand arrangement only by a rotation, but no rotation of the middle die would make it look like the right-hand die. The problem of counting the number of distinguishable arrangements belongs to the domain of combinatorics, and if you are good at systematic counting, you can solve it without groups. But the problem can also be solved with an appropriate group, and this provides a way of viewing the problem that is almost indispensable for more complicated problems. In each case a group is used to account systematically for the symmetry in the problem (such as the symmetry of a cube).

In each case, 1 is on the front
and 2 is on the back.

Figure 4

ALGEBRAIC EQUATIONS

The examples thus far have had to do with geometrical symmetry, but groups were first studied for a different reason. Much of the early history of modern algebra was tied closely to questions about equations. In beginning algebra we learn that each linear (first-degree) equation $ax + b = 0$ $(a \neq 0)$ has a unique solution $x = -b/a$, and each quadratic (second-degree) equation $ax^2 + bx + c = 0$ $(a \neq 0)$ has solutions

$$x = \frac{-b \pm \sqrt{b^2 - 4ac}}{2a}. \tag{1}$$

Methods for solving these equations were known by the sixteenth century. For example, particular types of quadratics had been handled by the ancient Egyptians, Babylonians, and Greeks, and by the Hindus and Arabs in the Middle Ages. But what about equations of degree higher than two? For instance, is there a general procedure or formula, like (1), for writing the solutions of a cubic (third-degree) equation $ax^3 + bx^2 + cx + d = 0$ in terms of the coefficients a, b, c, and d? Italian algebraists discovered in the sixteenth century that the answer is Yes, not only for cubics but also for quartic (fourth-degree) equations.

These solutions for cubics and quartics are fairly complicated, and their detailed form is not important here. What is important is that all of this leads to the following more general question: Can the solutions of each algebraic equation

$$a_n x^n + a_{n-1} x^{n-1} + \cdots + a_0 = 0 \tag{2}$$

be derived from the coefficients $a_n, a_{n-1}, \ldots, a_0$ by addition, subtraction, multiplication, division, and extraction of roots, each applied only finitely many times—or briefly, as we now say, is (2) *solvable by radicals*? By early in the nineteenth century mathematicians knew that the answer is *No*: for each $n > 4$ there are equations of degree n that are not solvable by radicals. However, *some* equations of each degree $n > 4$ are solvable by radicals, so there is the new problem of how to determine whether a given equation is or is not solvable in this way. With this problem we are brought back to groups: with each equation (2) we can associate a group, and the French mathematician Évariste Galois (1811–1832) discovered that properties of this group reveal whether the equation is solvable by radicals. The group associated with an equation measures an abstruse kind of symmetry involving the solutions of the equation. Thus the abstract idea of a group can be used to analyze both geometrical symmetry and solvability by radicals. Groups arise in other contexts as well, but we cannot pursue them all here.

RINGS AND FIELDS

The theory of groups was not the only part of algebra to be stimulated by questions about equations. One question that arises when first studying quadratic equations has to do with square roots of negative numbers: What can we say about the solutions in (1) if $b^2 - 4ac < 0$? Nowadays this creates little problem, since we know about complex numbers, and we generally feel just as comfortable with them as we do with integers and real numbers. But this is true only because earlier mathematicians worked out a clear understanding of the properties of all the familiar number systems. Two more ideas from modern algebra—*ring* and *field*—were essential for this. Roughly, a ring is a set with operations like addition, subtraction, and multiplication; and a field is a ring in which division is also possible. Precise definitions of ring and field are given at the appropriate places in the text, where we also see how these ideas can be used to characterize the familiar number systems. There are no surprises when we relate rings and fields to number systems, but the following application of fields to geometry should be more unexpected.

GEOMETRIC CONSTRUCTIONS

Among the geometric construction problems left unsolved by the ancient Greeks, three became especially famous. Each involved the construction of one geometrical segment from another, using only unmarked straightedge and compass (Figure 5):

I. Construct the edge of a cube having twice the volume of a given cube.

II. Show that every angle can be trisected.

III. Construct the side of a square having the same area as a circle of given radius.

These problems remained unsolved for more than 2000 years. Then, in the nineteenth century, it was proved that the constructions are impossible. What makes this interesting for us is that although the constructions were to be geometric, the proofs of their impossibility involve algebra. And the key algebraic concepts needed are the same as those used to analyze solvability by radicals; these include facts about fields that go far beyond what is necessary to characterize the familiar number systems.

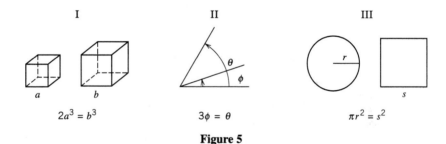

Figure 5

NUMBER THEORY

The motivation for studying some of the deeper properties of rings came from a source totally different from the applications already mentioned. *Pythagorean triples* are triples (x, y, z) of positive integers such that $x^2 + y^2 = z^2$. That is, they are the triples of integers that can occur as lengths of sides of right triangles (relative to an appropriate unit of length). Examples are $(3, 4, 5)$, $(5, 12, 13)$, $(8, 15, 17)$, and $(199, 19800, 19801)$. The Greek mathematician Diophantus derived a method for determining all such triples around A.D. 250. (The Babylonians had determined many Pythagorean triples by much the same method around 1500 B.C.) In reading about this problem in Diophantus's book *Arithmetica*, the French mathematician Pierre de Fermat (1601–1665) was led to introduce one of the most famous problems in mathematics: Are there nonzero integers x, y, z such that

$$x^n + y^n = z^n$$

for any integer $n > 2$? Actually, Fermat claimed that there are no such integers, and this claim eventually came to be known as *Fermat's Last Theorem*. But Fermat did not give a proof of his claim, and the problem of constructing a proof defied some of the world's best mathematicians for over 350 years.

Fermat's Last Theorem was finally proved in 1994 by Andrew Wiles of Princeton University. The proof by Wiles, who was born in Cambridge, England, and educated at Cambridge University, is extremely complicated and draws on ideas developed by other mathematicians over many years. Nothing in the statement of the theorem suggests the depth of the ideas required for its proof. For a hint at these ideas, see Section 41. For an interesting history of Fermat's Theorem and the work of Wiles, as well as an insight into mathematics as a creative process, see the book *Fermat's Enigma*, by Simon Singh, which is listed at the end of this Introduction.

In the book by Singh, Wiles is quoted as saying that "the definition of a good mathematical problem is the mathematics it generates rather then the problem itself." This brings out an important lesson from the history of mathematics, namely, that attempts to solve problems, both successful and unsuccessful, have been responsible for the development of the subject. In particular, attempts to solve Fermat's Last Theorem have had a profound effect on number theory and algebra.

ORDER

The three basic kinds of systems we have discussed—groups, rings, and fields—are examples of what are known as algebraic structures. Each such structure involves one or more operations like addition or multiplication of numbers. Some algebraic structures also involve a notion of order, like $\subseteq$ for sets and $\leq$ for numbers. For example, order must be taken into account in studying the familiar number systems. One formal idea that grew from questions about order is that of a *lattice*. Lattices can be represented by diagrams like those in Figure 6: the example on the left shows the subsets of $\{x, y, z\}$, with a sequence of segments connecting one set to another above it if the first set is contained in the second; the example on the right shows the positive factors of 30, with a sequence of segments connecting one integer to another above it if the first integer is a factor of the second. The similarity of these two diagrams suggests one of the purposes of lattice theory, just as the similarity of certain symmetric figures suggests one of the purposes of group theory. Lattice theory is concerned with analyzing the notion of order (subject to some definite rules), and with describing in abstract terms just what is behind the similarity of diagrams like those in Figure 6. Of course, there is more to this study of order than diagrams. Lattices were first

 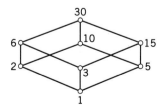

Figure 6

studied as natural generalizations of *Boolean algebras*, which were themselves introduced in the mid-nineteenth century by the British mathematician George Boole (1815–1864) for the purpose of giving an algebraic analysis of formal logic. The first significant use of lattices outside of this connection with logic was in ring theory and algebraic number theory; this interdependence of different branches of algebra is certainly not uncommon in modern mathematics—in fact, it is one of its characteristic features.

COMPUTER-RELATED ALGEBRA

A number of applications of modern algebra have grown with the advent of electronic computers and communication systems. These applications make use of many of the general ideas first introduced to handle much older problems. For example, one such application involves the use of Boolean algebras to study the design of computers and switching circuits. Another application is to algebraic coding (Chapter XV), which uses, among other things, finite fields—these are systems that have only finitely many elements but are otherwise much like the system of real numbers. Applications that use tools from modern algebra and combinatorics belong to the general area of discrete applied mathematics; this can be contrasted with classical applied mathematics, which uses tools from calculus and its extensions.

GENERAL REMARKS

Each topic discussed in this section will be touched on in the book, but they cannot all be treated thoroughly. It would take more than one volume to do that, and in any event there is even more to algebra than the topics introduced in this section might suggest. One of the methods used by the American Mathematical Society to classify current research divides mathematics into eight broad areas: algebra and the theory of numbers, analysis, applied mathematics, geometry, logic and foundations, statistics and probability, topology, and miscellaneous. Although the major branches represented in this list are in many ways interdependent, it is nonetheless true that each branch tends to have its own special outlook and its own special methods and techniques. The goal of this book is to go as far as possible in getting across the outlook and methods and techniques of algebra or, more precisely, that part of algebra devoted to the study of algebraic structures.

Most of the chapters end with notes that list other books, including some where more historical background can be found. Here are some general references that are concerned with history; the notes at the end of Chapter XI give a short list of more advanced general references on modern algebra.

Bell, E. T., *Development of Mathematics*, 2nd ed., Dover reprint, 1992.
____, *Men of Mathematics*, Touchstone Books, 1986. These two books by E. T. Bell are especially lively, though slightly romanticized.
Bourbaki, N., *Elements of the History of Mathematics*, Springer-Verlag, Berlin, New York, 1994.
Boyer, C. B., *History of Mathematics*, Wiley, New York, 1991. An excellent comprehensive survey.
Kline, M., *Mathematical Thought from Ancient to Modern Times*, Oxford University Press, London, 1990. Another excellent comprehensive history; more complete on modern topics than Boyer's book.

Nový, L., *Origins of Modern Algebra*, Noordhoff, Leyden, The Netherlands, 1973. A detailed account of algebra in the important period from 1770 to 1870.

Singh, S., *Fermat's Enigma*, Walker, New York, 1997.

Van der Waerden, B. L., *A History of Algebra: From al-Khowarizmi to Emmy Noether*, Springer-Verlag, New York, 1985.

The following web site is strongly recommended for references on all aspects of the history of mathematics: www-groups.dcs.st-and.ac.uk/~history/

CHAPTER I

MAPPINGS AND OPERATIONS

The most fundamental concept in modern algebra is that of an operation on a set. Addition and the other operations in the familiar number systems are examples, but we shall see that the general concept of operation is much broader than that. Before looking at operations, however, we devote several sections to some basic terminology and facts about mappings, which will be just as important for us as operations. The words *function* and *mapping* are synonymous; therefore, at least some of this material on mappings will be familiar from calculus and elsewhere. Our context for mappings will be more general than that in calculus, however, and one of the reasons we use *mapping* rather than *function* is to emphasize this generality. Notice that elementary facts about sets are collected in Appendix A; they will be used without explicit reference, and probably should be reviewed at the start. You are also urged to read Appendix B, which reviews some elements of logic and offers suggestions regarding proofs.

SECTION 1 MAPPINGS

Mappings are important throughout mathematics. In calculus, for instance, we study mappings (functions) that assign real numbers to real numbers. The mapping given by $f(x) = x^2$, for example, assigns to each real number x the real number x^2. The mapping given by $f(x) = \sin x$ assigns to each real number x the real number $\sin x$. The set $\mathbb{R}$ of real numbers plays two roles in these examples: first, $x \in \mathbb{R}$, and second, $f(x) \in \mathbb{R}$. In general these roles can be played by sets other than $\mathbb{R}$. Thus in the definition of mapping, which follows, S and T can denote any sets whatsoever.

Definition. A *mapping* from a set S to a set T is a relationship (rule, correspondence) that assigns to each element of S a uniquely determined element of T. The set S is called the *domain* of the mapping, and the set T is called the *codomain*.

An alternative definition of *mapping*, preferred by some, is given at the end of Appendix A.

Mappings will generally be denoted by Greek letters, and, to indicate that α is a mapping from S to T, we shall write $\alpha \colon S \to T$ or $S \xrightarrow{\alpha} T$. If x is an element of S, then $\alpha(x)$ will denote the unique element of T that is assigned to x; the element $\alpha(x)$ is called the *image* of x under the mapping α. Sometimes there will be a formula for $\alpha(x)$, as in the examples $f(x) = x^2$ and $f(x) = \sin x$. But that certainly need not be the case.

Example 1.1.   Let $S = \{x, y, z\}$ and $T = \{1, 2, 3\}$. Then α defined by $\alpha(x) = 2$, $\alpha(y) = 1$, and $\alpha(z) = 3$ is a mapping from S to T.

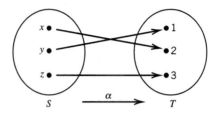

Another mapping, $\beta : S \to T$, is given by $\beta(x) = 1$, $\beta(y) = 3$, and $\beta(z) = 1$.

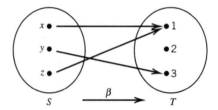

The following diagram does not represent a mapping from S to T for two reasons: first, it assigns two different elements to x, and second, it assigns no element to y.

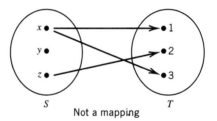

Not a mapping   ∎

Suppose that f and g are the mappings, each with the set of real numbers as domain, defined by $f(x) = (x + 1)^2$ and $g(x) = x^2 + 2x + 1$. Because $(x + 1)^2 = x^2 + 2x + 1$ for every real number x, it is reasonable to think of the mappings f and g as being *equal*. In general, any two mappings α and β are said to be *equal* if their domains are equal, their codomains are equal, and $\alpha(x) = \beta(x)$ for every x in their common domain.

Example 1.2.   In Example 1.1, $\alpha \neq \beta$. For example, $\alpha(y) = 1$ but $\beta(y) = 3$, so $\alpha(y) \neq \beta(y)$. Note that the definition of *equal* for mappings requires that $\alpha(x) = \beta(x)$ for *all* x in the common domain. If there is an x in the common domain such that $\alpha(x) \neq \beta(x)$, then $\alpha \neq \beta$. As explained in Appendix B, "all" is a *universal quantifier*, and the negation of a statement with a universal quantifier is a statement with an *existential quantifier*: "all" is universal, while "there is" and "there exists" are existential.

Thus Example 1.1 gives two different mappings (α and β) from S to T. There are, in fact, 27 different mappings from S to T, for S and T as in Example 1.1. The reason is that we have three independent choices (1, 2 or 3) for where to map each of x, y, and z. Thus the total number of choices is $3 \cdot 3 \cdot 3 = 3^3 = 27$ (3 choices for where to map x, and the same for each of y and z).   ∎

Example 1.3.   If S is any set, we shall use ι (iota) to denote the *identity mapping* from S to S, which is defined by $\iota(x) = x$ for each $x \in S$. If it is necessary to indicate which set S is being considered, ι_S can be written instead of ι. ∎

It is sometimes convenient to write $x \overset{\alpha}{\to} y$ or $x \mapsto y$ to indicate that y is the image of x under a mapping.

Example 1.4.   The rule $(s, t) \mapsto s + t$ for each ordered pair of real numbers s and t defines a mapping from the set of ordered pairs of real numbers to the set of real numbers. [That (s, t) is *ordered* means that it is to be distinguished from (t, s). The necessity for this distinction can be seen if addition is replaced by subtraction: $t - s \neq s - t$ unless $s = t$.] Mappings of this kind, assigning single elements of a set to pairs of elements from the same set, will be discussed at length in Section 3. ∎

If $\alpha : S \to T$ and A is a subset of S, then $\alpha(A)$ will denote the set of elements of T that are images of elements of A under the mapping α. In set-builder notation (described in Appendix A),

$$\alpha(A) = \{\alpha(x) : x \in A\}.$$

The set $\alpha(A)$ is called the *image* of A under the mapping α.

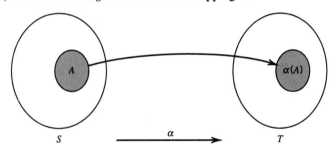

Example 1.5.   With α and β as in Example 1.1,

$$a(\{x, z\}) = \{2, 3\} \text{ and } \beta(\{x, z\}) = \{1\}.$$ ∎

If $\alpha : S \to T$ then $\alpha(S)$ will be called the *image* of α. If $\alpha : S \to T$ and $\alpha(S) = T$, then α is said to be *onto*. Thus α is onto if for each $y \in T$ there is at least one $x \in S$ such that $\alpha(x) = y$. (This use of the preposition *onto* as an adjective may be regrettable, but it is solidly established.)

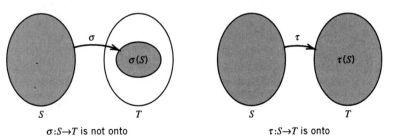

Example 1.6.   In terms of diagrams like those in Example 1.1, a mapping is onto provided each element of the codomain has at least one arrow pointing to it. Thus the mapping α in Example 1.1 is onto. However, the mapping β in Example 1.1 is not onto; its image is $\{1, 3\}$, which is a proper subset of $\{1, 2, 3\}$, the codomain of β.   ∎

Example 1.7.   With $f(x) = x^2$ and $g(x) = \sin x$ thought of as mappings from $\mathbb{R}$ to $\mathbb{R}$, neither is onto. The image of f is the set of nonnegative real numbers. The image of g is the set of real numbers between -1 and 1, inclusive.   ∎

A mapping $\alpha : S \to T$ is said to be *one-to-one* if

$$x_1 \neq x_2 \quad \text{implies} \quad \alpha(x_1) \neq \alpha(x_2) \quad (x_1, x_2 \in S),$$

that is, if unequal elements in the domain have unequal images in the codomain.

Example 1.8.   In terms of diagrams like those in Example 1.1, a mapping is one-to-one provided no two arrows point to the same element. The mapping α in Example 1.1 is one-to-one. However, the mapping β in Example 1.1 is not one-to-one, because $x \neq z$ but $\beta(x) = \beta(z)$.   ∎

Example 1.9.   The mapping $f(x) = x^2$, with domain $\mathbb{R}$, is not one-to-one, because $f(x) = x^2 = f(-x)$, although $x \neq -x$ when $x \neq 0$. The mapping $f(x) = \sin x$, with domain $\mathbb{R}$, is not one-to-one, because $f(x) = \sin x = \sin(x + 2n\pi) = f(x + 2n\pi)$ for each $x \in \mathbb{R}$ and each integer n.   ∎

The contrapositive of

$$x_1 \neq x_2 \quad \text{implies} \quad \alpha(x_1) \neq \alpha(x_2)$$

is

$$\alpha(x_1) = \alpha(x_2) \quad \text{implies} \quad x_1 = x_2.$$

(See Appendix B for a discussion of contrapositive statements.) Since a statement and its contrapositive are logically equivalent, we see that

$$\alpha : S \to T \qquad \text{is one-to-one}$$

iff[†]

$$\alpha(x_1) = \alpha(x_2) \quad \text{implies} \quad x_1 = x_2 \quad (x_1, x_2 \in S).$$

It is sometimes easier to work with this latter condition than with that given in the definition. With $\alpha : \mathbb{R} \to \mathbb{R}$ defined by $\alpha(x) = x - 1$, for instance, we have that $\alpha(x_1) = \alpha(x_2)$ implies $x_1 - 1 = x_2 - 1$, which implies $x_1 = x_2$; therefore, α is one-to-one.

Notice that the identity mapping on any set is both onto and one-to-one. Notice also that a mapping of a finite set to itself is onto iff it is one-to-one. In contrast, there are mappings of any infinite set to itself that are one-to-one but not onto, and also mappings that are onto but not one-to- one. Here are some examples.

[†]We follow the practice, now widely accepted by mathematicians, of using "iff" to denote "if and only if."

Example 1.10. Define mappings α and β from the set of natural numbers, $\{1, 2, 3, \ldots\}$, to itself, by

$$\alpha(n) = 2n$$

and

$$\beta(n) = \begin{cases} (n+1)/2 & \text{if } n \text{ is odd} \\ n/2 & \text{if } n \text{ is even.} \end{cases}$$

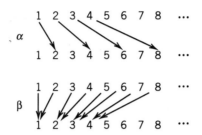

Then α is one-to-one but not onto; and β is onto but not one-to-one. The existence of such mappings is precisely what distinguishes infinite sets from finite sets. In fact, a set S can be defined as *infinite* if there exists a mapping from S to S that is one-to-one but not onto. ∎

We close this section with one remark on notation and another on terminology. The symbols $\mathbb{N}$, $\mathbb{Z}$, $\mathbb{Q}$, $\mathbb{R}$, and $\mathbb{C}$ will be reserved to denote the following sets:

$\mathbb{N}$	the set of all natural numbers, $\{1, 2, 3, \ldots\}$
$\mathbb{Z}$	the set of all integers, $\{\ldots, -2, -1, 0, 1, 2, \ldots\}$
$\mathbb{Q}$	the set of all rational numbers, that is, real numbers that can be expressed in the form a/b, with $a, b \in \mathbb{Z}$ and $b \neq 0$
$\mathbb{R}$	the set of all real numbers
$\mathbb{C}$	the set of all complex numbers

Familiarity with basic properties of $\mathbb{N}$, $\mathbb{Z}$, $\mathbb{Q}$ and $\mathbb{R}$ will be assumed throughout. All of these sets will be studied in Chapter VII.

Although we shall not use the following terminology, it should be mentioned because you may see it elsewhere. Sometimes, a one-to-one mapping is called an *injection*, an onto mapping is called a *surjection*, and a mapping that is both one-to-one and onto is called a *bijection* or a *one-to-one correspondence*. Also, what we are calling the codomain of a mapping is sometimes called the *range* of the mapping. Regrettably, *range* is also used for what we are calling the image of a mapping; this ambiguity over range is one reason for avoiding its use.

PROBLEMS _____

1.1. Let $S = \{w, x, y, z\}$ and $T = \{1, 2, 3, 4\}$, and define $\alpha : S \rightarrow T$ and $\beta : S \rightarrow T$ by $\alpha(w) = 2, \alpha(x) = 4, \alpha(y) = 1, \alpha(z) = 2$ and $\beta(w) = 4, \beta(x) = 2, \beta(y) = 3, \beta(z) = 1$.
(a) Is α one-to-one? Is β one-to-one? Is α onto? Is β onto?

(b) Let $A = \{w, y\}$ and $B = \{x, y, z\}$. Determine each of the following subsets of $T : \alpha\,(A)\,, \beta(B), \alpha(A \cap B), \beta(A \cup B)$.

1.2. Let α, β, and γ be mappings from $\mathbb{Z}$ to $\mathbb{Z}$ defined by $\alpha(n) = 2n$, $\beta(n) = n + 1$, and $\gamma(n) = n^3$ for each $n \in \mathbb{Z}$.
(a) Which of the three mappings are onto?
(b) Which of the three mappings are one-to-one?
(c) Determine $\alpha(\mathbb{N})$, $\beta\,(\mathbb{N})$, and $\gamma(\mathbb{N})$.

For Problems 1.3–1.6, assume $S = \{x, y, z\}$ and $T = \{1, 2, 3\}$. From Example 1.2, we know there are 27 mappings from S to T.

1.3. How many mappings are there from S onto T?

1.4. How many one-to-one mappings are there from S to T?

1.5. How many mappings are there from S to $\{1, 2\}$?

1.6. How many mappings are there from S onto $\{1, 2\}$?

For Problems 1.7–1.10, assume that S and T are sets, $\alpha : S \to T$, and $\beta : S \to T$. Complete each of the following statements. (The discussion of quantifiers in Appendix B may help.)

1.7. α is not onto iff for some . . . .   **1.8.** α is not one-to-one iff . . . .

1.9. $\alpha \neq \beta$ iff. . . .

1.10. β is one-to-one and onto iff for each $y \in T$ . . . .

Each f in Problems 1.11–1.16 defines a mapping from $\mathbb{R}$ (or a subset of $\mathbb{R}$) to $\mathbb{R}$. Determine which of these mappings are onto and which are one-to-one. Also describe $f(P)$ in each case, for P the set of positive real numbers.

1.11. $f(x) = 2x$ **1.12.** $f(x) = x - 4$ **1.13.** $f(x) = x^3$

1.14. $f(x) = x^2 + x$ **1.15.** $f(x) = e^x$ **1.16.** $f(x) = \tan x$

In Problems 1.17 and 1.18 let A denote the set of odd natural numbers, B the set of even natural numbers, and C the set of natural numbers that are multiples of 4.

1.17. With α as in Example 1.10, describe $\alpha\,(A)\,, \alpha(B)$, and $\alpha(C)$.

1.18. With β as in Example 1.10, describe $\beta(A), \beta(B)$, and $\beta(C)$.

In Problems 1.19 and 1.20, for each $n \in \mathbb{Z}$ the mapping $f_n : \mathbb{Z} \to \mathbb{Z}$ is defined by $f_n(x) = nx$.

1.19. For which values of n is f_n onto?

1.20. For which values of n is f_n one-to-one?

———————— †

1.21. Assume that S and T are finite sets containing m and n elements, respectively.
(a) How many mappings are there from S to T?
(b) How many one-to-one mappings are there from S to T? (Consider two cases: $m > n$ and $m \leq n$.)

1.22. (a) How many mappings are there from a 2-element set onto a 2-element set?
(b) from a 3-element set onto a 2-element set?
(c) from an n-element set onto a 2-element set?

1.23. A mapping $f : \mathbb{R} \to \mathbb{R}$ is onto iff each horizontal line (line parallel to the x-axis) intersects the graph of f at least once.
(a) Formulate a similar condition for $f : \mathbb{R} \to \mathbb{R}$ to be one-to-one.

———————————

†For each set of problems, solutions for most of the odd-numbered problems preceding the double line can be found at the back of the book.

(b) Formulate a similar condition for $f : \mathbb{R} \to \mathbb{R}$ to be both one-to-one and onto.

1.24. For each ordered pair (a, b) of integers define a mapping $a_{a,b} : \mathbb{Z} \to \mathbb{Z}$ by $a_{a,b}(n) = an + b$.
   (a) For which pairs (a, b) is $a_{a,b}$ onto?
   (b) For which pairs (a, b) is $a_{a,b}$ one-to-one?

1.25. With β as defined in Example 1.10, for each $n \in \mathbb{N}$ the equation $\beta(x) = n$ has exactly two solutions. (The solutions of $\beta(x) = 2$ are $x = 3$ and $x = 4$, for example.)
   (a) Define a mapping $\gamma : \mathbb{N} \to \mathbb{N}$ such that for each $n \in \mathbb{N}$ the equation $\gamma(x) = n$ has exactly three solutions.
   (b) Define a mapping $\gamma : \mathbb{N} \to \mathbb{N}$ such that for each $n \in \mathbb{N}$ the equation $\gamma(x) = n$ has exactly n solutions. (It suffices to describe γ in words.)
   (c) Define a mapping $\gamma : \mathbb{N} \to \mathbb{N}$ such that for each $n \in \mathbb{N}$ the equation $\gamma(x) = n$ has infinitely many solutions.

1.26. Prove that there is a mapping from a set to itself that is one-to-one but not onto iff there is a mapping from the set to itself that is onto but not one-to-one. (Compare Example 1.10.)

1.27. Prove that if $\alpha : S \to T$ and A and B are subsets of S, then $\alpha(A \cup B) = \alpha(A) \cup \alpha(B)$. (See Appendix B for half of the proof.)

1.28. (a) Prove that if $\alpha : S \to T$, and A and B are subsets of S, then $\alpha(A \cap B) \subseteq \alpha(A) \cap \alpha(B)$.
   (b) Give an example (specific S, T, A, B, and α) to show that equality need not hold in part (a). (For the simplest examples S will have two elements.)

1.29. Prove that a mapping $\alpha : S \to T$ is one-to-one iff $\alpha(A \cap B) = \alpha(A) \cap \alpha(B)$ for every pair of subsets A and B of S. (Compare Problem 1.28.)

SECTION 2   COMPOSITION. INVERTIBLE MAPPINGS

Assume that $\alpha : S \to T$ and $\beta : T \to U$. Then $\alpha(x) \in T$ for each $x \in S$, so it makes sense to write $\beta(\alpha(x))$, which is an element of U.

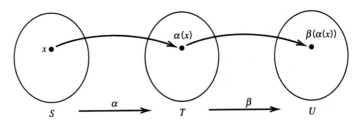

Application of α followed by β in this way is seen to yield a mapping from S to U. This mapping is called the *composition* (or *composite*) of α and β; it is denoted by $\beta \circ \alpha$. Thus, by definition,

$$(\beta \circ \alpha)(x) = \beta(\alpha(x))$$

for each $x \in S$. Note carefully: in $\beta \circ \alpha$, it is α, the mapping on the right, that is applied first.

Example 2.1.   Let $S = \{x, y, z\}$, $T = \{1, 2, 3\}$, and $U = \{a, b, c\}$. Define $\alpha : S \to T$ by

$$\alpha(x) = 2, \quad \alpha(y) = 1, \quad \text{and} \quad \alpha(z) = 3,$$

as shown in the diagram below. Also, define $\beta : T \to U$ by

$$\beta(1) = b, \quad \beta(2) = c, \quad \text{and} \quad \beta(3) = a.$$

Then

$$(\beta \circ \alpha)(x) = \beta(\alpha(x)) = \beta(2) = c$$
$$(\beta \circ \alpha)(y) = \beta(\alpha(y)) = \beta(1) = b$$
$$(\beta \circ \alpha)(z) = \beta(\alpha(z)) = \beta(3) = a.$$

Example 2.2.   Let α and β denote the mappings, each with the set of real numbers as both domain and codomain, defined by

$$\alpha(x) = x^2 + 2 \quad \text{and} \quad \beta(x) = x - 1.$$

Then

$$(\alpha \circ \beta)(x) = \alpha(\beta(x))$$
$$= \alpha(x - 1)$$
$$= (x - 1)^2 + 2$$
$$= x^2 - 2x + 3,$$

while

$$(\beta \circ \alpha)(x) = \beta(\alpha(x))$$
$$= \beta(x^2 + 2)$$
$$= (x^2 + 2) - 1$$
$$= x^2 + 1.$$

In particular, for example, $(\alpha \circ \beta)(0) = 3$ but $(\beta \circ \alpha)(0) = 1$. This example shows that $\beta \circ \alpha$ and $\alpha \circ \beta$ need not be equal, even if both are defined.   ■

It will be important to know which compositions are onto and which are one-to-one. The following theorem provides the answer.

Theorem 2.1.   *Assume that* $\alpha : S \to T$ *and* $\beta : T \to U$.

(a) *If* α *and* β *are onto, then* $\beta \circ \alpha$ *is onto.*

(b) *If* $\beta \circ \alpha$ *is onto, then* β *is onto.*

(c) *If* α *and* β *are one-to-one, then* $\beta \circ \alpha$ *is one-to-one.*

(d) *If* $\beta \circ \alpha$ *is one-to-one, then* α *is one-to-one.*

PROOF.   (a) Assume that both α and β are onto. To prove that $\beta \circ \alpha$ is onto, we must establish that if $z \in U$, then there is an element $x \in S$ such that $(\beta \circ \alpha)(x) = z$.

Let $z \in U$. Because β is onto, there exists $y \in T$ such that $\beta(y) = z$. Since α is also onto, there exists $x \in S$ such that $\alpha(x) = y$. We now have

$$(\beta \circ \alpha)(x) = \beta(\alpha(x)) = \beta(y) = z.$$

Therefore, $\beta \circ \alpha$ is onto.

(b) Assume that $\beta \circ \alpha$ is onto and that $z \in U$. Then there exists $x \in S$ such that $(\beta \circ \alpha)(x) = z$. But then $\beta(\alpha(x)) = z$ with $\alpha(x) \in T$. Hence, β is onto.

(c) Assume that both α and β are one-to-one. To prove that $\beta \circ \alpha$ is one-to-one, we shall prove that if $x_1, x_2 \in S$ and $(\beta \circ \alpha)(x_1) = (\beta \circ \alpha)(x_2)$, then $x_1 = x_2$.

If $(\beta \circ \alpha)(x_1) = (\beta \circ \alpha)(x_2)$, then $\alpha(x_1) = \alpha(x_2)$ since β is one-to-one. Therefore, $x_1 = x_2$ since α is one-to-one. This proves that $\beta \circ \alpha$ is one-to-one.

(d) Assume that $\beta \circ \alpha$ is one-to-one. If $x_1, x_2 \in S$ and $\alpha(x_1) = \alpha(x_2)$, then $\beta(\alpha(x_1)) = \beta(\alpha(x_2))$; that is, $(\beta \circ \alpha)(x_1) = (\beta \circ \alpha)(x_2)$. This implies $x_1 = x_2$ because $\beta \circ \alpha$ is one-to-one. Therefore, α is one-to-one.   ∎

A mapping $\beta : T \to S$ is an *inverse* of $\alpha : S \to T$ if both $\beta \circ \alpha = \iota_S$ and $\alpha \circ \beta = \iota_T$. A mapping is said to be *invertible* if it has an inverse. If a mapping is invertible, then its inverse is unique (this is Problem 4.13).

Example 2.3.   The mapping α in Example 2.1 is invertible. Its inverse is the mapping γ defined by

$$\gamma(1) = y, \quad \gamma(2) = x, \quad \text{and} \quad \gamma(3) = z.$$

For example,

$$(\gamma \circ \alpha)(x) = \gamma(2) = x = \iota_S(x)$$

and

$$(\alpha \circ \gamma)(1) = \alpha(y) = 1 = \iota_T(1).$$

There are also four other equations to be checked, those involving $(\gamma \circ \alpha)(y)$, $(\gamma \circ \alpha)(z)$, $(\alpha \circ \gamma)(2)$, and $(\alpha \circ \gamma)(3)$. In terms of the diagram in Example 2.1, γ is gotten by reversing the direction of the arrows under α.   ∎

Example 2.4.   The mapping $\alpha : \mathbb{R} \to \mathbb{R}$ defined by $\alpha(x) = x^2$ is not invertible. It fails on two counts, by the following theorem.   ∎

Theorem 2.2.   *A mapping is invertible iff it is both one-to-one and onto.*

Remark.   This theorem has the form p iff q, so we must prove both if p then q, and if q then p (Appendix B). In the following proof we first prove if p then q, so we begin by assuming that the given mapping is invertible.

PROOF.   First assume that $\alpha : S \to T$ is invertible, with inverse β. Then $\beta \circ \alpha$, being the identity mapping on S, is one-to-one; therefore, α must be one-to-one by Theorem 2.1(d). On the other hand, $\alpha \circ \beta$, being the identity mapping on T, is onto. Therefore, α

must be onto by Theorem 2.1(b). This proves that if α is invertible then it is both one-to-one and onto.

Now assume that $\alpha : S \to T$ is both one-to-one and onto. We shall show that α is invertible by describing an inverse. Assume $t \in T$. Then, because α is onto, there is at least one element $s \in S$ such that $\alpha(s) = t$. But α is also one-to-one, so this element s must be unique; let $\beta(t) = s$. This can be done for each element $t \in T$, and in this way we obtain a mapping $\beta : T \to S$. Moreover, from the way in which β is constructed it can be seen that $\beta \circ \alpha = \iota_S$ and $\alpha \circ \beta = \iota_T$, so that β is an inverse of α. Thus α is invertible.   ■

Warning: Some authors write mappings on the right rather than on the left. Our $\alpha(x)$ becomes, for them, $x\alpha$. Then in $\beta \circ \alpha$ it is β, the mapping on the left, that is applied first, because $x(\beta \circ \alpha) = (x\beta)\alpha$. Although we shall consistently write mappings on the left, it is important when reading other sources to take note of which convention is being followed.

PROBLEMS

Let α, β, and γ be mappings from $\mathbb{Z}$ to $\mathbb{Z}$ defined by $\alpha(n) = 2n$, $\beta(n) = n + 1$ and $\gamma(n) = n^2$. Write a formula for each of the compositions in 2.1–2.6. Also determine the image in each case.

2.1. $\alpha \circ \alpha$ **2.2.** $\gamma \circ \alpha$ **2.3.** $\alpha \circ \beta$

2.4. $\beta \circ \beta$ **2.5.** $\beta \circ \gamma$ **2.6.** $\gamma \circ \gamma$

2.7. Prove that if $\alpha : S \to T$, then $\alpha \circ \iota_S = \alpha$ and $\iota_T \circ \alpha = \alpha$.

2.8. Describe the inverse of the mapping β in Example 2.1.

Each mapping f in Problems 2.9 and 2.10 defines an invertible mapping from $\mathbb{R}$ (or a subset of $\mathbb{R}$) to $\mathbb{R}$. Write a formula for the inverse (call it g) in each case.

2.9. (a) $f(x) = 5x$ (b) $f(x) = x - 4$ (c) $f(x) = 10^x$

2.10. (a) $f(x) = -x/2$ (b) $f(x) = x^3$ (c) $f(x) = \log_e x$ $(x > 0)$

2.11. For α a mapping, decide whether each of the following is true or false. (Appendix B gives examples involving *if, only if, necessary conditions,* and *sufficient conditions.*)
   (a) α is invertible if α is one-to-one.
   (b) α is invertible only if α is one-to-one.
   (c) A necessary condition for α to be invertible is that it be one-to-one.
   (d) A sufficient condition for α to be invertible is that it be one-to-one.

2.12. For α a mapping, decide whether each of the following is true or false.
   (a) α is invertible only if α is onto.
   (b) α is invertible if α is onto.
   (c) A sufficient condition for α to be onto is that it be invertible.
   (d) A necessary condition for α to be onto is that it be invertible.

2.13. Consider f and g, mappings from $\mathbb{R}$ to $\mathbb{R}$, defined by $f(x) = \sin x$ and $g(x) = 2x$. Is $f \circ g$ equal to $g \circ f$?

2.14. Which of the functions sine, cosine, and tangent, as mappings from $\mathbb{R}$ to $\mathbb{R}$, are invertible?

2.15. Complete the following statement: A mapping $\alpha : S \to T$ is not invertible iff α is not one-to-one or . . . .

2.16. Assume $\alpha : S \to T$ and $\beta : T \to S$. Complete the following statement: β is not an inverse of α iff $\beta \circ \alpha \neq \iota_S$ or . . . .

2.17. Assume that $\alpha : S \rightarrow T$ and $\beta : T \rightarrow U$. Consider the following statement.

$$\text{If } \alpha \text{ is one-to-one and } \beta \text{ is onto, then} \tag{2.1}$$
$$\beta \circ \alpha \text{ is one-to-one and onto.}$$

(a) Is Statement (2.1) true? Justify your answer.
(b) Write the converse of Statement (2.1). Is this converse true? Justify your answer.
(c) Write the contrapositive of Statement (2.1). Is this contrapositive true? Justify your answer.

2.18. Assume that $\alpha : S \rightarrow T$ and $\beta : T \rightarrow U$. Consider the following statement.

$$\text{If } \alpha \text{ is not one-to-one, then} \tag{2.2}$$
$$\beta \circ \alpha \text{ is not invertible.}$$

(a) Is Statement (2.2) true? Justify your answer.
(b) Write the converse of Statement (2.2). Is this converse true? Justify your answer.
(c) Write the contrapositive of Statement (2.2). Is this contrapositive true? Justify your answer.

2.19. For α and β as in Example 1.10, determine both $\beta \circ \alpha$ and $\alpha \circ \beta$. Is either α or β invertible?

2.20. Prove that the inverse of an invertible mapping is invertible.

2.21. Give an example of sets S, T, and U and mappings $\alpha : S \rightarrow T$ and $\beta : T \rightarrow U$ such that $\beta \circ \alpha$ is onto, but α is not onto. [Compare Theorem 2.1(b).]

2.22. Give an example of sets S, T, and U and mappings $\alpha : S \rightarrow T$ and $\beta : T \rightarrow U$ such that $\beta \circ \alpha$ is one-to-one, but β is not one-to-one. [Compare Theorem 2.1(d).]

2.23. Prove that if $\alpha : S \rightarrow T$, $\beta : T \rightarrow U$, $\gamma : T \rightarrow U$, α is onto, and $\beta \circ \alpha = \gamma \circ \alpha$, then $\beta = \gamma$.

2.24. Prove that if $\beta : S \rightarrow T$, $\gamma : S \rightarrow T$, $\alpha : T \rightarrow U$, α is one-to-one, and $\alpha \circ \beta = \alpha \circ \gamma$, then $\beta = \gamma$.

2.25. Give an example to show that the condition "α is onto" cannot be omitted from Problem 2.23.

2.26. Give an example to show that the condition "α is one-to-one" cannot be omitted from Problem 2.24.

2.27. Assume that $\alpha : S \rightarrow T$ and $\beta : T \rightarrow U$. Use Theorems 2.1 and 2.2 to prove each of the following statements.
(a) If α and β are invertible, then $\beta \circ \alpha$ is invertible.
(b) If $\beta \circ \alpha$ is invertible, then β is onto and α is one-to-one.

SECTION 3   OPERATIONS

If one integer is added to another, the result is an integer. If one integer is subtracted from another, the result is also an integer. These examples—addition and subtraction of integers—are special cases of what are known as *operations*. In each case there is a set (here the integers), and a relationship that assigns to each ordered pair (a, b) of elements of that set another element of the same set: $a + b$ in one case, $a - b$ in the other. The general definition of operation is as follows.

Definition.   An *operation* on a set S is a relationship (rule, correspondence) that assigns to each ordered pair of elements of S a uniquely determined element of S.

Thus an operation is a special kind of mapping: First, $S \times S$, the *Cartesian product* of S with S, is the set of all ordered pairs (a, b) with $a \in S$ and $b \in S$ (Appendix A). Then an operation on S is simply a mapping from $S \times S$ to S. In the case of addition as an operation on the integers, $(a, b) \mapsto a + b$.

Example 3.1. On the set of positive integers, multiplication is an operation: $(m, n) \mapsto mn$, where mn has the usual meaning, m times n. Division is not an operation on the set of positive integers, because $m \div n$ is not necessarily a positive integer ($1 \div 2 = 1/2$, for instance). ■

The last example illustrates a point worth stressing. To have an operation on a set S, it is essential that if $a, b \in S$, then the image of the ordered pair (a, b) be in S. This property of an operation is referred to as *closure*, or we say that S is *closed* with respect to the operation.

If there is an established symbol to denote the image of a pair under an operation, as in the case of $a + b$ for addition of numbers, then that symbol is used. Otherwise some other symbol is adopted, such as $(a, b) \mapsto a * b$ or just $(a, b) \mapsto ab$, for instance, where it must be specified what $a * b$ or ab is to mean in each case. We often say "operation $*$" when we mean "operation denoted by $*$."

Example 3.2. If $*$ is defined by $m * n = m^n$ for all positive integers m and n, the result is an operation on the set of positive integers. Notice that $3 * 2 = 3^2 = 9$, while $2 * 3 = 2^3 = 8$. Thus $3 * 2 \neq 2 * 3$, so that, just as with subtraction, the order makes a difference. ■

Example 3.3. Let S denote any nonempty set, and let $M(S)$ denote the set of all mappings from S to S. Suppose that $\alpha \in M(S)$ and $\beta \in M(S)$. Then $\alpha : S \to S$, $\beta : S \to S$, and $\beta \circ \alpha : S \to S$, so that $\beta \circ \alpha \in M(S)$. Thus $\circ$ is an operation on $M(S)$. We shall return to this in Section 4. ■

Example 3.4. If S is a finite set, then we can specify an operation on S by means of a table, similar to the addition and multiplication tables used in beginning arithmetic. We first form a square, and then list the elements of S across the top and also down the left-hand side. For an operation $*$, we put $a * b$ at the intersection of the (horizontal) row with a at the left and the (vertical) column with b at the top. For Table 3.1, $u * v = w$, $v * u = v$, and so forth. Any way of filling in the nine spaces in the square, with entries chosen from the set $\{u, v, w\}$, will define an operation on $\{u, v, w\}$. Changing one or more of the nine entries will give a different operation. (If the nine entries are left unchanged, but $*$ is changed to some other symbol, the result would not be considered a different operation.) Tables defining operations in this way are called *Cayley* tables, after the British mathematician Arthur Cayley (1821–1895).

Table 3.1

$*$	u	v	w
u	u	w	w
v	v	v	v
w	w	u	v

■

Example 3.5. Let $M(2, \mathbb{R})$ denote the set of all 2×2 matrices with real numbers as entries. Addition and multiplicatioin are defined on $M(2, \mathbb{R})$ as follows.

$$\begin{bmatrix} a & b \\ c & d \end{bmatrix} + \begin{bmatrix} w & x \\ y & z \end{bmatrix} = \begin{bmatrix} a+w & b+x \\ c+y & d+z \end{bmatrix}$$

$$\begin{bmatrix} a & b \\ c & d \end{bmatrix} \begin{bmatrix} w & x \\ y & z \end{bmatrix} = \begin{bmatrix} aw+by & ax+bz \\ cw+dy & cx+dz \end{bmatrix}$$

More generally, for any positive integer n, both matrix multiplication and matrix addition are operations on the set $M(n, \mathbb{R})$ of all $n \times n$ matrices with real numbers as entries. These examples are discussed more fully in Appendix D. ∎

What we are calling *operations* are often called *binary operations*, to emphasize that they are mappings of ordered *pairs*, rather than mappings of single elements or ordered triples or such. [Examples of operations that are not binary are $a \mapsto -a$ and $(a, b, c) \mapsto \alpha(b + c)$, where $a, b, c \in \mathbb{R}$.] We shall have no occasion to discuss explicitly any operations other than binary operations, so we shall not carry along the extra word *binary*.

The notion of operation is so fundamental in algebra that algebra could almost be defined as the study of operations (with binary operations being the most important). But this would be something like defining mathematics as the study of sets and mappings—there is no question of the importance of these concepts, but at the same time they are too general to be of real interest. In calculus, for example, it is not *all* functions $f : \mathbb{R} \to \mathbb{R}$ that are of interest, but only functions that have some property such as continuity or differentiability. In the same way, in algebra the operations that are of interest usually possess certain special properties. We now introduce some of the most important of these properties.

Definition. An operation $*$ on a set S is said to be *associative* if it satisfies the condition

$$a * (b * c) = (a * b) * c \qquad\qquad \textit{associative law}$$

for all $a, b, c \in S$.

For example, addition of real numbers is associative: $a + (b + c) = (a + b) + c$ for all $a, b, c \in \mathbb{R}$. Subtraction of real numbers, however, is not associative: $2 - (3 - 4) = 2 - (-1) = 3$, but $(2 - 3) - 4 = (-1) - 4 = -5$. Notice that if the equation in the associative law fails for even one triple (a, b, c), then the operation is not associative. (See the discussion in Appendix B on the negation of statements with quantifiers.)

Multiplication of real numbers is associative: $a(bc) = (ab)c$. But the operation defined in Example 3.2 is not associative: for example, $2 * (3 * 2) = 2 * (3^2) = 2 * 9 = 2^9 = 512$ but $(2 * 3) * 2 = 2^3 * 2 = 8 * 2 = 8^2 = 64$.

To motivate the next definition, think of the following properties of the integers 0 and 1: $m + 0 = 0 + m = m$ and $m \cdot 1 = 1 \cdot m = m$, for every integer m.

Definition. An element e in a set S is an *identity* (or *identity element*) for an operation $*$ on S if

$$e * a = a * e = a \quad \text{for each } a \in S.$$

Thus 0 is an identity for addition of integers, and 1 is an identity for multiplication of integers. Note that the definition requires *both* $e * a = a$ and $a * e = a$, for each $a \in S$. (See Problems 3.11 and 3.12.) A similar remark applies to the next definition.

Definition.   Assume that $*$ is an operation on S, with identity e, and that $a \in S$. An element b in S is an *inverse* of a relative to $*$ if

$$a * b = b * a = e.$$

Example 3.6.   Relative to addition as an operation on the set of integers, each integer has an inverse, its negative: $a + (-a) = (-a) + a = 0$ for each integer a. It is important to notice that the inverse of an element must be in the set under consideration. Relative to addition as an operation on the set of nonnegative integers, no element other than 0 has an inverse: the negative of a positive integer is negative.   ∎

Example 3.7.   Relative to multiplication as an operation on the set of real numbers, each real number different from 0 has an inverse, its reciprocal: $a \cdot (1/a) = (1/a) \cdot a = 1$. Multiplication is also an operation on the set of integers (with identity 1), but in this case only 1 and -1 have inverses.   ∎

We have seen that it is possible to have an operation $*$ and elements a and b such that $a * b \neq b * a$ (subtraction of integers, or Example 3.2, for instance). Operations for which this cannot happen are numerous enough and important enough to deserve a special name.

Definition.   An operation $*$ on a set S is said to be *commutative* if

$$a * b = b * a \qquad\qquad \textit{commutative law}$$

for all $a, b \in S$.

Addition and multiplication of integers are commutative. Other examples will occur in the problems and elsewhere.

PROBLEMS

Which of the following equations define operations on the set of integers? Of those that do, which are associative? Which are commutative? Which have identity elements?

3.1. $m * n = mn + 1$

3.2. $m * n = (m + n)/2$

3.3. $m * n = m$

3.4. $m * n = mn^2$

3.5. $m * n = m^2 + n^2$

3.6. $m * n = 2^{mn}$

3.7. $m * n = 3$

3.8. $m * n = \sqrt{mn}$

3.9. Does $(m, n) \mapsto m^n$ define an operation on the set of all integers? (Compare Example 3.2.)

3.10. There is an identity for the operation in Example 3.3. What is it?

3.11. Prove that the operation in Example 3.2 has no identity element.

3.12. (a) Prove that the operation in Example 3.4 has no identity element.
(b) Change one entry in the table in Example 3.4 so that u becomes an identity element.

3.13. If $*$ is an operation on S, T is a subset of S, and T is closed with respect to $*$, then two of the following three statements are necessarily true, but one may be false. Which two are true?
  (a) If $*$ is associative on S, then $*$ is associative on T.
  (b) If there is an identity element for $*$ on S, then there is an identity element for $*$ on T.
  (c) If $*$ is commutative on S, then $*$ is commutative on T.

Assume that $*$ is an operation on S. Complete each of the following statements.

3.14. $*$ is not associative iff $a * (b * c) \neq (a * b) * c$ ....

3.15. $*$ is not commutative iff $a * b \neq b * a$ ....

3.16. $e \in S$ is not an identity element for $*$ iff ....

3.17. There is no identity element for $*$ iff for each $e \in S$ there is an element $a \in S$ such that ....

3.18. Complete the following table in a way that makes $*$ commutative.

$*$	a	b	c	d
a	a	b		d
b		c		
c	c	d	a	b
d		a		c

3.19. Determine the smallest subset A of $\mathbb{Z}$ such that $2 \in A$ and A is closed with respect to addition.

3.20. Determine the smallest subset B of $\mathbb{Q}$ such that $2 \in B$ and B is closed with respect to addition and division.

3.21. How many different operations are there on a 1-element set? 2-element set? 3-element set? n-element set? (See the remarks in Example 3.4.)

3.22. How many different commutative operations are there on a 1-element set? 2-element set? 3-element set? n-element set? (The Cayley table for a commutative operation must have a special kind of symmetry.)

3.23. (a) Complete the following Cayley table in such a way that u becomes an identity element. In how many ways can this be done?

$*$	u	v
u		
v		

  (b) Can the table be completed in such a way that u and v both become identity elements? Why or why not?
  (c) Prove: An operation $*$ on a set S (any S) can have at most one identity element.

3.24. Complete the following table in such a way that $*$ is commutative and has an identity element, and each element has an inverse. (There is only one correct solution. First explain why y must be the identity element.)

$*$	w	x	y	z
w	y			x
x	z	w		
y				
z				w

3.25. Prove that addition is commutative on the set $M(2, \mathbb{R})$ of all 2×2 matrices with real numbers as entries (Example 3.5). Which property of addition of real numbers do you need?

3.26. Prove that matrix multiplication (Example 3.5) is not commutative as an operation on the set of all 2×2 matrices with real numbers as entries.

3.27. Assume that $*$ is an operation on S with identity element e. Prove that to check whether $*$ is associative on S it suffices to check the associativity condition for the nonidentity elements of S; that is,

$$a * (b * c) = (a * b) * c$$

is automatically true if any one of a, b, or c is equal to e.

3.28. Assume that $*$ is an associative operation on S and that $a \in S$. Let

$$C(a) = \{x \colon x \in S \text{ and } a * x = x * a\}.$$

Prove that $C(a)$ is closed with respect to $*$.

3.29. Prove: If $*$ is an associative operation on S, then

$$
\begin{aligned}
a * (b * (c * d)) &= a * ((b * c) * d) \\
&= (a * b) * (c * d) \\
&= (a * (b * c)) * d \\
&= ((a * b) * c) * d
\end{aligned}
$$

for all $a, b, c, d \in S$. (See the second paragraph in Section 14.)

3.30. Assume that $*$ is an operation on S with identity element e and that

$$x * (y * z) = (x * z) * y$$

for all $x, y, z \in S$. Prove that $*$ is commutative and associative.

3.31. Assume that e is an identity element for an operation $*$ on a set S. If $a, b \in S$ and $a * b = e$, then a is said to be a *left inverse* of b and b is said to be a *right inverse* of a. Prove that if $*$ is associative, b is a left inverse of a, and c is a right inverse of a, then $b = c$.

SECTION 4   COMPOSITION AS AN OPERATION

In Example 3.3 we saw that if S is any nonempty set, then composition is an operation on $M(S)$, the set of all mappings from S to S. It is worthwhile to look more closely at this operation, for its importance is matched only by that of addition and the other operations on the familiar number systems. The most general properties are summarized in the following theorem.

Theorem 4.1.   *Let S denote any nonempty set.*

(a)   *Composition is an associative operation on $M(S)$, with identity element ι_S.*

(b)   *Composition is an associative operation on the set of all invertible mappings in $M(S)$, with identity ι_S.*

PROOF.   Associativity means that $\gamma \circ (\beta \circ \alpha) = (\gamma \circ \beta) \circ \alpha$ for all $\alpha, \beta, \gamma \in M(S)$. By the definition of equality for mappings, this means that

$$[\gamma \circ (\beta \circ \alpha)](x) = [(\gamma \circ \beta) \circ \alpha](x)$$

for each $x \in S$. To verify this, we can write

$$[\gamma \circ (\beta \circ \alpha)](x) = \gamma ((\beta \circ \alpha)(x))$$
$$= \gamma (\beta (\alpha (x)))$$
$$= (\gamma \circ \beta) (\alpha (x))$$
$$= [(\gamma \circ \beta) \circ \alpha](x).$$

(As in all proofs, it is important to understand the justification for each step. Remember this test: Could you explain it to someone else?)

It is easy to verify that $\iota_S \circ \alpha = \alpha \circ \iota_S = \alpha$ for each $\alpha \in M(S)$, which proves that ι_S is an identity element. (Compare Problem 2.7 with $S = T$.)

Now move to part (b) of the theorem. Notice that the question of whether $\gamma \in M(S)$ is an inverse of $\alpha \in M(S)$ means the same whether taken in the sense of Section 2 (preceding Example 2.3), or in the sense of Section 3 (preceding Example 3.6): $\gamma \circ \alpha = \alpha \circ \gamma = \iota_S$. To prove that composition is an operation on the set of all invertible mappings in $M(S)$, assume that $\alpha, \beta \in M(S)$ and that both α and β are invertible. Then α and β are one-to-one and onto by Theorem 2.2. Therefore, $\beta \circ \alpha$ is both one-to-one and onto by Theorem 2.1, parts (c) and (a). But this implies that $\beta \circ \alpha$ is invertible, again by Theorem 2.2.

Since composition is associative as an operation on all of $M(S)$, it is certainly associative when restricted to the invertible elements of $M(S)$. The proof is now finished by the observation that ι_S is invertible. ■

Example 2.2 shows that composition as an operation on $M(\mathbb{R})$ is not commutative. Problem 4.10 gives a more general statement about commutativity.

Notice that composition is an operation on a subset of $M(S)$ iff that subset is closed with respect to composition. Also, whenever a subset of $M(S)$ is closed with respect to composition, then composition is necessarily associative as an operation on that subset; the first paragraph of the proof of Theorem 4.1 provides a proof. Many important operations involve composition on special sets of invertible mappings. We conclude this section by giving two examples of this; more examples will come later.

Example 4.1.   Let p denote a fixed point in a plane P, and let G denote the set of all rotations of the plane about the point p. Each element of G represents an element of $M(P)$. Agree that two rotations are equal if they differ by a multiple of $360°$. Then composition is an operation of G: if α and β are rotations about p, then $\beta \circ \alpha$ is the rotation obtained by first applying α and then β (Figure 4.1). For example, if α denotes clockwise rotation through $70°$, and β clockwise rotation through $345°$, then $\beta \circ \alpha$ is clockwise rotation through $415°$

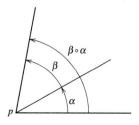

Figure 4.1

or, equivalently, $55°$, since we can ignore multiples of $360°$. This operation is associative by Theorem 4.1.

An identity element is rotation through $0°$, and each rotation has an inverse: rotation of the same magnitude in the opposite direction. Finally, as an operation on G, composition is commutative.   ∎

Example 4.2.   For each ordered pair (a, b) of real numbers with $a \neq 0$, let $\alpha_{a,b} : \mathbb{R} \to \mathbb{R}$ be defined by $\alpha_{a,b}(x) = ax + b$. Let A denote the set of all such mappings. Then composition is an operation on A. To verify this, let (a, b) and (c, d) be ordered pairs of real numbers with $a \neq 0$ and $c \neq 0$. Then $ac \neq 0$, and

$$
\begin{aligned}
(\alpha_{a,b} \circ \alpha_{c,d})(x) &= \alpha_{a,b}(\alpha_{c,d}(x)) \\
&= \alpha_{a,b}(cx + d) \\
&= a(cx + d) + b \\
&= acx + ad + b \\
&= \alpha_{ac,ad+b}(x).
\end{aligned}
$$

Thus $\alpha_{a,b} \circ \alpha_{c,d} = \alpha_{ac,ad+b}$. Notice that A is a subset of $M(\mathbb{R})$, and that composition is, as always, associative. Further properties of this example are brought out in the problems.

Each mapping $\alpha_{a,b}$ can be interpreted geometrically by considering what it does to the points on a real line. If $a > 1$, for instance, then $\alpha_{a,0}$ magnifies the distance of each point from the origin by a factor of $a : \alpha_{a,0}(x) = ax$. Also, if $b > 0$, then $\alpha_{1,b}$ translates each point b units to the right (assuming the real line is directed to the right): $\alpha_{1,b}(x) = x + b$. Now observe that $\alpha_{a,b} = \alpha_{1,b} \circ \alpha_{a,0}$. It follows from these observations that if $a > 1$ and $b > 0$, then $\alpha_{a,b}$ corresponds to $\alpha_{a,0}$ (magnification) followed by $\alpha_{1,b}$ (translation), as shown in Figure 4.2. Problem 4.7 asks what happens in other cases.   ∎

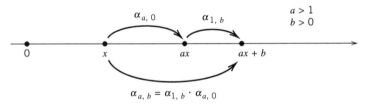

Figure 4.2

4.1. With $S = \{a, b\}$, the set $M(S)$ contains four elements; denote these by π, ρ, σ, τ, defined as follows:

$$
\begin{array}{llll}
\pi(a) = a & \rho(a) = a & \sigma(a) = b & \tau(a) = b \\
\pi(b) = a & \rho(b) = b & \sigma(b) = a & \tau(b) = b
\end{array}
$$

(a) Construct the Cayley table for composition ($\circ$) as an operation on $M(S) = \{\pi, \rho, \sigma, \tau\}$. (As a start, $\rho \circ \tau = \tau$ and $\sigma \circ \tau = \pi$.)

(b) Which is the identity element?

(c) Is $\circ$ commutative as an operation on $M(S)$?

(d) Which elements of $M(S)$ are invertible?

(e) Is $\circ$ commutative as an operation on the set of invertible elements in $M(S)$?

4.2. Let $S = \{a, b, c\}$ and let $A = \{\alpha, \beta, \gamma, \delta\}$, where α, β, γ, and δ are the elements of $M(S)$ defined as follows.

$$\begin{array}{lll} \alpha(a) = a & \alpha(b) = b & \alpha(c) = c \\ \beta(a) = b & \beta(b) = a & \beta(c) = c \\ \gamma(a) = a & \gamma(b) = a & \gamma(c) = a \\ \delta(a) = b & \delta(b) = b & \delta(c) = b \end{array}$$

(a) Construct a Cayley table for A with respect to $\circ$. This table will show, in particular, that A is closed with respect to $\circ$.

(b) The operation $\circ$ is associative on A. Give a reason. (No calculations are necessary.)

(c) Is $\circ$ commutative as an operation on A?

(d) The identity element of $M(S)$ is in A. Which element is it?

4.3. Consider Example 4.1, and let ρ_1, ρ_2, ρ_3, and ρ_4 denote clockwise rotation through $0°$, $90°$, $180°$, and $270°$, respectively. Composition is an operation on $\{\rho_1, \rho_2, \rho_3, \rho_4\}$ ($\rho_4 \circ \rho_2 = \rho_1$, for instance). Construct the corresponding Cayley table. Is there an identity element? Does each element have an inverse?

4.4. Let S denote a nonempty set, and let $N(S)$ denote the set of one-to-one mappings from S to S.

(a) Which theorems prove that composition is an associative operation on $N(S)$? (The answer should be specific and complete.)

(b) Is there an identity element for composition as an operation on $N(S)$?

(c) For which sets S is $N(S)$ the same as the set of all invertible mappings from S to S?

4.5. (a) Consider Example 4.1, and let α denote clockwise rotation through $\pi/2$ radians. Let H denote the smallest subset of G such that $\alpha \in H$ and H is closed with respect to $\circ$. Determine H. (*Suggestion:* Denote $\alpha \circ \alpha$ by α^2, $\alpha \circ \alpha \circ \alpha$ by α^3, and so on. H contains four distinct elements.)

(b) Same as (a) with $\pi/2$ replaced by $\pi/6$.

(c) Same as (a) with $\pi/2$ replaced by π/k ($k \in \mathbb{N}$). How many distinct elements (rotations) does H contain? (Treat rotations as indistinct if they differ by an integral multiple of 2π.)

4.6. Consider the operation $\circ$ on the set A in Example 4.2.

(a) Verify that $\alpha_{1,0}$ is an identity element.

(b) Prove that each $\alpha_{a,b} \in A$ is invertible by verifying that it is one-to-one and onto. (Remember that $a \neq 0$.)

(c) Determine (c, d) so that $\alpha_{c,d}$ is an inverse of $\alpha_{a,b}$.

4.7. Give a geometric interpretation of $\alpha_{a,b}$ in Example 4.2 under each of the following conditions. (There can be shrinking rather than magnification, and translation to the left as well as to the right.)

(a) $0 < a < 1$ and $b = 0$.   (b) $a < 0$ and $b = 0$.   (c) $a = 1$ and $b < 0$.

4.8. Let B and C denote the following subsets of A in Example 4.2:

$$B = \left\{\alpha_{a,0} : a \in \mathbb{R} \text{ and } a \neq 0\right\}$$

$$C = \left\{\alpha_{1,b} : b \in \mathbb{R}\right\}$$

(a) Verify that $\circ$ is an operation on B. (That is, verify that B is closed with respect to $\circ$.) Is it associative? Commutative? Is there an identity element?

(b) Verify that $\circ$ is an operation on C. Is it associative? Commutative? Is there an identity element?

(c) Verify that each mapping in A is the composition of a mapping in B and a mapping in C.

4.9. Consider Example 4.2 and let D denote the smallest subset of A such that $\alpha_{1,1} \in D$ and D is closed with respect to $\circ$. Determine D.

4.10. Verify that if S contains more than one element, then composition is not a commutative operation on $M(S)$. (Try "constant" mappings, like π and τ in Problem 4.1.)

4.11. Let S denote the set of all real numbers except 0 and 1. Consider the six mappings from S to S defined as follows:

$$\alpha_1(x) = x \qquad \alpha_2(x) = \frac{1}{x} \qquad \alpha_3(x) = 1 - x$$

$$\alpha_4(x) = 1 - \frac{1}{x} \qquad \alpha_5(x) = \frac{1}{1-x} \qquad \alpha_6(x) = \frac{x}{x-1}.$$

(a) Verify that composition is an operation on $\{\alpha_1, \ldots, \alpha_6\}$ by constructing a Cayley table. [As a start,

$$(\alpha_6 \circ \alpha_2)(x) = \alpha_6\left(\frac{1}{x}\right) = \frac{\frac{1}{x}}{\frac{1}{x}-1} = \frac{1}{1-x} = \alpha_5(x),$$

so $\alpha_6 \circ \alpha_2 = \alpha_5$.]
(b) There is an identity element. What is it?
(c) Show that each of the six elements has an inverse.
(d) Is $\circ$ commutative as an operation on $\{\alpha_1, \ldots, \alpha_6\}$?
(e) The operation $\circ$ on $\{\alpha_1, \ldots, \alpha_6\}$ is associative. Why? (You may refer to the text.)

4.12. Prove that if $\alpha : S \to T$, $\beta : T \to U$, and $\gamma : U \to V$ are any mappings, then $\gamma \circ (\beta \circ \alpha) = (\gamma \circ \beta) \circ \alpha$. (The associativity in Theorem 4.1 is the special case of this with $S = T = U = V$.)

4.13. Prove that each invertible mapping has a unique inverse. [Assume that $\alpha : S \to T$, $\beta : T \to S$, $\gamma : T \to S$, $\beta \circ \alpha = \gamma \circ \alpha = \iota_S$, and $\alpha \circ \beta = \alpha \circ \gamma = \iota_T$. Show that $\beta = \gamma$ by using $\beta \circ (\alpha \circ \gamma) = (\beta \circ \alpha) \circ \gamma$, which is a consequence of Problem 4.12.]

4.14. Is composition an operation on the set of all continuous mappings from $\mathbb{R}$ to $\mathbb{R}$? (The answer requires either a theorem or an example from calculus.)

4.15. Prove that if V is a vector space, then composition is an operation on the set of all linear transformations from V to V. (This requires some knowledge of linear algebra. The basic ideas are reviewed in Appendix D.)

4.16. Assume that S is a nonempty set and that $\alpha \in M(S)$. Then left and right inverses for α relative to $\circ$ are defined as in Problem 3.31.
(a) Prove that α has a left inverse relative to $\circ$ iff α is one-to-one.
(b) Prove that α has a right inverse relative to $\circ$ iff α is onto.

CHAPTER II

INTRODUCTION TO GROUPS

We are now ready for one of the central ideas of modern mathematics, that of a group. In the first chapter we met a number of examples of sets with operations, and we observed that such operations may or may not possess any of several special properties such as associativity or the existence of an identity element. We shall see that a group is a set together with an operation such that certain specified properties are required to hold. These properties have been singled out because they arise naturally in many important special cases. By studying groups we can arrive at a clearer understanding of each of those special cases.

This chapter gives an introduction to groups through examples and a connection with symmetry. Later chapters treat the general theory and applications of groups.

SECTION 5   DEFINITION AND EXAMPLES

It takes patience to appreciate the diverse ways in which groups arise, but one of these ways is so familiar that we can use it to ease our way into the basic definition. To this end, recall the following three things about the set of integers with respect to addition. First, addition is associative. Second, 0 is an identity element. And third, relative to 0, each integer has an inverse (its negative). Much more can be said about the integers, of course, but these are the properties that are important at the moment; they show that the integers with addition form a group, in the sense of the following definition.

Definition.   A *group* is a set G together with an operation $*$ on G such that each of the following axioms is satisfied:

associativity
$$a * (b * c) = (a * b) * c \qquad \text{for all} \quad a, b, c \in G,$$

existence of an identity element
there is an element $e \in G$ such that $a * e = e * a = a$ for each $a \in G$, and

existence of inverse elements
for each $a \in G$ there is an element $b \in G$ such that $a * b = b * a = e$.

Notice that a group consists of a pair of things, a set and an operation on that set. In particular, the set must be closed with respect to the operation. Very often, a group is referred to by naming only the underlying set, but that is safe only if it is clear what operation is intended. As a special case, whenever we refer to "the group of integers" the operation is meant to be addition.

Example 5.1.   The set of even integers together with addition is a group. Addition is an operation because the sum of two even integers is an even integer. The associative law is true for all integers, so it is certainly true for the subset of all even integers. The identity element is 0 (an even integer), and the inverse of an even integer x is $-x$ (again an even integer).   ∎

Example 5.2.   The set of positive integers with addition is not a group, because there is no identity element. Even if we considered the positive integers along with 0 we would not get a group, because no element other than 0 would have an inverse.   ∎

Example 5.3.   The set {0} together with addition is a group. Notice that because a group must contain an identity element, the set underlying a group must always contain at least one element.   ∎

Example 5.4.   The set of positive rational numbers with multiplication is a group. If r/s and u/v are positive (where none of the integers r, s, u, or v is 0), then $(r/s)(u/v) = ru/sv$ is also positive. The associative law we take to be a generally known fact from arithmetic. (We shall have more to say about this in Chapter VII.) The identity element is 1 and the inverse of r/s is s/r ($r \neq 0$, $s \neq 0$).   ∎

Example 5.5.   Tables 5.1 and 5.2 define operations on the set {a, b, c} that yield groups. In Table 5.1 ($*$), a is an identity element and the inverses of a, b, and c are a, c, and b, respectively. In Table 5.2 (#), b is an identity element and the inverses of a, b, and c are c, b, and a, respectively. The verification of associativity is Problem 5.18. This example illustrates why, in general, we should specify the operation, not just the set, when talking about a group.   ∎

Example 5.6.   If S is any nonempty set, then the set of all invertible mappings in $M(S)$ is a group with composition as the operation. This is merely a restatement of Theorem 4.1(b). We shall return to groups of this type in the next section.   ∎

Table 5.1

$*$	a	b	c
a	a	b	c
b	b	c	a
c	c	a	b

Table 5.2

#	a	b	c
a	c	a	b
b	a	b	c
c	b	c	a

Example 5.7.   Let p denote a fixed point in a plane, and let G denote the set of all rotations of the plane about the point p. In Example 4.1 we observed that composition is an operation on this set G, and we also verified everything needed to show that this gives a group.   ∎

Example 5.8.   Let A denote the set of all mappings $\alpha_{a,b}\colon \mathbb{R} \to \mathbb{R}$, as defined in Example 4.2. Recall that $a, b, \in \mathbb{R}$, $a \neq 0$, and $\alpha_{a,b}(x) = ax + b$ for each $x \in \mathbb{R}$. With composition of mappings as the operation, this yields a group. The rule for composition was worked out in Example 4.2. The identity element is $\alpha_{1,0}$ [Problem 4.6(a)]. The inverse of $\alpha_{a,b}$ is $\alpha_{a^{-1},-a^{-1}b}$ (Problem 5.19).   ∎

Example 5.9.   Examples of groups of matrices are given in Problems 5.16 and 5.17, and in Appendix D. Such groups are very important.   ∎

Examination of the groups given thus far will reveal that in each case there is only one identity element. The definition requires that there be one; the point now is that there cannot be more than one. Similarly, each element in each group has only one inverse element. Here is a formal statement and proof of these two facts.

Theorem 5.1.   *Assume that G together with $*$ is a group.*

(a)   *The identity element of G is unique. That is, if e and f are elements of G such that*

$$e * a = a * e = a \quad \text{for each } a \in G,$$

and

$$f * a = a * f = a \quad \text{for each } a \in G,$$

then $e = f$.

(b)   *Each element in a group has a unique inverse. That is, if a, x, and y are elements of G, e is the identity element of G, and*

$$a * x = x * a = e$$

and

$$a * y = y * a = e,$$

then $x = y$.

PROOF.   (a) Assume that e and f are as stated. Then $e * a = a$ for each $a \in G$, and so, in particular, $e * f = f$. Similarly, using $a = e$ in $a * f = a$, we have $e * f = e$. Thus $f = e * f = e$, and so $e = f$, as claimed.

(b) With a, x, and y as stated, write

$$
\begin{aligned}
x &= x * e & &(e \text{ is the identity}) \\
&= x * (a * y) & &(a * y = e) \\
&= (x * a) * y & &(\text{associativity}) \\
&= e * y & &(x * a = e) \\
&= y & &(e \text{ is the identity}).
\end{aligned}
$$
    ∎

Because of Theorem 5.1, it makes sense to speak of *the* identity element of a group, and *the* inverse of a group element. It is customary to use a^{-1} for the inverse of a group element a

when there is no conflicting natural notation (such as $-a$ in the case of the integers with addition). Thus $a * a^{-1} = a^{-1} * a = e$. This is in accordance with the notation for inverses relative to multiplication of real numbers, and with notation for other exponents in groups, to be introduced in Section 14. If several groups are being considered at once, e_G can be used in place of e to denote the identity of a group G.

The number of elements in the set underlying a group is called the *order* of the group; we denote this $|G|$. (More generally, $|S|$ denotes the number of elements in the set S.) A group is said to be *finite* or *infinite* depending on whether its order is finite or infinite. The group in Example 5.3 has order 1. The groups in Example 5.5 each have order 3. We shall compute the orders of the groups in Example 5.6, for S finite, in the next section. All other groups considered thus far are infinite.

PROBLEMS

In Problems 5.1–5.10, decide whether the given set of numbers forms a group with respect to the given operation. If it does, give the identity element and the inverse of each element. If it does not, give a reason. In each case, be sure to check whether the given operation is an operation on the given set.

5.1. $\{1\}$, multiplication.

5.2. All nonzero rational numbers, multiplication.

5.3. All rational numbers, addition.

5.4. All rational numbers, multiplication.

5.5. $\{-1, 1\}$, multiplication.

5.6. $\{-1, 0, 1\}$, addition.

5.7. All integers, multiplication.

5.8. $\{n : n = 10k$ for some $k \in \mathbb{Z}\}$, addition.

5.9. All nonzero rational numbers, division.

5.10. All integers, subtraction.

5.11. Verify that $\{2^m : m \in \mathbb{Z}\}$ is a group with respect to multiplication. Identify clearly the properties of $\mathbb{Z}$ and $\mathbb{R}$ that you use.

5.12. Verify that $\{2^m 3^n : m, n \in \mathbb{Z}\}$ is a group with respect to multiplication. Identify clearly the properties of $\mathbb{Z}$ and $\mathbb{R}$ that you use.

5.13. Let F denote $M(\mathbb{R})$, the set of all mappings from $\mathbb{R}$ to $\mathbb{R}$. For $f, g \in F$ define $f + g$ by $(f + g)(x) = f(x) + g(x)$ for all $x \in \mathbb{R}$. Then $f + g \in F$. Verify that with this operation F is a group.

5.14. Let $H = \{f : f : \mathbb{R} \to \mathbb{R}$ and $f(x) \neq 0$ for all $x \in \mathbb{R}\}$. For $f, g \in H$, define fg by $(fg)(x) = f(x)g(x)$ for all $x \in \mathbb{R}$. Then $fg \in H$. Verify that with this operation H is a group. How does this group differ from the group of invertible mappings in $M(\mathbb{R})$ (Example 5.6)?

5.15. If $|S| > 1$, then $M(S)$ is not a group with respect to composition. Why?

5.16. Verify that $M(2, \mathbb{Z})$, the set of all 2×2 matrices with integers as entries, forms a group with respect to matrix addition, which is defined in Example 3.5.

5.17. (For those who have studied linear algebra.) Verify that the set of all invertible (nonsingular) 2×2 matrices with real numbers as entries forms a group with respect to matrix multiplication.

5.18. Verify the associative law for the operation $*$ in Example 5.5. (Notice that each time the identity is involved there is really no problem. See Problem 3.27.)

5.19. For Example 5.8, verify the claim that the inverse of $\alpha_{a,b}$ is $\alpha_{a^{-1}, -a^{-1}b}$.

5.20. If $\{a, b\}$ with operation $*$ is to be a group, with a the identity element, then what must the Cayley table be?

5.21. If $\{x, y, z\}$ with operation $*$ is to be a group, with x the identity element, then what must the Cayley table be?

5.22. Prove that if G is a group, $a \in G$, and $a * b = b$ for some $b \in G$, then a must be the identity element of G.

5.23. There are four assumptions in Theorem 5.1(a):

$$e * a = a \text{ for each } a \in G, \qquad f * a = a \text{ for each } a \in G,$$
$$a * e = a \text{ for each } a \in G, \qquad a * f = a \text{ for each } a \in G,$$

The proof of Theorem 5.1(a) uses only two of these assumptions. Which two? Which of the three axioms for a group are used?

5.24. Formulate questions like those in Problem 5.23 with Theorem 5.1(b) in place of Theorem 5.1(a). Answer them.

SECTION 6   PERMUTATIONS

A *permutation* of a nonempty set S is a one-to-one mapping from S onto S. Because a mapping from S to S is one-to-one and onto iff it is invertible, the permutations of S are the same as the invertible elements in $M(S)$. We observed in Example 5.6 that with composition as the operation these invertible elements form a group. Such groups are of sufficient interest for us to state this as a theorem.

Theorem 6.1.   *The set of all permutations of a nonempty set S is a group with respect to composition. This group is called the* symmetric group *on S, and will be denoted Sym(S).*

   Any group whose elements are permutations, with composition as the operation, is called a *permutation group*; or, if we want to specify the underlying set S, a *permutation group on S*. Therefore Sym(S) is a permutation group on S. But, in general, a permutation group on S need not contain *all* of the permutations of S. Examples are the group of rotations of a plane about a fixed point p (Example 5.7), and the group of mappings $\alpha_{a,b} : \mathbb{R} \to \mathbb{R}$ ($a \neq 0$) (Example 5.8). We shall see that permutation groups have a number of important applications. At the moment, we concentrate on some elementary facts about the groups of the form Sym(S), especially for S finite.

   When S is the set $\{1, 2, \ldots, n\}$, consisting of the first n positive integers, the group Sym(S) is commonly denoted S_n. An element α of S_n can be conveniently represented in *two-row form* as follows. First write $1, 2, \ldots, n$, and then below each number k write its image $\alpha(k)$. Thus

$$\begin{pmatrix} 1 & 2 & 3 & 4 \\ 2 & 4 & 3 & 1 \end{pmatrix}$$

represents the permutation in S_4 defined by $1 \mapsto 2, 2 \mapsto 4, 3 \mapsto 3$, and $4 \mapsto 1$. The identity element of S_n is

$$\begin{pmatrix} 1 & 2 & \cdots & n \\ 1 & 2 & \cdots & n \end{pmatrix}.$$

The inverse of an element is obtained by reading from the bottom entry to the top entry rather than from top to bottom: if 1 appears beneath 4 in α, then 4 will appear beneath 1 in α^{-1}. Thus

$$\begin{pmatrix} 1 & 2 & 3 & 4 \\ 2 & 4 & 3 & 1 \end{pmatrix}^{-1} = \begin{pmatrix} 1 & 2 & 3 & 4 \\ 4 & 1 & 3 & 2 \end{pmatrix}.$$

In composing permutations we always follow the same convention we use in composing any other mappings: read from right to left. Thus

$$\begin{pmatrix} 1 & 2 & 3 & 4 \\ 2 & 4 & 1 & 3 \end{pmatrix} \circ \begin{pmatrix} 1 & 2 & 3 & 4 \\ 3 & 4 & 1 & 2 \end{pmatrix} = \begin{pmatrix} 1 & 2 & 3 & 4 \\ 1 & 3 & 2 & 4 \end{pmatrix}$$

but

$$\begin{pmatrix} 1 & 2 & 3 & 4 \\ 3 & 4 & 1 & 2 \end{pmatrix} \circ \begin{pmatrix} 1 & 2 & 3 & 4 \\ 2 & 4 & 1 & 3 \end{pmatrix} = \begin{pmatrix} 1 & 2 & 3 & 4 \\ 4 & 2 & 3 & 1 \end{pmatrix}.$$

Warning: Some authors compose permutations from left to right. One must check in each case to see what convention is being followed.

Example 6.1.   The elements of S_3 are

$$\begin{pmatrix} 1 & 2 & 3 \\ 1 & 2 & 3 \end{pmatrix} \quad \begin{pmatrix} 1 & 2 & 3 \\ 2 & 3 & 1 \end{pmatrix} \quad \begin{pmatrix} 1 & 2 & 3 \\ 3 & 1 & 2 \end{pmatrix}$$

$$\begin{pmatrix} 1 & 2 & 3 \\ 2 & 1 & 3 \end{pmatrix} \quad \begin{pmatrix} 1 & 2 & 3 \\ 3 & 2 & 1 \end{pmatrix} \quad \begin{pmatrix} 1 & 2 & 3 \\ 1 & 3 & 2 \end{pmatrix}.$$

Here are two representative computations.

$$\begin{pmatrix} 1 & 2 & 3 \\ 3 & 1 & 2 \end{pmatrix} \circ \begin{pmatrix} 1 & 2 & 3 \\ 3 & 2 & 1 \end{pmatrix} = \begin{pmatrix} 1 & 2 & 3 \\ 2 & 1 & 3 \end{pmatrix}$$

$$\begin{pmatrix} 1 & 2 & 3 \\ 2 & 3 & 1 \end{pmatrix}^{-1} = \begin{pmatrix} 1 & 2 & 3 \\ 3 & 1 & 2 \end{pmatrix} \qquad \blacksquare$$

The group S_1 has order 1; S_2 has order 2; and, as we have just seen, S_3 has order 6. To give a general formula for the order of S_n, we first recall that if n is a positive integer, then $n!$ (read *n factorial*) is defined by $n! = 1 \cdot 2 \cdots \cdot n$, the product of all positive integers up to and including n. Thus $1! = 1, 2! = 2, 3! = 6, 5! = 120, 20! = 2,432,902,008,176,640,000$.

Theorem 6.2. *The order of S_n is $n!$.*

PROOF. The problem of computing the number of elements in S_n is the same as that of computing the number of different ways the integers $1, 2, \ldots, n$ can be placed in the n blanks indicated (using each integer just once):

$$\begin{pmatrix} 1 & 2 & \cdots & n \\ \underline{} & \underline{} & \cdots & \underline{} \end{pmatrix}.$$

If we begin filling these blanks from the left, there are n possibilities for the first blank. Once that choice has been made, there remain $n - 1$ possibilities for the second blank. Then there are $n - 2$ possibilities for the third blank. And so on. The theorem follows by repeated application of this basic counting principle: If one thing can be done in r different ways, and after that a second thing can be done in s different ways, then the two things can be done together in rs different ways. ∎

A group is said to be *Abelian* if its operation is commutative ($a * b = b * a$ for all a, b). This is in honor of the Norwegian mathematician Niels Henrik Abel (1802–1829), whose contributions will be discussed in Chapter XI. *Non-Abelian* means not Abelian. Here is one more fact about the groups S_n.

Theorem 6.3. *If $n \geq 3$, then S_n is non-Abelian.*

PROOF. If α and β in S_n are defined by

$$\alpha = \begin{pmatrix} 1 & 2 & 3 & \cdots \\ 1 & 3 & 2 & \cdots \end{pmatrix} \quad \text{and} \quad \beta = \begin{pmatrix} 1 & 2 & 3 & \cdots \\ 3 & 2 & 1 & \cdots \end{pmatrix},$$

with each number after 3 mapped to itself in each case, then

$$\beta \circ \alpha = \begin{pmatrix} 1 & 2 & 3 & \cdots \\ 3 & 1 & 2 & \cdots \end{pmatrix} \quad \text{but} \quad \alpha \circ \beta = \begin{pmatrix} 1 & 2 & 3 & \cdots \\ 2 & 3 & 1 & \cdots \end{pmatrix}.$$

Thus $\beta \circ \alpha \neq \alpha \circ \beta$, and the group is non-Abelian. ∎

It is easily seen that S_1 and S_2 are Abelian. Indeed, if S is *any* set containing only 1 or 2 elements, then Sym(S) is Abelian. On the other hand, if S contains more than 2 elements, then Sym(S) is non-Abelian (Problem 6.11).

Elements of S_n are frequently written using *cycle notation*. If S is a set, and $a_1, a_2, \ldots, a_k \in S$, then $(a_1 a_2 \cdots a_k)$ denotes the permutation of S for which

$$a_1 \mapsto a_2, \quad a_2 \mapsto a_3, \quad \cdots, \quad a_{k-1} \mapsto a_k, \quad a_k \mapsto a_1,$$

and

$$x \mapsto x \quad \text{for all other } x \in S.$$

Such a permutation is called a *cycle*, or a *k-cycle*. If a is any element of S, then the 1-cycle (a) is the identity permutation of S. Cycles are composed just as any other permutations (except that the symbol $\circ$ is usually omitted). We shall occasionally follow the common practice of referring to a composition of cycles, or of other permutations, as a *product*. If necessary for clarity, $(a_1, a_2, \ldots, a_k)$ can be written in place of $(a_1 a_2 \cdots a_k)$.

Example 6.2. In checking the following equations, remember that, in this book, in composing permutations the one on the right is applied first.

$$(1 \quad 2 \quad 4) = \begin{pmatrix} 1 & 2 & 3 & 4 & 5 \\ 2 & 4 & 3 & 1 & 5 \end{pmatrix}$$

$$(1 \quad 2 \quad 4)(3 \quad 5) = \begin{pmatrix} 1 & 2 & 3 & 4 & 5 \\ 2 & 4 & 3 & 1 & 5 \end{pmatrix} \circ \begin{pmatrix} 1 & 2 & 3 & 4 & 5 \\ 1 & 2 & 5 & 4 & 3 \end{pmatrix}$$

$$= \begin{pmatrix} 1 & 2 & 3 & 4 & 5 \\ 2 & 4 & 5 & 1 & 3 \end{pmatrix}$$

$$(1) = (2) = (3) = (1)(2)(3)$$

$$(1 \quad 2 \quad 4)(3 \quad 4) = (1 \quad 2 \quad 4 \quad 3)$$

$$(3 \quad 4)(1 \quad 2 \quad 4) = (1 \quad 2 \quad 3 \quad 4)$$

$$(1 \quad 2 \quad 4)(3)(5) = (1 \quad 2 \quad 4)$$

$$(1 \quad 2 \quad 3 \quad 4) = (2 \quad 3 \quad 4 \quad 1) = (3 \quad 4 \quad 1 \quad 2) = (4 \quad 1 \quad 2 \quad 3)$$

Notice that $(1 \quad 2 \quad 4)$ can denote an element of S_n for any $n \geq 4$. For instance, in S_4 it is the same as $(1 \quad 2 \quad 4)(3)$, and in S_5 it is the same as $(1 \quad 2 \quad 4)(3)(5)$. This ambiguity seldom causes trouble. ∎

Cycles $(a_1 \, a_2 \cdots a_m)$ and $(b_1 b_2 \cdots b_n)$ are *disjoint* if $a_i \neq b_j$ for all i, j. For example, $(1 \quad 2 \quad 4)$ and $(3 \quad 5 \quad 6)$ are disjoint, but $(1 \quad 2 \quad 4)$ and $(3 \quad 4 \quad 6)$ are not. Disjoint cycles commute; that is, if α and β represent disjoint cycles, then $\alpha\beta = \beta\alpha$ (Problem 6.15).

> Any permutation of a finite set is either a cycle or can be written as a product of pairwise disjoint cycles; and, except for the order in which the cycles are written, and the inclusion or omission of 1-cycles, this can be done in only one way.

We shall omit the proof of that statement, but it is illustrated in the following example. In each case, we simply start the first cycle with 1, continue until we get back to 1, and then close the first cycle. Then start the second cycle with the smallest number not in the first cycle, continue until we get back to that number, and then close the second cycle. And so on, never repeating a number that has already appeared. Other facts about cycles are given in the problems.

Example 6.3. In each of the following equations the cycles on the right are pairwise disjoint.

$$\begin{pmatrix} 1 & 2 & 3 & 4 & 5 \\ 3 & 4 & 1 & 5 & 2 \end{pmatrix} = (1 \quad 3)(2 \quad 4 \quad 5) = (2 \quad 4 \quad 5)(1 \quad 3)$$

$$(1 \quad 4 \quad 5)(2 \quad 3 \quad 5) = (1 \quad 4 \quad 5 \quad 2 \quad 3)$$

$$(2 \quad 4)(1 \quad 3 \quad 2)(2 \quad 5 \quad 4) = (1 \quad 3 \quad 4)(2 \quad 5) = (2 \quad 5)(1 \quad 3 \quad 4)$$

$$(1 \quad 6)(1 \quad 5)(1 \quad 4)(1 \quad 3)(1 \quad 2) = (1 \quad 2 \quad 3 \quad 4 \quad 5 \quad 6)$$

$$(1 \quad 2 \quad 3 \quad 4)^{-1} = (4 \quad 3 \quad 2 \quad 1) = (1 \quad 4 \quad 3 \quad 2)$$

$$(1 \quad 5 \quad 4 \quad 6 \quad 3 \quad 2)(4 \quad 3 \quad 6)(2 \quad 5) = (1 \quad 5)(2 \quad 4)(3)(6)$$

$$= (1 \quad 5)(2 \quad 4) \qquad ∎$$

Example 6.4. Table 6.1 is the Cayley table for S_3. Here is a reminder of two of our conventions: across from α and below β is $\alpha \circ \beta$, which is the permutation obtained by first applying β and then α.

Table 6.1

	(1)	(1 2 3)	(1 3 2)	(1 2)	(1 3)	(2 3)
(1)	(1)	(1 2 3)	(1 3 2)	(1 2)	(1 3)	(2 3)
(1 2 3)	(1 2 3)	(1 3 2)	(1)	(1 3)	(2 3)	(1 2)
(1 3 2)	(1 3 2)	(1)	(1 2 3)	(2 3)	(1 2)	(1 3)
(1 2)	(1 2)	(2 3)	(1 3)	(1)	(1 3 2)	(1 2 3)
(1 3)	(1 3)	(1 2)	(2 3)	(1 2 3)	(1)	(1 3 2)
(2 3)	(2 3)	(1 3)	(1 2)	(1 3 2)	(1 2 3)	(1)

■

PROBLEMS

6.1. Assume $\alpha = \begin{pmatrix} 1 & 2 & 3 & 4 \\ 1 & 4 & 3 & 2 \end{pmatrix}$ and $\beta = \begin{pmatrix} 1 & 2 & 3 & 4 \\ 3 & 1 & 4 & 2 \end{pmatrix}$.

Compute each of the following.
(a) $\beta \circ \alpha$ (b) $\alpha \circ \beta$ (c) α^{-1} (d) β^{-1}
(e) $\beta^{-1} \circ \alpha^{-1}$ (f) $\alpha^{-1} \circ \beta^{-1}$ (g) $(\beta \circ \alpha)^{-1}$ (h) $(\alpha \circ \beta)^{-1}$

6.2. Repeat Problem 6.1 using $\alpha = \begin{pmatrix} 1 & 2 & 3 & 4 \\ 3 & 4 & 1 & 2 \end{pmatrix}$ and $\beta = \begin{pmatrix} 1 & 2 & 3 & 4 \\ 4 & 3 & 1 & 2 \end{pmatrix}$.

6.3. Write each of the following as a single cycle or a product of disjoint cycles.
(a) $\begin{pmatrix} 1 & 2 & 3 & 4 & 5 & 6 \\ 3 & 5 & 6 & 4 & 2 & 1 \end{pmatrix}$ (b) $(1 \ \ 2)(1 \ \ 3)(1 \ \ 4)$
(c) $(1 \ \ 3)^{-1}(2 \ \ 4)(2 \ \ 3 \ \ 5)^{-1}$ (d) $(1 \ \ 4 \ \ 5)(1 \ \ 2 \ \ 3 \ \ 5)(1 \ \ 3)$

6.4. Write each of the following as a single cycle or a product of disjoint cycles.
(a) $\begin{pmatrix} 1 & 2 & 3 & 4 & 5 & 6 \\ 1 & 6 & 4 & 5 & 3 & 2 \end{pmatrix}$ (b) $(1 \ \ 4)(1 \ \ 3)(1 \ \ 2)$
(c) $(1 \ \ 2 \ \ 3)^{-1}(2 \ \ 3)(1 \ \ 2 \ \ 3)$ (d) $(2 \ \ 4 \ \ 5)(1 \ \ 3 \ \ 5 \ \ 4)(1 \ \ 2 \ \ 5)$

6.5. (a) Write all of the elements of S_4 both in two-row form and using cycle notation.
(b) Which elements of S_4 are their own inverse?

6.6. (a) How many elements of S_3 map 3 to 3?
(b) How many elements of S_n map n to n?

6.7. (a) Write $(a_1 a_2 \cdots a_k)^{-1}$ in cycle notation (without the symbol for inverse).
(b) For which values of k will every k-cycle be its own inverse?

6.8. Rewrite the proof of Theorem 6.3 using cycle notation.

6.9. Show that every element of S_n is a 2-cycle or can be written as a product of 2-cycles. [*Suggestion*: $(a_1 a_2 \cdots a_k) = (a_1 a_k) \cdots (a_1 a_3)(a_1 a_2)$. Two-cycles, which we'll return to in Section 7, are also called *transpositions*.]

6.10. Complete: If G (with operation $*$) is a group, then G is non-Abelian iff $a * b \neq b * a$ ....

6.11. Prove that if S contains at least three elements, then $\text{Sym}(S)$ is non-Abelian. (The main idea is already in the proof of Theorem 6.3.)

6.12. By Problem 4.2, if $|S| = 2$, then composition, as an operation on $M(S)$, is not commutative. However, $\text{Sym}(S)$ is Abelian. Explain.

6.13. Prove that the group in Example 5.8 is non-Abelian.

6.14. Consider the problem of completing the following Cayley table.

$*$	u	v	w	x	y	z
u						
v						
w						
x						
y						
z						

(a) In how many ways can the table be completed (so that $*$ is an operation)?
(b) Same as (a), but with u required to be an identity element.
(c) Make use of the Cayley table for S_3 (Example 6.4) to complete this table in such a way as to obtain a group. [Consider replacement of (1) by u, (1 2 3) by v, and so on.]

6.15. Assume that α and β are disjoint cycles representing elements of S_n, say $\alpha = (a_1 a_2 \cdots a_s)$ and $\beta = (b_1 b_2 \cdots b_t)$ with $a_i \neq b_j$ for all i and j.
(a) Compute $(\alpha \circ \beta)(a_k)$ and $(\beta \circ \alpha)(a_k)$ for $1 \leq k \leq s$. [Here, $(\alpha \circ \beta)(a_k)$ denotes the image of a_k under the mapping $\alpha \circ \beta$; that is, (a_k) is not a 1-cycle.]
(b) Compute $(\alpha \circ \beta)(b_k)$ and $(\beta \circ \alpha)(b_k)$ for $1 \leq k \leq t$.
(c) Compute $(\alpha \circ \beta)(m)$ and $(\beta \circ \alpha)(m)$ for $1 \leq m \leq n$ with $m \neq a_i$ and $m \neq b_j$ for all i and j.
(d) What do parts (a), (b), and (c), taken together, prove about the relationship between $\alpha \circ \beta$ and $\beta \circ \alpha$?

SECTION 7   SUBGROUPS

The set of even integers is a subset of the set of all integers, and both sets are groups with respect to addition. Thus the even integers form a subgroup of the group of all integers, according to the following definition.

Definition.   A subset H of a group G is a *subgroup* of G if H is itself a group with respect to the operation on G.

Notice that if G is a group with operation $*$, H is a subgroup of G, and $a, b \in H$, then $a * b \in H$. That is, H must be closed with respect to the operation $*$. In particular, $a * a \in H$ for each $a \in H$.

Example 7.1.
   (a)   The group of integers with addition is a subgroup of the group of real numbers with addition.

(b)   With multiplication, $\{1, -1\}$ is a subgroup of the group of nonzero real numbers.

(c)   Any group is a subgroup of itself.

(d)   If e is the identity of a group G, then $\{e\}$ is a subgroup of G. ∎

Theorem 7.1 will provide a convenient way to decide whether a subset of a group is a subgroup. But first we need the following preliminary result.

Lemma 7.1.   Let G be a group with operation $*$, and let H be a subgroup of G.

(a)   If f is the identity of H and e is the identity of G, then $f = e$.

(b)   If $a \in H$, then the inverse of a in H is the same as the inverse of a in G.

PROOF.   (a) If f is the identity of H, then $f * f = f$. Therefore, if f^{-1} denotes the inverse of f in G, then

$$f^{-1} * (f * f) = f^{-1} * f$$
$$\left(f^{-1} * f\right) * f = e$$
$$e * f = e$$
$$f = e.$$

(b) Assume $a \in H$. Let a^{-1} denote the inverse of a in G and let c denote the inverse of a in H. Then $a * c = c * a = f$, so $a * c = c * a = e$ by part (a) of the proof. However, Theorem 5.1(b) implies that a^{-1} is the unique element x in G satisfying $a * x = x * a = e$. Therefore, $c = a^{-1}$. ∎

Theorem 7.1.   *Let G be a group with operation $*$, and let H be a subset of G. Then H is a subgroup of G iff*

(a)   *H is nonempty,*

(b)   *if $a \in H$ and $b \in H$, then $a * b \in H$, and*

(c)   *if $a \in H$, then $a^{-1} \in H$.*

PROOF.   Assume H to be a subgroup. Then, being a group, it must contain at least an identity element and thus be nonempty, confirming condition (a). The necessity of closure for a group, condition (b), has already been pointed out. Now consider condition (c). If $a \in H$, then a must have an inverse in the set H. By Lemma 7.1(b), this inverse is a^{-1}, the inverse of a in G. Thus $a^{-1} \in H$. We have now proved that if H is a subgroup, then conditions (a), (b), and (c) must be satisfied.

Assume now that H is a subset satisfying (a), (b), and (c). To verify that H is a group we shall verify that with respect to $*$, H satisfies the conditions in the definition of a group in Section 5. Property (b) ensures that $*$ is an operation on H. The associative law is satisfied automatically: If $a * (b * c) = (a * b) * c$ is true for all elements in G, then it is certainly true for all elements in H, a subset of G. To show that H contains e, the identity element of G, let x denote any element of H; there is such an element by condition (a). By condition (c), $x^{-1} \in H$. Therefore, by condition (b), $e = x * x^{-1} \in H$. Thus H is a subgroup. ∎

Problem 7.22 contains a variation on Theorem 7.1, showing how (b) and (c) can be combined into a single condition. If H is known to be a finite set, then condition (c) of Theorem 7.1 can be omitted altogether (Problem 14.35).

Example 7.2.   If k is an integer, the set of all integral multiples of k satisfies the conditions of Theorem 7.1 and is therefore a subgroup of $\mathbb{Z}$ (with respect to $+$). In this case, the inverse of an element is the negative of the element. The special case $k = 2$ gives the subgroup of all even integers.   ∎

Example 7.3.   Table 7.1 shows that if $H = \{(1), (1 \quad 2 \quad 3), (1 \quad 3 \quad 2)\}$, then H is a subgroup of S_3. Checking the conditions of Theorem 7.1, we see first that H is nonempty. Closure is fulfilled because nothing appears in the table except (1), $(1 \quad 2 \quad 3)$, and $(1 \quad 3 \quad 2)$. And condition (c) is satisfied because $(1)^{-1} = (1)$, $(1 \quad 2 \quad 3)^{-1} = (1 \quad 3 \quad 2)$, and $(1 \quad 3 \quad 2)^{-1} = (1 \quad 2 \quad 3)$.

Table 7.1

	(1)	(1 2 3)	(1 3 2)
(1)	(1)	(1 2 3)	(1 3 2)
(1 2 3)	(1 2 3)	(1 3 2)	(1)
(1 3 2)	(1 3 2)	(1)	(1 2 3)

∎

We now consider a general class of subgroups of which Example 7.3 is a special case. First, we define a *transposition* to be a 2-cycle in S_n (for any n). For example, the transpositions in S_3 are $(1 \quad 2)$, $(1 \quad 3)$, and $(2 \quad 3)$. It can be verified that every element of S_n is a transposition or a product of transpositions (Problem 6.9). For instance, $(1 \quad 2 \quad 3) = (1 \quad 3)(1 \quad 2)$.

When we write an element of S_n as a product of transpositions we may need either an even number of factors, as in

$$(1 \quad 2 \quad 3 \quad 4 \quad 5) = (1 \quad 5)(1 \quad 4)(1 \quad 3)(1 \quad 2),$$

or an odd number of factors, as in

$$(1 \quad 2 \quad 5)(3 \quad 4) = (1 \quad 5)(1 \quad 2)(3 \quad 4).$$

We define a permutation to be *even* or *odd* according to whether it can be written as a product of an even or an odd number of transpositions, respectively. Thus the preceding equations show that $(1 \quad 2 \quad 3 \quad 4 \quad 5)$ is even and $(1 \quad 2 \quad 5)(3 \quad 4)$ is odd.

We must be careful of one thing with this definition. A permutation can be written as a product of transpositions in more than one way, such as $(1 \quad 2 \quad 3) = (1 \quad 3)(1 \quad 2) = (2 \quad 3)(1 \quad 2)(1 \quad 3)(2 \quad 3)$. It must be proved that the number of transpositions needed is necessarily either even or odd, depending *only* on the given permutation. For example, one of our representations of $(1 \quad 2 \quad 3)$ has two factors and the other has four; any other representation of $(1 \quad 2 \quad 3)$ as a product of transpositions also will have an even number of factors. We shall not prove this, but a proof can be found in many books on modern algebra or in almost any book on basic group theory. Thus both of the terms "even" and "odd" are well-defined for permutations. We can now state the following result.

Theorem 7.2.   *The set of all even permutations in S_n forms a subgroup of S_n for each $n \geq 2$. This subgroup is called the* alternating group *of degree n, and will be denoted by A_n. The order of A_n is $\frac{1}{2}(n!)$.*

PROOF.   Use Theorem 7.1. The identity permutation is in A_n because $(1) = (12)(12)$. If $a, b \in A_n$, then $ab \in A_n$ because a product of an even number of transpositions and an even number of transpositions will give an even number of transpositions. Finally, the inverse of $a = (a_1 a_2)(a_3 a_4) \cdots (a_{k-1} a_k)$ is $a^{-1} = (a_{k-1} a_k) \cdots (a_3 a_4)(a_1 a_2)$, so that a^{-1} can be written using the same number of transpositions as a. Thus $a \in A_n$ implies $a^{-1} \in A_n$.

To prove that the order of A_n is $\frac{1}{2}(n!)$, it suffices to prove that S_n has the same number of even permutations as odd permutations, since S_n has order $n!$. To do this, it suffices to prove that the mapping $\theta \colon A_n \to S_n$ defined by $\theta(a) = a(1 \quad 2)$ is one-to-one and that $\theta(A_n)$ is the set of all odd permutations in S_n. This is left to Problem 7.7.   ∎

The group H in Example 7.3 is A_3. Problem 7.8 asks you to find the elements in A_4.

We close this section with two other types of subgroups of permutation groups. One type will be used in Section 8 in studying symmetry, and the other will be used in Chapter XIII in applications to combinatorics.

Assume that G is a permutation group on a set S, and that T is a subset of S. Let

$$G_T = \{\alpha \in G : \alpha(t) = t \quad \text{for each} \quad t \in T\}. \tag{7.1}$$

We say that the elements of G_T leave T *elementwise invariant.*

Example 7.4.   Let $S = \{1, 2, 3, 4\}$, $G = \mathrm{Sym}(S) = S_4$, and $T = \{1, 2\}$. Then

$$G_T = \{(1)(2)(3)(4), (1)(2)(3 \quad 4)\} = \{(1), (3 \quad 4)\}$$

(see Figure 7.1). This is a subgroup of G.

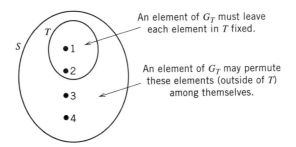

An element of G_T must leave each element in T fixed.

An element of G_T may permute these elements (outside of T) among themselves.

Figure 7.1   ∎

Just as in the example, G_T is always a subgroup. Before proving that, however, we introduce another subset of G closely related to G_T. If α is a permutation of S, and T is a subset of S, then $\alpha(T)$ denotes the set of all elements $\alpha(t)$ for $t \in T$ Let

$$G_{(T)} = \{\alpha \in G : \alpha(T) = T\}. \tag{7.2}$$

Thus, if $\alpha \in G_{(T)}$ then α may permute the elements of T among themselves, but it sends no element of T outside T: We say that the elements of $G_{(T)}$ leave T *invariant.*

Example 7.5.   With S, G, and T as in Example 7.4,

$$G_{(T)} = \{(1)(2)(3)(4), (1 \quad 2)(3)(4), (1)(2)(3 \quad 4), (1 \quad 2)(3 \quad 4)\}$$
$$= \{(1), (1 \quad 2), (3 \quad 4), (1 \quad 2)(3 \quad 4)\}$$

(see Figure 7.2). This also is a subgroup of S_4.

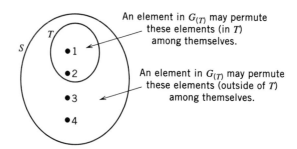

An element in $G_{(T)}$ may permute these elements (in T) among themselves.

An element in $G_{(T)}$ may permute these elements (outside of T) among themselves.

Figure 7.2   ∎

Theorem 7.3.   *If G is a permutation group on S, and T is a subset of S, then G_T and $G_{(T)}$ are subgroups of G. Also, G_T is a subgroup of $G_{(T)}$.*

PROOF.   Apply Theorem 7.1, first to G_T. Since ι, the identity mapping of S, is in G_T, the set G_T is nonempty. If $\alpha, \beta \in G_T$, then

$$(\alpha \circ \beta)(t) = \alpha(\beta(t)) = \alpha(t) = t$$

for each $t \in T$ so $\alpha \circ \beta \in G_T$. Finally, if $\alpha \in G_T$ and $t \in T$ then

$$\alpha(t) = t$$
$$\alpha^{-1}(\alpha(t)) = \alpha^{-1}(t)$$
$$(\alpha^{-1} \circ \alpha)(t) = \alpha^{-1}(t)$$
$$t = \alpha^{-1}(t)$$

so $\alpha^{-1} \in G_T$. The proof that $G_{(T)}$ is a group is similar; simply replace t by T in the obvious places (Problem 7.15).

To prove that G_T is a subgroup of $G_{(T)}$, assume $\alpha \in G_T$. Then $\alpha(t) = t$ for each $t \in T$ so $\alpha(T) = T$. That is, $\alpha \in G_{(T)}$.   ∎

PROBLEMS

7.1. Decide in each case whether the given subset is a subgroup of S_4. Justify your answers.
  (a) $\{(1), (1\ \ 3\ \ 4), (1\ \ 4\ \ 3\}$
  (b) $\{(1), (1\ \ 2\ \ 3), (2\ \ 3\ \ 4)\}$
  (c) $\{(1), (1\ \ 2)(3\ \ 4)\}$
  (d) $\{(1), (1\ \ 2\ \ 3\ \ 4), (1\ \ 4\ \ 3\ \ 2)\}$

7.2. Decide in each case whether the given subset is a subgroup of S_5. Justify your answers.
  (a) $\{(1), (1\ \ 3\ \ 5), (1\ \ 5\ \ 3)\}$
  (b) $\{(1), (1\ \ 3), (2\ \ 4), (1\ \ 3)(2\ \ 4)\}$
  (c) $\{(1), (1\ \ 2\ \ 3\ \ 4\ \ 5), (1\ \ 5\ \ 4\ \ 3\ \ 2)\}$
  (d) $\{(1), (1\ \ 2\ \ 3), (1\ \ 3\ \ 2), (4\ \ 5), (1\ \ 2\ \ 3)(4\ \ 5), (1\ \ 3\ \ 2)(4\ \ 5)\}$

7.3. Let $S = \{1, 2, 3\}$ and $G = S_3$. Write all of the elements of G_T and $G_{(T)}$ in each case.
  (a) $T = \{1\}$
  (b) $T = \{2, 3\}$

7.4. Let $S = \{1, 2, 3, 4\}$ and $G = S_4$. Write all of the elements of G_T and $G_{(T)}$ in each case.
  (a) $T = \{1\}$
  (b) $T = \{1, 2, 3\}$

7.5. Verify that $\{\alpha_{a,0} : a \in \mathbb{R}, a \neq 0\}$ is a subgroup of the group in Example 5.8. Characterize the group in geometric terms. (See Example 4.2.)

7.6. Repeat Problem 7.5 using $\{\alpha_{1,b} : b \in \mathbb{R}\}$.

7.7. Prove the claims about θ in the proof of Theorem 7.2.

7.8. Find the elements in A_4.

7.9. Which of the following are subgroups of the group in Example 5.8?
(a) $\{\alpha_{a,0} : a \in \mathbb{Q}, a \neq 0\}$ (b) $\{\alpha_{a,0} : a \in \mathbb{Z}, a \neq 0\}$
(c) $\{\alpha_{1,b} : b \in \mathbb{Z}\}$ (d) $\{\alpha_{1,b} : b \in \mathbb{N}\}$

7.10. Let F denote $M(\mathbb{R})$, the set of all mappings $f: \mathbb{R} \to \mathbb{R}$, made into a group as in Problem 5.13. Let

$$H = \{f \in F : f(x) \in \mathbb{Z} \quad \text{for each} \quad x \in \mathbb{R}\}.$$

Prove that H is a subgroup of F. State clearly the properties of $\mathbb{Z}$ that you use.

7.11. Consider the group $M(2, \mathbb{Z})$, the set of all 2×2 matrices with integers as entries, with matrix addition as the operation (Problem 5.16).
(a) Prove that the set of all diagonal matrices (those with zeroes in the upper right-hand and lower left-hand corners) forms a subgroup.
(b) Find a subgroup of $M(2, \mathbb{Z})$ besides the group itself, the subgroup containing only the zero matrix [Example 7.1(d)], and the subgroup in part (a) of this problem.

7.12. Find a subgroup of $\mathbb{Q}$ (operation $+$) that contains $\mathbb{Z}$ but is different from both $\mathbb{Z}$ and $\mathbb{Q}$.

7.13. Prove that if H and K are subgroups of a group G (with operation $*$), then $H \cap K$ is a subgroup of G. (Compare Problem 7.14.)

7.14. Let $H = \{(1), (1\ \ 2)\}$ and $K = \{(1), (1\ \ 2\ \ 3), (1\ \ 3\ \ 2)\}$. Both H and K are subgroups of S_3. Show that $H \cup K$ is not a subgroup of S_3. It follows that a union of subgroups is not necessarily a subgroup. (Compare Problem 7.13. Also see Problem 14.38.)

7.15. Prove in detail that $G_{(T)}$ is a subgroup of G. (That is, complete the proof of Theorem 7.3.)

7.16. For $G = S_n$, state necessary and sufficient conditions on T, a subset of $S = \{1, 2, \ldots, n\}$, for $G_{(T)} = G_T$.

7.17. For a subset T of S, let T' denote the complement of T in S, that is,

$$T' = \{x : x \in S \quad \text{and} \quad x \notin T\}.$$

Prove that if $S = \{1, 2, \ldots, n\}$ and $G = S_n$, then $G_{(T)} = G_{(T')}$.

7.18. With S and G as in Problem 7.17, find necessary and sufficient conditions on n and T for $G_T = G_{T'}$.

7.19. With S and G as in Problem 7.17, find necessary and sufficient conditions on n and T for $G_T \subseteq G_{T'}$.

7.20. Let $S = \{1, 2, \ldots, n\}$, $T = \{1, 2, \ldots, k\}$ $(0 \leq k \leq n)$, and $G = S_n$.
(a) What is $|G_T|$ (the order of G_T)?   (b) What is $|G_{(T)}|$?

7.21. Prove that if G is a group with identity e, and $x \in G$ and $x * x = x$, then $x = e$. (*Question:* How would you prove this in $\mathbb{Z}$ with $+$ in place of $*$, and 0 in place of e?)

7.22. Prove that if G is a group with operation $*$, and H is a subset of G, then H is a subgroup of G iff
(a) H is nonempty, and
(b) if $a \in H$ and $b \in H$, then $a * b^{-1} \in H$.

7.23. Assume that G is a group with operation $*$ and that $a \in G$. Let

$$C(a) = \{x \in G: a * x = x * a\}.$$

Prove that $C(a)$ is a subgroup of G. [$C(a)$ is called the *centralizer* of a in G.]

7.24. Assume that G is a group with operation $*$ and let

$$Z(G) = \{x \in G: a * x = x * a \quad \text{for every } a \in G\}.$$

Prove that $Z(G)$ is a subgroup of G. [$Z(G)$ is called the *center* of G.]

7.25. Assume that G is a permutation group on a set S, and that T is a subset of S. Let

$$G_{[T]} = \{\alpha \in G: \alpha (T) \subseteq T\}.$$

Then $G_{(T)} \subseteq G_{[T]}$.

(a) Give an example of a permutation α of a set S and a subset T of S such that $\alpha(T) \subsetneqq T$. (Necessarily, S and T will be infinite.)

(b) Give an example to show that $G_{[T]}$ need not be a subgroup of G. (Compare Theorem 7.3.)

SECTION 8   GROUPS AND SYMMETRY

Much of the importance of groups comes from their connection with symmetry. Just as numbers can be used to measure size (once a unit of measurement has been chosen), groups can be used to measure symmetry. With each figure we associate a group, and this group characterizes the symmetry of the figure. This application of groups extends from geometry to crystallography, and is introduced in this section and then discussed more fully in Chapter XIV. Another connection with symmetry—more abstract and not geometrical—arises in the study of algebraic equations. A group is associated with each equation, and this group characterizes a type of symmetry involving the solutions of the equation; questions about the solvability of an equation can be answered by studying the group associated with the equation. This application is discussed in Chapter XI.

We now look at how to associate a group with each figure in a plane. Let P denote the set of all points in a plane, and let M denote the set of all permutations of P that preserve distance between points. Thus, if p and q are in P, and μ is in M, then the distance between $\mu(p)$ and $\mu(q)$ is equal to the distance between p and q. The permutations in M are called *motions* or *isometries* of the plane. We shall prove in Theorem 8.1 that M, with composition, is a group. But first we consider three types of motions: rotations, reflections (through lines), and translations.

Rotations   If p is a fixed point in P, then any rotation of the plane about p is a motion of the plane. (Rotations were discussed in Examples 4.1 and 5.7.)

Reflections   The reflection of the plane P through a line L in P is the mapping that sends each point p in P to the point q such that L is the perpendicular bisector of the segment pq (Figure 8.1).

Translations   A translation of P is a mapping that sends all points the same distance in the same direction. For instance, the translation sending p_1 to q_1 in Figure 8.2 would send p_2 to q_2 and p_3 to q_3.

A third type of motion of a plane, *glide-reflection*, is discussed in Section 54. In this section we need only rotations and reflections.

Figure 8.1

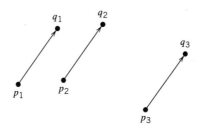

Figure 8.2

Theorem 8.1.   *The set M of all motions (isometries) of a plane P forms a subgroup of $\mathrm{Sym}(P)$.*

PROOF.   For $p, q \in P$ let $d(p, q)$ denote the distance between p and q. With this notation, if $\alpha \in \mathrm{Sym}(P)$ then

$$\alpha \in M \quad \text{iff} \quad d(\alpha(p), \alpha(q)) = d(p, q).$$

To prove that M is a subgroup we shall verify the conditions in Theorem 7.1. (Remember that the operation on $\mathrm{Sym}(P)$ is composition.)

First, ι, the identity of $\mathrm{Sym}(P)$, is clearly in M, so M is nonempty.

Assume $\alpha \in M$ and $\beta \in M$. If $p, q \in P$ then

$$
\begin{aligned}
d((\alpha \circ \beta)(p), (\alpha \circ \beta)(q)) & \\
= d(\alpha(\beta(p)), \alpha(\beta(q))) & \\
= d(\beta(p), \beta(q)) & \quad \text{since } \alpha \in M, \beta(p) \in P, \text{ and } \beta(q) \in P \\
= d(p, q) & \quad \text{since } \beta \in M.
\end{aligned}
$$

Thus $\alpha \circ \beta \in M$.

Finally, assume $\alpha \in M$ and $p, q \in P$ Then

$$
\begin{aligned}
d(p, q) = d(\iota(p), \iota(q)) & \quad \text{since } \iota \in M \\
= d(\alpha(\alpha^{-1}(p)), \alpha(\alpha^{-1}(q))) & \quad \text{since } \alpha \circ \alpha^{-1} = \iota \\
= d(\alpha^{-1}(p), \alpha^{-1}(q)) & \quad \text{since } \alpha \in M, \alpha^{-1}(p) \in P, \text{ and } \alpha^{-1}(q) \in P.
\end{aligned}
$$

Thus $\alpha^{-1} \in M$.   ∎

Now let T denote any subset of P. Since M is a permutation group on P [that is, a subgroup of $\mathrm{Sym}(P)$], we can use M in place of G in the defining equation for $G_{(T)}$ in Equation (7.2):

$$M_{(T)} = \{a \in M : \alpha(T) = T\}.$$

By Theorem 7.3, $M_{(T)}$ is a subgroup of M. Thus we can make the following definition.

Definition.   If T is a set of points in a plane, then $M_{(T)}$, the group of all motions leaving T invariant, is called the *group of symmetries* (or *symmetry group*) of T.[†]

Example 8.1.   Consider a square, a rectangle, and a parallelogram (Figure 8.3). Any motion of one of the figures will permute the vertices of the figure among themselves and the sides of the figure among themselves. Moreover, any motion will be completely determined by the way it permutes the vertices. It follows that in each case the group of symmetries will correspond to a subgroup of Sym$\{a, b, c, d\}$, and thus will have order at most $4! = 24$ (Theorem 6.2). In fact, the order must be less than 24, because some permutations of the vertices clearly cannot arise from motions of the plane (Problem 8.7). It turns out that the groups of the three figures have orders 8, 4, and 2, respectively. Their elements are listed below. The lines V (for vertical), H (for horizontal), and D_1 and D_2 (for diagonal) are not affected by the different motions. For example, rotation 90° clockwise around p will change the positions of a, b, c, and d, but not of V, H, D_1, and D_2. Notice that the more symmetric the figure, the larger its group of symmetries.

Group of symmetries of the square in Figure 8.3a

$$\mu_1 = \text{identity permutation}$$
$$\mu_2 = \text{rotation 90° clockwise around } p$$
$$\mu_3 = \text{rotation 180° clockwise around } p$$
$$\mu_4 = \text{rotation 270° clockwise around } p$$
$$\mu_5 = \text{reflection through } H$$
$$\mu_6 = \text{reflection through } V$$
$$\mu_7 = \text{reflection through } D_1$$
$$\mu_8 = \text{reflection through } D_2$$

Group of symmetries of the rectangle in Figure 8.3b

$$\mu_1 = \text{identity permutation}$$
$$\mu_3 = \text{rotation 180° clockwise around } p$$
$$\mu_5 = \text{reflection through } H$$
$$\mu_6 = \text{reflection through } V$$

Group of symmetries of the parallelogram in Figure 8.3c

$$\mu_1 = \text{identity permutation}$$
$$\mu_3 = \text{rotation 180° clockwise around } p$$

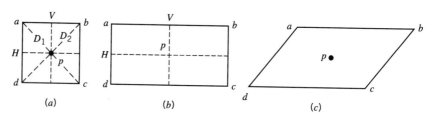

(a)   (b)   (c)

Figure 8.3

[†]Notice the difference between "symmetric" group, as used in Section 6, and "symmetry" group, as used here.

Figure 8.4 illustrates how to compute entries for the Cayley tables of these groups. It shows that the result of $\mu_5 \circ \mu_2$ is the same as μ_8, reflection through D_2; and $\mu_2 \circ \mu_5$ is the same as μ_7, reflection through D_1. Notice again that when we make such calculations we assume H, V, D_1, and D_2 to be fixed.

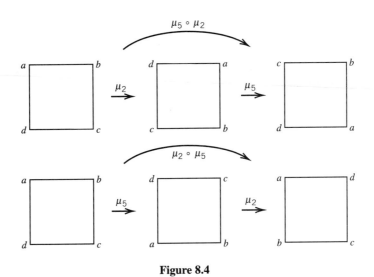

Figure 8.4   ■

Table 8.1 is the Cayley table for the group of the square, which has each of the other two groups as a subgroup.

Table 8.1

$\circ$	μ_1	μ_2	μ_3	μ_4	μ_5	μ_6	μ_7	μ_8
μ_1	μ_1	μ_2	μ_3	μ_4	μ_5	μ_6	μ_7	μ_8
μ_2	μ_2	μ_3	μ_4	μ_1	μ_7	μ_8	μ_6	μ_5
μ_3	μ_3	μ_4	μ_1	μ_2	μ_6	μ_5	μ_8	μ_7
μ_4	μ_4	μ_1	μ_2	μ_3	μ_8	μ_7	μ_5	μ_6
μ_5	μ_5	μ_8	μ_6	μ_7	μ_1	μ_3	μ_4	μ_2
μ_6	μ_6	μ_7	μ_5	μ_8	μ_3	μ_1	μ_2	μ_4
μ_7	μ_7	μ_5	μ_8	μ_6	μ_2	μ_4	μ_1	μ_3
μ_8	μ_8	μ_6	μ_7	μ_5	μ_4	μ_2	μ_3	μ_1

The symmetry groups for the figures in the following exercises contain only rotations and reflections. The amounts of the rotations may differ from those in Example 8.1, however.

PROBLEMS

8.1. Draw figures like those in Figure 8.4 to verify the entries for $\mu_3 \circ \mu_6$ and $\mu_6 \circ \mu_3$ in Table 8.1.

8.2. Draw figures like those in Figure 8.4 to verify the entries for $\mu_4 \circ \mu_8$ and $\mu_8 \circ \mu_4$ in Table 8.1.

8.3. Determine the group of symmetries of an equilateral triangle.

8.4. Determine the group of symmetries of an isosceles triangle.

8.5. Determine the group of symmetries of a regular pentagon. (It will have order 10.)

8.6. Determine the permutation of the vertices of the square $abcd$ corresponding to each motion μ_i, $(1 \le i \le 8)$ in Table 8.1. [Example: μ_2 corresponds to $(a \quad b \quad c \quad d)$.]

8.7. The permutation $(ab)(c)(d)$ of the vertices of the square $abcd$ (Figure 8.3a) does not correspond to any motion of the plane. Why?

8.8. Consider the mapping $T \mapsto M_{(T)}$ from the set of subsets of a plane to the set of symmetry groups. Is it one-to-one? Explain.

8.9. Using the notation of Example 8.1, determine the group of symmetries of a rhombus (Figure 8.3c, with $ab = ad$).

8.10. Considered as geometric objects, the 26 capital letters of the alphabet fall into five sets, with letters in each set having the same group of symmetries. Determine the five sets. (*Suggestion:* **A, B, N, H,** and **F** belong to different sets.)

Determine the group of symmetries of each of the following figures. (It suffices in each case to use motions similar to those in Example 8.1.)

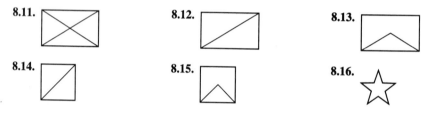

8.11.   **8.12.**   **8.13.**

8.14.   **8.15.**   **8.16.**

NOTES ON CHAPTER II

The origins of group theory can be found primarily in the theory of equations, number theory, and the study of geometrical transformations. The earliest connection was with the theory of equations in the late eighteenth and early nineteenth centuries, and will be discussed in Sections 46 and 47. The connection with number theory was related to work that will be discussed in Section 41. The symmetry groups in Section 8 are examples of geometrical transformation groups; although we shall return to symmetry in Chapter XIV, there are other kinds of transformation groups that we shall not be able to consider.

The following references discuss the history of group theory. G. A. Miller (1865–1951) was the first distinguished American group theorist.

1.   Kleiner, I., The evolution of group theory: A brief survey, *Mathematics Magazine*, **59** (1986), 195–215.
2.   Kline, M., *Mathematical Thought from Ancient to Modern Times*, Oxford University Press, London, 1990.
3.   *The Collected Works of G. A. Miller*, 5 vols., University of Illinois, Urbana, 1935–1959.
4.   Novy, L., *Origins of Modern Algebra*, Noordhoff, Leyden, The Netherlands, 1973.
5.   Wussing, H., *The Genesis of the Abstract Group Concept*, MIT Press, Cambridge, Mass., 1984.

For many more references, check the web site www-groups.dcs.st-and.ac.uk/~history/

CHAPTER III

EQUIVALENCE. CONGRUENCE. DIVISIBILITY

The first section in this chapter is devoted to equivalence relations, which occur often not only in algebra but throughout mathematics. The other sections are devoted to elementary facts about the integers. These facts are used to construct examples of groups and of other algebraic systems yet to be introduced. They will also help us understand some of the elementary facts about groups to be proved in the next chapter.

SECTION 9   EQUIVALENCE RELATIONS

Consider the following statements:

1. If $x, y \in \mathbb{R}$, then either

$$x = y \quad \text{or} \quad x \neq y.$$

2. If $x, y \in \mathbb{R}$, then either

$$x \leq y \quad \text{or} \quad x \nleq y.$$

3. If ABC and DEF are triangles, and $\cong$ denotes congruence, then either

$$ABC \cong DEF \quad \text{or} \quad ABC \ncong DEF.$$

In each statement, there is a set ($\mathbb{R}$, $\mathbb{R}$, and all triangles, respectively) and a relation on that set ($=$, $\leq$, and $\cong$, respectively). The relationship may or may not hold between ordered pairs of elements from the set. We are concerned now with relations of this type that satisfy three specific conditions. (The symbol $\sim$ in the following definition is read *tilde*.)

Definition.   A relation $\sim$ on a nonempty set S is an *equivalence relation* on S if it satisfies the following three properties:

If $a \in S$, then $a \sim a$.	*reflexive*
If $a, b \in S$ and $a \sim b$, then $b \sim a$.	*symmetric*
If $a, b, c \in S$ and $a \sim b$ and $b \sim c$, then $a \sim c$.	*transitive*

Of the relations in **1**, **2**, and **3**, the first and third are equivalence relations, but the second is not (because it is not symmetric). In the first relation, $\mathbb{R}$ can be replaced by any nonempty set, and the result will still be an equivalence relation; that is, for any nonempty set S, equality ($=$) is an equivalence relation on S.

If $\sim$ is an equivalence relation and $a \sim b$, we say that a and b are *equivalent*, or we use the specific term involved if there is one (such as equal or congruent).

Example 9.1.   Let p denote a fixed point in a plane P, and for points x and y in P let $x \sim y$ mean that x and y are equidistant from p. This is an equivalence relation on the set of points in P. A point x will be equivalent to a point q iff x lies on the circle through q with center p. (The point p is equivalent only to itself; think of $\{p\}$ as a circle with radius 0.)   ∎

Example 9.2.   Let L denote the set of all lines in a plane with rectangular coordinate system. For $l_1, l_2 \in L$, let $l_1 \sim l_2$ mean that l_1 and l_2 have equal slopes or that both slopes are undefined. This is an equivalence relation on L. The set of lines equivalent to a line l consists of l and all lines in L that are parallel to l.   ∎

Example 9.3.   Let $\alpha: S \to T$ be a mapping. For $x, y \in S$, let $x \sim y$ mean that $\alpha(x) = \alpha(y)$. This is an equivalence relation on S. Here are two special cases.

(a)   Let $S = \{u, v, w\}$, $T = \{1, 2, 3\}$, and define $\alpha: S \to T$ by

$$\alpha(u) = 3, \quad \alpha(v) = 1, \quad \text{and} \quad \alpha(w) = 3.$$

Then $u \sim w$ because $\alpha(u) = \alpha(w)$. But $u \nsim v$ because $\alpha(u) \neq \alpha(v)$. Here is a complete list of the equivalences between elements of S:

$$u \sim u, \quad v \sim v, \quad w \sim w, \quad u \sim w, \quad w \sim u$$

(b)   Let $S = T = \mathbb{R}$, and let α be the sine function. Then $x_1 \sim x_2$ iff $\sin x_1 = \sin x_2$. Thus, for example, the set of real numbers equivalent to π in this case is

$$\{x: \sin x = 0\} = \{0, \pm \pi, \pm 2\pi, \ldots\}.$$
   ∎

Now return to Example 9.1. A different way to define the equivalence relation in that example is to say that $x \sim y$ iff x and y lie on a common circle with center p. These circles form a partition of the set of points in the plane, in the sense of the following definition.

Definition.   A collection $\mathcal{P}$ of nonempty subsets of a nonempty set S forms a *partition* of S provided

(a) S is the union of the sets in $\mathcal{P}$, and
(b)   if A and B are in $\mathcal{P}$ and $A \neq B$, then $A \cap B = \varnothing$.

Alternatively, the collection $\mathcal{P}$ forms a partition of S if each element of S is contained in one and only one of the sets in $\mathcal{P}$. Notice that each element of $\mathcal{P}$ is a subset of S. Figure 9.1 shows an example.

We have observed that the equivalence relation in Example 9.1 induces a partition of the underlying set. In fact, every equivalence relation induces a partition and, conversely, every partition induces an equivalence relation. Before proving this we make the following definition.

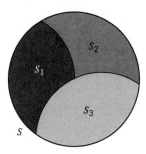

Figure 9.1 $\mathcal{P} = \{S_1, S_2, S_3\}$ forms a partition of S.

Definition.   Let $\sim$ be an equivalence relation on a set S, let $a \in S$, and let $[a] = \{x \in S : a \sim x\}$. This subset $[a]$ of S is called the *equivalence class* of a (relative to $\sim$).

In Example 9.1, the equivalence class of a point q is the set of all points on the circle with center at p and passing through q. In Example 9.3(a),

$$[u] = \{u, w\}, \quad [v] = \{v\}, \quad \text{and} \quad [w] = \{u, w\}.$$

Notice that it is always true that $a \in [a]$, because $\sim$ is reflexive. And if $b \in [a]$ then $a \in [b]$, because $\sim$ is symmetric.

Theorem 9.1.   *If $\sim$ is an equivalence relation on a set S, then the set of equivalence classes of $\sim$ forms a partition of S. Conversely, let $\mathcal{P}$ be a partition of S, and define a relation $\sim$ on S by $a \sim b$ iff there is a set in $\mathcal{P}$ that contains both a and b; then $\sim$ is an equivalence relation on S. Thus there is a natural one-to-one correspondence between the equivalence relations on a set and the partitions of the set.*

PROOF.   Let $\sim$ be an equivalence relation on S. If $a \in S$, then a belongs to at least one equivalence class, namely $[a]$, and thus S is indeed the union of the equivalence classes. It remains to be proved that if two equivalence classes are unequal then they are disjoint; or, alternatively, if they are not disjoint, then they are equal. To this end, assume that $[a] \cap [b] \neq \varnothing$, and let c denote an element in the intersection. If x denotes any element in $[a]$, then we have both $a \sim c$ and $a \sim x$; thus $c \sim a$ and $a \sim x$, so $c \sim x$. But we also know that $b \sim c$; hence we can conclude that $b \sim x$, that is, $x \in [b]$. This shows that $[a] \subseteq [b]$. In the same way, it can be shown that $[a] \supseteq [b]$. Therefore $[a] = [b]$, which is what we were to prove.

Now assume $\sim$ to be defined as in the converse statement. If $a \in S$, then $a \sim a$ because there is some set in the partition containing a. If there is a set containing both a and b, then it contains both b and a, so that the symmetry of $\sim$ is trivial. Finally, assume that $a \sim b$ and $b \sim c$. Then there is a set in $\mathcal{P}$ containing both a and b, call it A; there is also a set in $\mathcal{P}$ containing both b and c, call it B. Since $b \in A \cap B$, we have $A \cap B \neq \varnothing$; thus, $A = B$ because $\mathcal{P}$ is a partition. Both a and c belong to this set; therefore $a \sim c$. This proves that $\sim$ is transitive. ∎

Example 9.4.   Let E denote the set of even integers and O the set of odd integers. Then $\{E, O\}$ forms a partition of the set of all integers. For the corresponding equivalence relation, $a \sim b$ iff a and b are both even or both odd. Alternatively, $a \sim b$ iff $a - b$ is even. ■

In working with an equivalence relation on a set S, it is often useful to have a *complete set of equivalence class representatives*—that is, a subset of S containing precisely one element from each equivalence class.

Example 9.5.

(a) In Example 9.4, $\{0, 1\}$ is a complete set of equivalence class representatives. Each integer is equivalent to either 0 or 1, but no integer is equivalent to both. More generally, any set $\{a, b\}$ of integers with a even and b odd is a complete set of equivalence class representatives in this case.

(b) In Example 9.3, $\{u, v\}$ is a complete set of equivalence class representatives. Another complete set of equivalence class representatives in this case is $\{v, w\}$.

(c) In Example 9.1, the set of all points on any ray (half-line) with endpoint p is a complete set of equivalence class representatives, because such a ray intersects each equivalence class (circle centered at p) in precisely one point. ■

The type of equivalence relation in the following theorem is important in applications to combinatorics. Chapter XIII will give a generalization and more examples.

Theorem 9.2.   *Let G be a permutation group on S and define a relation $\sim$ on S by*

$$a \sim b \quad \textit{iff} \quad \alpha(a) = b \quad \textit{for some } \alpha \in G$$

Then $\sim$ is an equivalence relation on S.

PROOF.   *Reflexive*: If $a \in S$, then $\iota(a) = a$, where ι is the identity element of G. Thus $a \sim a$.

Symmetric: If $a, b \in S$ and $a \sim b$, then there exists $\alpha \in G$ such that $\alpha(a) = b$. Since G is a group, $\alpha^{-1} \in G$. And $\alpha^{-1}(b) = a$, so $b \sim a$.

Transitive: Assume that $a, b, c \in S, a \sim b$, and $b \sim c$. Then $\alpha(a) = b$ and $\beta(b) = c$ for some $\alpha, \beta \in G$. Because G is a group, $\beta \circ \alpha \in G$. From $(\beta \circ \alpha)(a) = \beta(\alpha(a)) = \beta(b) = c$, we see that $a \sim c$. ■

Example 9.6.   Consider Theorem 9.2 for $S = \{1, 2, 3, 4, 5\}$ and G the group $\{(1), (1 \quad 2 \quad 5), (1 \quad 5 \quad 2)\}$. In this case, the equivalence classes are

$$\{1, 2, 5\}, \qquad \{3\}, \quad \text{and} \quad \{4\}.$$

A complete set of equivalence class representatives is $\{1, 3, 4\}$. ■

PROBLEMS

9.1. Assume that $S = \{w, x, y, z\}$ and that $w \sim y$ and $z \sim y$. Which of the following must also be true if $\sim$ is to be an equivalence relation on S?

(a) $y \sim y$ (b) $y \sim z$ (c) $w \sim z$ (d) $y \sim x$

9.2. Assume that $T = \{v, w, x, y, z\}$ and that $w \sim x$ and $x \sim y$. Which of the following must also be true if $\sim$ is to be an equivalence relation on T?

(a) $z \sim z$ (b) $x \sim w$ (c) $v \sim z$ (d) $y \sim w$

9.3. If $\mathcal{P} = \{\{1, 3\}, \{2\}, \{4, 5\}\}$, then $\mathcal{P}$ is a partition of $\{1, 2, 3, 4, 5\}$. For the corresponding equivalence relation $\sim$ (see Theorem 9.1), which of the following are true?

(a) $4 \sim 5$ (b) $3 \sim 3$ (c) $1 \sim 2$ (d) $5 \sim 1$

9.4. Repeat Problem 9.3 with $\mathcal{P} = \{\{1, 4, 5\}, \{2, 3\}\}$.

9.5. For points with coordinates (x_1, y_1) and (x_2, y_2) in a plane with rectangular coordinate system, let $(x_1, y_1) \sim (x_2, y_2)$ mean that $y_1 = y_2$.
(a) Prove that $\sim$ is an equivalence relation on the set of points in the plane. State clearly the properties of the relation $=$ on $\mathbb{R}$ that are used in the proof.
(b) Describe the equivalence classes geometrically.
(c) Give a complete set of equivalence class representatives.

9.6. For points (x_1, y_1) and (x_2, y_2) in a plane with rectangular coordinate system, let $(x_1, y_1) \sim (x_2, y_2)$ mean that either $x_1 = x_2$ or $y_1 = y_2$. Explain why $\sim$ is not an equivalence relation on the set of points in the plane.

9.7. Define a relation $\sim$ on $\mathbb{R}$ by

$$a \sim b \quad \text{iff} \quad |a| = |b|.$$

(a) Prove that $\sim$ is an equivalence relation on $\mathbb{R}$. State clearly the properties of the relation $=$ on $\mathbb{R}$ that are used in the proof.
(b) Give a complete set of equivalence class representatives.

9.8. Define a relation $\sim$ on the set $\mathbb{N}$ of natural numbers by

$$a \sim b \quad \text{iff} \quad a = b \cdot 10^k \quad \text{for some } k \in \mathbb{Z}.$$

(a) Prove that $\sim$ is an equivalence relation on $\mathbb{N}$. State clearly the properties of $\mathbb{Z}$ that are used in the proof.
(b) Give a complete set of equivalence class representatives.

9.9. For $x, y \in \mathbb{R}$, let $x \sim y$ mean that $xy > 0$. Which properties of an equivalence relation does $\sim$ satisfy? Answer the same question with $xy \geq 0$ in place of $xy > 0$.

9.10. Give a complete set of equivalence class representatives for the equivalence relation in Example 9.2.

9.11. Determine a complete set of equivalence class representatives for the equivalence relation induced on $\mathbb{R}$ by the sine function in Example 9.3(b). What does this have to do with the inverse sine function?

9.12. Repeat Problem 9.11 with the tangent function in place of the sine function.

9.13. (a) Find all of the partitions of $\{x, y, z\}$.
(b) How many different equivalence relations are there on a 3-element set? (Two equivalence relations are different if they induce different partitions.)

9.14. How many different equivalence relations are there on a 4-element set? (Compare Problem 9.13.)

9.15. For $x, y \in \mathbb{R}$, let $x \sim y$ mean that $x \sim y$ is an integer. Verify that $\sim$ is an equivalence relation. Describe the equivalence classes geometrically, with the elements of $\mathbb{R}$ identified with the points on a line in the usual way. Give a complete set of equivalence class representatives.

9.16. For points (x_1, y_1) and (x_2, y_2) in a plane with rectangular coordinate system, let $(x_1, y_1) \sim (x_2, y_2)$ mean that $x_1 - x_2$ is an integer.
(a) Prove that $\sim$ is an equivalence relation.

(b) Give a geometric description of the equivalence class to which $(0, 0)$ belongs.

(c) Give a complete set of equivalence class representatives.

9.17. Repeat Problem 9.16, but let $(x_1, y_1) \sim (x_2, y_2)$ mean that both $x_1 - x_2$ and $y_1 - y_2$ are integers.

9.18. For sets S and T, let $S \sim T$ mean that there is an invertible mapping of S onto T. Prove that $\sim$ is reflexive, symmetric, and transitive.

9.19. Consider the equivalence relation in Theorem 9.2. Find the equivalence classes and a complete set of equivalence class representatives in each of the following special cases.
(a) $S = \{1, 2, \ldots, n\}$ and $G = S_n$.
(b) $S = \{1, 2, 3, 4\}$ and $G = \{(1), (2 \ 3)\}$.
(c) $S = \{1, 2, 3, 4, 5\}$ and $G = \{(1), (1 \ 2), (1 \ 3), (2 \ 3), (1 \ 2 \ 3), (1 \ 3 \ 2)\}$.

9.20. For polynomials $f(x)$ and $g(x)$ with real coefficients, let $f(x) \sim g(x)$ mean that $f'(x) = g'(x)$ (where the primes denote derivatives). Prove that $\sim$ is an equivalence relation, and give a complete set of equivalence class representatives. (A polynomial with real coefficients is an expression of the form $a_0 + a_1 x + \cdots + a_n x^n$, where $a_0, a_1, \ldots, a_n \in \mathbb{R}$.)

9.21. Let $\sim$ be a relation on a set S. Complete each of the following statements.
(a) $\sim$ is not reflexive iff ....
(b) $\sim$ is not symmetric iff ....
(c) $\sim$ is not transitive iff ....

9.22. Find a flaw in the following "proof" that a relation on a set S is reflexive if it is both symmetric and transitive: Let $x \in S$. From $x \sim y$, by symmetry, we have $y \sim x$. By transitivity, $x \sim y$ and $y \sim x$ imply $x \sim x$. Therefore, $\sim$ is reflexive.

SECTION 10   CONGRUENCE. THE DIVISION ALGORITHM

We get an equivalence relation on the set of integers by agreeing that two integers are equivalent iff either both are even or both are odd (Example 9.4). Another way to say this is to say that two integers are equivalent iff their difference is even. The notion of congruence of integers generalizes this example. Congruence was first treated systematically at the beginning of the nineteenth century by the eminent German mathematician Carl Friedrich Gauss (1777–1855); it has played a crucial role in the theory of numbers ever since. We shall see that congruence is also a fruitful source for examples in modern algebra. In fact, many concepts in modern algebra first arose in work relating to congruence.

Before defining congruence we need some elementary facts about divisibility. An integer m is *divisible* by an integer n if there is an integer q (for quotient) such that $m = nq$. Thus 6 is divisible by 3 because $6 = 3 \cdot 2$. But 6 is not divisible by 4 or 5. If m is divisible by n, we also say that n *divides* m, and that m is a *multiple* of n, and we write $n \mid m$. So $3 \mid 6$ but $4 \nmid 6$. If $n \mid m$, we also say that n if a *factor* of m. An integer p is a *prime* if $p > 1$ and p is divisible by no positive integer other than 1 and p itself.

Notice that if $n \mid m$, then $n \mid (-m)$. (Why?) Also, if $n \mid a$ and $n \mid b$, then $n \mid (a + b)$. [*Proof:* If $a = nq_1$ and $b = nq_2$, then $a + b = n(q_1 + q_2)$, and $q_1 + q_2$ is an integer if q_1 and q_2 are integers.]

Definition.   Let n be a positive integer. Integers a and b are said to be *congruent modulo n* if $a - b$ is divisible by n. This is written $a \equiv b \pmod{n}$.

Two integers are congruent modulo 2 iff either both are even or both are odd. That is the example from the introductory paragraph. Here are other examples: $17 \equiv 3 \pmod 7$ because 7 divides $17 - 3 = 14$; $4 \equiv 22 \pmod 9$ because 9 divides $4 - 22 = -18$; $19 \equiv 19 \pmod{11}$; $17 \not\equiv 3 \pmod 8$.

When working with congruences it helps to be able to move easily among the following equivalent statements.

$$a \equiv b \pmod n$$
$$n \mid (a - b)$$
$$a - b = un \quad \text{for some } u \in \mathbb{Z}$$
$$a = b + un \quad \text{for some } u \in \mathbb{Z}$$

(See Problem 10.21.)

Theorem 10.1.   *Congruence modulo n is an equivalence relation on the set of integers, for each positive integer n.*

PROOF.   *Reflexive*: If a is an integer, then $a \equiv a \pmod n$ because $n \mid (a - a) = 0$.
Symmetric: If $a \equiv b \pmod n$, then $n \mid (a - b)$, so $n \mid (b - a)$ and $b \equiv a \pmod n$.
Transitive: If $a \equiv b \pmod n$ and $b \equiv c \pmod n$, then $n \mid (a - b)$ and $n \mid (b - c)$; but then $n \mid [(a - b) + (b - c)] = a - c$, so $a \equiv c \pmod n$. ∎

The equivalence classes for this equivalence relation are called *congruence classes mod n*, or simply *congruence classes* if n is clear from the context. (These classes are sometimes called *residue classes*, but we will not use this term.)

Example 10.1.
   (a)   There are two congruence classes mod 2: the even integers and the odd integers.
   (b)   There are four congruence classes mod 4:

$$\{\ldots, -8, -4, 0, 4, 8, \ldots\}$$
$$\{\ldots, -7, -3, 1, 5, 9, \ldots\}$$
$$\{\ldots, -6, -2, 2, 6, 10, \ldots\}$$
$$\{\ldots, -5, -1, 3, 7, 11, \ldots\}.$$

∎

Notice that in the last example there are four congruence classes and each integer is congruent to either 0, 1, 2, or 3 (mod 4). In the language of Section 9, $\{0, 1, 2, 3\}$ is a complete set of equivalence class representatives. We are now going to show that this is typical, by showing that there are always n congruence classes modulo n and that each integer is congruent to either $0, 1, 2, \ldots$, or $n - 1 \pmod n$. But first we need more information about the integers—information that is important far beyond our immediate need, by the way. We start from the following principle.

Least Integer Principle.   *Every nonempty set of positive integers contains a least element.*

The Least Integer Principle is really an axiom, whose role will be clarified in Section 29. (See also Appendix C.) At the moment we need it to prove the Division Algorithm, which will follow an example.

If 11 is divided by 4, there is a quotient of 2 and a remainder of 3:

$$\tfrac{11}{4} = 2 + \tfrac{3}{4}, \quad \text{or} \quad 11 = 4 \cdot 2 + 3.$$

This illustrates the following result.

Division Algorithm.   *If a and b are integers with b > 0, then there exist unique integers q and r such that*

$$a = bq + r, \quad 0 \le r < b.$$

Before giving the proof, we shall look at the idea behind it in the special case $11 = 4 \cdot 2 + 3$ ($a = 11$, $b = 4$, $q = 2$, and $r = 3$). Consider the display in Example 10.1(b): $\mathbb{Z}$ has been partitioned into $b = 4$ rows (congruence classes); $r = 3$ is the smallest positive number in the row containing $a = 11$; and $q = 2$ is the number of positions (multiples of $b = 4$) that we must move to the right to get from $r = 3$ to $a = 11$.

Here is another illustration: $-6 = 4 \cdot (-2) + 2$. Again, $b = 4$; $r = 2$ is the smallest positive number in the row containing $a = -6$; $q = -2$ is the number of positions that we must move (regarding left as negative) to get from $r = 2$ to $a = -6$. In terms of such a display, with the integers partitioned into b rows (congruence classes), the set S in the proof that follows consists of the elements in the row to which a belongs.

PROOF.   We shall prove first that q and r exist, and then that they are unique. Consider the set $S = \{a - bt : t \text{ is an integer}\}$. Let S' denote the set of nonnegative elements of S. Then $S' \ne \varnothing$, which can be seen as follows. If $a \ge 0$, then $t = 0$ yields $a \in S'$. If $a < 0$, then with $t = a$ we find $a - ba \in S$; but $a - ba = a(1 - b) \ge 0$ because $a < 0$ and $1 - b \le 0$ (recall $1 \le b$), and the product of two nonpositive integers is nonnegative.

Let r denote the least integer in S' (if $0 \in S'$ then $r = 0$; otherwise, apply the Least Integer Principle). Let q denote the corresponding value of t, so that $a - bq = r$ and $a = bq + r$. Then $0 \le r$ by choice; therefore, it suffices to show that $r < b$. Assume, on the contrary, that $r \ge b$. Then

$$a - b(q + 1) = a - bq - b = r - b \ge 0,$$

and thus $a - b(q + 1) \in S'$. But

$$a - b(q + 1) = a - bq - b < a - bq = r$$

because $b > 0$, and this contradicts the choice of r as the least element in S'. Thus we do have

$$a = bq + r, \quad 0 \le r < b.$$

To prove uniqueness, suppose that

$$a = bq_1 + r_1, \quad 0 \le r_1 < b$$

and

$$a = bq_2 + r_2, \quad 0 \le r_2 < b.$$

We must show that $q_1 = q_2$ and $r_1 = r_2$. We have

$$bq_1 + r_1 = bq_2 + r_2$$
$$b(q_1 - q_2) = r_2 - r_1.$$

Thus $b \mid (r_2 - r_1)$. But $0 \leq r_1 < b$ and $0 \leq r_2 < b$, so $-b < r_2 - r_1 < b$. The only multiple of b strictly between $-b$ and b is 0. Therefore, $r_2 - r_1 = 0$, and $r_2 = r_1$. But then $b(q_1 - q_2) = 0$ with $b \neq 0$ so that $q_1 = q_2$. ∎

We can now return to congruences.

Theorem 10.2.   *Let n be a positive integer. Then each integer is congruent modulo n to precisely one of the integers $0, 1, 2, \ldots, n - 1$.*

PROOF.   If a is an integer, then by the Division Algorithm there are unique integers q and r such that

$$a = nq + r, \quad 0 \leq r < n.$$

From this, $a - r = nq$ so that $n \mid (a - r)$ and $a \equiv r \pmod{n}$. Thus a is congruent to *at least* one of the integers $0, 1, 2, \ldots, n - 1$. To show that r is unique, assume that $a \equiv s \pmod{n}$ with $0 \leq s < n$. Then $a - s = nt$ (for some integer t), and

$$a = nt + s, \quad 0 \leq s < n.$$

Thus $s = r$ by the uniqueness of r in the Division Algorithm. ∎

PROBLEMS

10.1. List all positive divisors of each of the following integers.
   (a) 20
   (b) 63
   (c) −101

10.2. There are 25 primes less than 100. What are they?

10.3. For each of the following integers, find the smallest nonnegative integer to which it is congruent modulo 7.
   (a) 12
   (b) 100
   (c) −25

10.4. Example 10.1(b) shows the four congruence classes modulo 4. Make a similar array for the congruence classes modulo 3.

10.5. Find all x such that $2x \equiv x \pmod{5}$.

10.6. There are 10 integers x such that $-25 < x < 25$ and $x \equiv 3 \pmod{5}$. Find them all.

10.7. Find all x such that $0 \leq x < 6$ and $2x \equiv 4 \pmod{6}$.

10.8. For which n is $25 \equiv 4 \pmod{n}$?

10.9. For a, $b \in \mathbb{N}$, let $a \sim b$ mean that the decimal representations of a and b have the same last (units) digit. This is an equivalence relation on $\mathbb{N}$. How does it relate to congruence?

10.10. Let k denote a positive integer. For a, $b \in \mathbb{N}$, let $a \sim b$ mean that the decimal representations of a and b have the same digits in each of the last k positions. (Example: If $k = 3$, then $4587 \sim 30, 587$.) This is an equivalence relation on $\mathbb{N}$. How does it relate to congruence? (Compare Problem 10.9.)

For each pair a, b in Problems 10.11 and 10.12, find the unique integers q and r such that $a = bq + r$ with $0 \leq r < b$.

10.11. (a) $a = 19, b = 5$
   (b) $a = -7, b = 5$
   (c) $a = 11, b = 17$

10.12. (a) $a = 50, b = 6$
   (b) $a = 13, b = 20$
   (c) $a = 30, b = 1$

10.13. Prove that if $a \mid b$ and $b \mid c$, then $a \mid c$.

10.14. Prove that if $a \mid b$ and $b \mid a$, then $a = \pm b$. (You may assume that the only divisors of 1 are ± 1.)

In Problems 10.15 and 10.16, assume that $a \equiv b \pmod{n}$ and $c \equiv d \pmod{n}$.

10.15. Prove that $a + c \equiv b + d \pmod{n}$.

10.16. Prove that $ac \equiv bd \pmod{n}$. (*Suggestion:* $a = b + un, c = d + vn$.)

10.17. Disprove with a counterexample: If $a^2 \equiv b^2 \pmod{n}$, then $a \equiv b \pmod{n}$.

10.18. Consider the following statement: If $a \equiv b \pmod{n}$, then $a^2 \equiv b^2 \pmod{n^2}$. If the statement is true, give a proof. If it is false, give a counterexample.

_____

10.19. Prove that if $a \equiv b \pmod{n}$ and $n \mid a$, then $n \mid b$.

10.20. Prove that if $m \mid n$ and $a \equiv b \pmod{n}$, then $a \equiv b \pmod{m}$.

10.21. Prove the equivalence of the four statements just preceding Theorem 10.1. Also prove that $a \equiv b \pmod{n}$ iff a and b leave the same remainder on division by n.

10.22. Prove that if n is a positive integer and $a, b \in \mathbb{Z}$, then there is an integer x such that $a + x \equiv b \pmod{n}$.

10.23. Disprove with a counterexample: If n is a positive integer and $a, b \in \mathbb{Z}$, then there is an integer x such that $ax \equiv b \pmod{n}$.

10.24. Prove that if a is an odd integer, then $a^2 \equiv 1 \pmod{8}$.

10.25. How many positive integers divide (a) 3? (b) 9? (c) 27? (d) 3^k, if k is a positive integer?

10.26. Assume that p is a prime and that k is a positive integer. How many positive integers divide (a) p? (b) p^2? (c) p^k?

10.27. Verify that each of the following statements is false. (Compare the Least Integer Principle.)
(a) Every nonempty set of integers contains a least element.
(b) Every nonempty set of positive rational numbers contains a least element.

10.28. Use the Division Algorithm to prove that if a and b are integers, with $b \neq 0$, then there exist unique integers q and r such that
$$a = bq = r, \quad 0 \leq r < |b|.$$
(*Suggestion:* If $b > 0$, this is the Division Algorithm. Otherwise $|b| = -b$.)

SECTION 11   INTEGERS MODULO n

We have seen that if n is a positive integer, then there are n congruence classes modulo n. With n fixed and k an integer, let $[k]$ denote the congruence class to which k belongs $\pmod{n}$. With $n = 5$, for example,
$$[2] = [7] = [-33] = \{\ldots, -8, -3, 2, 7, 12, \ldots\}.^{\dagger}$$

By Theorem 10.2, $\{[0], [1], \ldots, [n-1]\}$ is a complete set of congruence classes modulo n, in the sense that each integer is in precisely one of these classes. Let $\mathbb{Z}_n$ denote the set

_____

†Notice that $[k]$ is ambiguous unless n has been specified. For example, $[3]$ means one thing if $n = 5$, and something else if $n = 6$. With reasonable attention to context there should be no confusion, however. In case of doubt, $[k]_n$ can be used in place of $[k]$.

$\{[0], [1], \ldots, [n-1]\}$. We shall show that there is a natural operation on this set that makes it a group.

Definition.   For $[a] \in \mathbb{Z}_n$ and $[b] \in \mathbb{Z}_n$, define $[a] \oplus [b]$ by

$$[a] \oplus [b] = [a + b].$$

Example 11.1.   Choose $n = 5$. Then $[3] \oplus [4] = [3 + 4] = [7] = [2]$, and $[-29] \oplus [7] = [-22] = [3]$. ■

There is a question about the definition of $\oplus$: Is it really an operation on $\mathbb{Z}_n$? Or, as it is sometimes expressed, is $\oplus$ *well-defined*? Notice that $[a] \oplus [b]$ has been defined in terms of a $a + b$. What if representatives other than a and b are chosen from $[a]$ and $[b]$? For example, with $n = 5$ again, $[3] = [18]$ and $[4] = [-1]$; therefore, it should be true that $[3] \oplus [4] = [18] \oplus [-1]$. Is that true? Yes, because $[3] \oplus [4] = [7] = [2]$ and $[18] \oplus [-1] = [17] = [2]$. The following lemma settles the question in general.

Lemma 11.1.   In $\mathbb{Z}_n$, if $[a_1] = [a_2]$ and $[b_1] = [b_2]$, then $[a_1 + b_1] = [a_2 + b_2]$.

PROOF.   If $[a_1] = [a_2]$ and $[b_1] = [b_2]$, then for some integers u and v

$$a_1 = a_2 + un \quad \text{and} \quad b_1 = b_2 + vn.$$

Addition yields

$$a_1 + b_1 = (a_2 + un) + (b_2 + vn)$$
$$= a_2 + b_2 + (u + v)n.$$

Thus $(a_1 + b_1) - (a_2 + b_2)$ is n times an integer, $u + v$, and hence $a_1 + b_1 \equiv a_2 + b_2$ (mod n). Therefore, $[a_1 + b_1] = [a_2 + b_2]$. ■

Problem 11.18 is designed to help remove doubts as to whether Lemma 11.1 is really necessary.

Theorem 11.1. $\mathbb{Z}_n$ is an Abelian group with respect to the operation $\oplus$.

PROOF.   Associativity:

$$\begin{aligned}
[a] \oplus ([b] \oplus [c]) &= [a] \oplus [b + c] &&\text{definition of } \oplus \\
&= [a + (b + c)] &&\text{definition of } \oplus \\
&= [(a + b) + c] &&\text{associativity of } + \\
&= [a + b] \oplus [c] &&\text{definition of } \oplus \\
&= ([a] \oplus [b]) \oplus)[c] &&\text{definition of } \oplus .
\end{aligned}$$

The identity is $[0]$:

$$[0] \oplus [a] = [0 + a] = [a]$$
$$[a] \oplus [0] = [a + 0] = [a].$$

Problem 11.10 asks you to prove that the inverse of $[a]$ is $[-a]$. Problem 11.7 asks you to prove that the group is Abelian, that is, $[a] \oplus [b] = [b] \oplus [a]$ for all $[a], [b] \in \mathbb{Z}_n$. ■

The first parts of the preceding proof are typical of proofs of properties of $\oplus$, in that first the definition of $\oplus$ is used, then the corresponding property of $+$, and finally the definition of $\oplus$ again. Whenever $\mathbb{Z}_n$ is referred to as a group the operation is understood to be $\oplus$. This group is called the *group of integers modulo n* (or *mod n*).

Corollary.   *There is a group of order n for each positive integer n.*

PROOF. $\mathbb{Z}_n$ contains n elements, $[0], [1], \ldots,$ and $[n-1]$.   ∎

To appreciate this corollary, try to find another way to construct a group of order 20, for instance. Even for small orders, the associative law is especially hard to verify for an operation that is not "natural" in some sense.

Example 11.2.   Table 11.1 is the Cayley table for $\mathbb{Z}_6$.

Table 11.1

$\oplus$	[0]	[1]	[2]	[3]	[4]	[5]
[0]	[0]	[1]	[2]	[3]	[4]	[5]
[1]	[1]	[2]	[3]	[4]	[5]	[0]
[2]	[2]	[3]	[4]	[5]	[0]	[1]
[3]	[3]	[4]	[5]	[0]	[1]	[2]
[4]	[4]	[5]	[0]	[1]	[2]	[3]
[5]	[5]	[0]	[1]	[2]	[3]	[4]

Using multiplication rather than addition, we obtain another operation on $\mathbb{Z}_n$ as follows:

$$[a] \odot [b] = [ab].$$ ∎

Example 11.3.   Choose $n = 6$. Then

$$[2] \odot [5] = [10] = [4]$$
$$[3] \odot [-4] = [-12] = [0].$$

As with $\oplus$, we must verify that $\odot$ is well-defined. Lemma 11.2 does that.   ∎

Lemma 11.2.   In $\mathbb{Z}_n$, if $[a_1] = [a_2]$ and $[b_1] = [b_2]$, then $[a_1 b_1] = [a_2 b_2]$.

PROOF.   If $[a_1] = [a_2]$ and $[b_1] = [b_2]$, then for some integers u and v

$$a_1 = a_2 + un \quad \text{and} \quad b_1 = b_2 + vn.$$

Therefore,

$$a_1 b_1 = (a_2 + un)(b_2 + vn)$$
$$= a_2 b_2 + (a_2 v + ub_2 + uvn)n.$$

Thus $a_1 b_1 - a_2 b_2$ is an integer $(a_2 v + ub_2 + uvn)$ times n. Therefore $[a_1 b_1] = [a_2 b_2]$.   ∎

In contrast to $\mathbb{Z}_n$ with $\oplus$, $\mathbb{Z}_n$ with $\odot$ is not a group. (See Problem 11.11, for example.) The operation $\odot$ does have some important properties, however; the next lemma gives two of these, and Chapter VI will give more.

Lemma 11.3.   The operation $\odot$ on $\mathbb{Z}_n$ is associative and commutative and has [1] as an identity element.

PROOF.   Make the obvious changes in the proof of Theorem 11.1.   ∎

Let $\mathbb{Z}_n^{\#}$ denote the set $\{[1], [2], \ldots, [n-1]\}$, that is, $\mathbb{Z}_n$ with [0] deleted. Although $\mathbb{Z}_n$ with $\odot$ is never a group, the next example shows that $\mathbb{Z}_n^{\#}$ with $\odot$ can be a group.

Example 11.4. $\mathbb{Z}_5^{\#}$ is a group with respect to the operation $\odot$. Associativity is a consequence of Lemma 11.3. Table 11.2 shows closure, and also that [1] is an identity and that the inverses of [1], [2], [3], and [4], are [1], [3], [2], and [4], respectively.

Table 11.2

$\odot$	[1]	[2]	[3]	[4]
[1]	[1]	[2]	[3]	[4]
[2]	[2]	[4]	[1]	[3]
[3]	[3]	[1]	[4]	[2]
[4]	[4]	[3]	[2]	[1]

∎

Example 11.5. $\mathbb{Z}_6^{\#}$ is not a group with respect to $\odot$. For example, since $[2] \odot [3] = [6] = [0]$, $\mathbb{Z}_6^{\#}$ is not even closed with respect to $\odot$. In Section 26 we shall uncover a general principle that will explain why $\mathbb{Z}_5^{\#}$ (the previous example) is a group, but $\mathbb{Z}_6^{\#}$ is not a group.   ∎

PROBLEMS

11.1. Give five integers in [3] as an element of $\mathbb{Z}_5$, that is, if $n = 5$.

11.2. Give five integers in [3] as an element of $\mathbb{Z}_6$, that is, if $n = 6$.

11.3. Simplify each of the following expressions in $\mathbb{Z}_5$. Write each answer as [0], [1], [2], [3], or [4].
   (a) $[3] \oplus [4]$
   (b) $[2] \oplus [-7]$
   (c) $[17] \oplus [76]$
   (d) $[3] \odot [4]$
   (e) $[2] \odot [-7]$
   (f) $[17] \odot [76]$
   (g) $[3] \odot ([2] \oplus [4])$
   (h) $([3] \odot [2]) \oplus ([3] \odot [4])$

11.4. (a) to (h). Simplify each of the expressions in Problem 11.3 after interpreting it in $\mathbb{Z}_6$ rather than $\mathbb{Z}_5$, and write each answer as [0], [1], [2], [3], [4], or [5].

11.5. Construct the Cayley table for the group $\mathbb{Z}_3$.

11.6. Construct the Cayley table for the group $\mathbb{Z}_4$.

11.7. Prove that each group $\mathbb{Z}_n$ is Abelian.

11.8. Prove that the operation $\odot$ on $\mathbb{Z}_n$ is commutative.

11.9. Prove that $[a] \odot ([b] \oplus [c]) = ([a] \odot [b]) \oplus ([a] \odot [c])$ for all $[a], [b], [c] \in \mathbb{Z}_n$.

11.10. Prove that $[-a]$ is an inverse for $[a]$ in $\mathbb{Z}_n$.

11.11. There is no inverse for $[0]$ relative to the operation $\odot$ on $\mathbb{Z}_n$. Why? (The element $[1]$ is an identity element for $\odot$.)

11.12. Write the proof of Lemma 11.3 in detail.

11.13. Prove or disprove that $\mathbb{Z}_3^{\#}$ is a group with respect to $\odot$.

11.14. Prove or disprove that $\mathbb{Z}_4^{\#}$ is a group with respect to $\odot$.

11.15. Prove that $\{[0], [2], [4]\}$ is a subgroup of $\mathbb{Z}_6$. Construct the Cayley table for the subgroup.

11.16. Prove that $\{[0], [3], [6], [9]\}$ is a subgroup of $\mathbb{Z}_{12}$. Construct the Cayley table for the subgroup.

11.17. (a) Prove that if n is even, then exactly one nonidentity element of $\mathbb{Z}_n$ is its own inverse.
   (b) Prove that if n is odd, then no nonidentity element of $\mathbb{Z}_n$ is its own inverse.
   (c) Prove that $[0] \oplus [1] \oplus \cdots \oplus [n-1]$ equals either $[0]$ or $[n/2]$ in $\mathbb{Z}_n$.
   (d) What does part (c) imply about $0 + 1 + \cdots + (n-1)$ modulo n?

11.18. Define an equivalence relation on the set of integers by letting $a \sim b$ mean that either both a and b are negative or both a and b are nonnegative. There are two equivalence classes: $[-1]$, consisting of the negative integers; and $[0]$, consisting of the nonnegative integers. Attempt to define an operation $\boxplus$ on the set $\{[-1], [0]\}$ of equivalence classes by

$$[a] \boxplus [b] = [a + b]$$

for $a, b \in \mathbb{Z}$ in analogy with the definition of $\oplus$ on $\mathbb{Z}_n$. Show that $\boxplus$ is not well-defined.

SECTION 12   GREATEST COMMON DIVISORS. THE EUCLIDEAN ALGORITHM

There is a close relationship between divisibility properties of the integers and some of the elementary properties of groups. In this section and the next we consider properties of divisibility that will be useful when we return to groups in the next chapter.

Theorem 12.1.   *If a and b are integers, not both zero, then there is a unique positive integer d such that*

(a) $d \mid a$ and $d \mid b$, *and*

(b)   *if c is an integer such that $c \mid a$ and $c \mid b$, then $c \mid d$.*

Property (a) states that d is a *common divisor* of a and b; property (b) ensures that d is the greatest such divisor. Therefore, the integer d in the theorem is called the *greatest common divisor* of a and b. It is denoted (a, b). (The context will usually make it clear whether this or some other interpretation of the ordered pair notation is intended.) Examples are $(4, -6) = 2$, $(-7, 0) = 7$, and $(25, 33) = 1$.

The following proof of Theorem 12.1 shows how to compute (a, b) by a systematic procedure known as the *Euclidean Algorithm*. Another proof, which shows the existence of (a, b), but not how to compute it, is outlined in Problem 12.23.

PROOF.   First consider the case $b > 0$. By the Division Algorithm (Section 10), there are unique integers q_1 and r_1 such that

$$a = bq_1 + r_1, \quad 0 \leq r_1 < b.$$

If $r_1 = 0$, then $b \mid a$, and b will satisfy the conditions for d in parts (a) and (b). If $r_1 \neq 0$, we can apply the Division Algorithm again, getting integers q_2 and r_2 such that

$$b = r_1 q_2 + r_2, \quad 0 \leq r_2 < r_1.$$

Repeated application of the Division Algorithm in this way produces a sequence of pairs of integers $q_1, r_1; q_2, r_2; q_3, r_3; \ldots$ such that

$$
\begin{aligned}
a &= bq_1 + r_1, & 0 \leq r_1 < b \\
b &= r_1 q_2 + r_2, & 0 \leq r_2 < r_1 \\
r_1 &= r_2 q_3 + r_3, & 0 \leq r_3 < r_2 \\
& \cdots &
\end{aligned}
\tag{12.1}
$$

Because each remainder is nonnegative, and $r_1 > r_2 > r_3 > \ldots$, we must eventually reach a remainder that is zero. If r_{k+1} denotes the first zero remainder, then the process terminates with

$$
\begin{aligned}
r_{k-2} &= r_{k-1}q_k + r_k, \quad 0 < r_k < r_{k-1} \\
r_{k-1} &= r_k q_{k+1}.
\end{aligned}
$$

We shall show that r_k, the *last nonzero remainder*, satisfies requirements (a) and (b) for d in the theorem.

Notice first that $r_k \mid r_{k-1}$, because $r_{k-1} = r_k q_{k+1}$. But then $r_k \mid r_{k-2}$ because $r_{k-2} = r_{k-1}q_k + r_k$ and $r_k \mid r_{k-1}$ and $r_k \mid r_k$. Continuing in this way, we can work through the equations in (12.1), from the end, and obtain $r_k \mid r_{k-1}, r_k \mid r_{k-2}, r_k \mid r_{k-3}, \cdots$, until we arrive at $r_k \mid r_1, r_k \mid b$, and finally $r_k \mid a$. Thus r_k is a common divisor of a and b.

Now, moving to condition (b), assume that $c \mid a$ and $c \mid b$. Then $c \mid r_1$ because $r_1 = a - bq_1$. But if $c \mid b$ and $c \mid r_1$, then $c \mid r_2$, because $r_2 = b - r_1 q_2$. Continuing in this way, we can work through the equations in (12.1), from the beginning, and obtain $c \mid r_1, c \mid r_2, c \mid r_3, \ldots$, and finally $c \mid r_k$. This verifies condition (b).

If $b < 0$, we simply go through the same process starting with a and $-b$, rather than a and b; then observe that since b and $-b$ have the same divisors, a greatest common divisor of a and $-b$ will also be a greatest common divisor of a and b.

If $b = 0$, then $|a|$ satisfies the requirements (a) and (b) for d in the theorem.

To prove the uniqueness of (a, b), assume that d_1 and d_2 are integers, each satisfying both of the requirements (a) and (b) for d. Then $d_1 \mid d_2$ and $d_2 \mid d_1$. Therefore, since both d_1 and d_2 are positive, $d_1 = d_2$.   ∎

Example 12.1.   Here is the Euclidean Algorithm applied to compute $(1001, 357)$.

$$
\begin{aligned}
1001 &= 357 \cdot 2 + 287 \\
357 &= 287 \cdot 1 + 70 \\
287 &= 70 \cdot 4 + 7 \\
70 &= 7 \cdot 10.
\end{aligned}
$$

Therefore $(1001, 357) = 7$.   ∎

If a and b are integers, then any integer that is equal to $am + bn$ for some integers m and n is said to be a *linear combination* of a and b. The equations that arise in the Euclidean Algorithm can be used to express (a, b) as a linear combination of a and b. Before proving this we illustrate the idea with an example.

Example 12.2.   From Example 12.1, $(1001, 357) = 7$. To express 7 as a linear combination of 1001 and 357, we use the equations in Example 12.1, beginning with $287 = 70 \cdot 4 + 7$ and working backward one step at a time.

$$7 = 287 - 70 \cdot 4. \tag{12.2}$$

Solve the equation $357 = 287 \cdot 1 + 70$ for 70 (the remainder), substitute in (12.2), and simplify.

$$7 = 287 - (357 - 287 \cdot 1)4$$
$$7 = 287 \cdot 5 - 357 \cdot 4. \tag{12.3}$$

Solve the equation $1001 = 357 \cdot 2 + 287$ for 287 (the remainder), substitute in (12.3), and simplify.

$$7 = (1001 - 357 \cdot 2)5 - 357 \cdot 4$$
$$7 = 1001 \cdot 5 - 357 \cdot 14$$
$$7 = 1001(5) + 357(-14).$$

Thus $(1001, 357) = 7 = 1001m + 357n$ for $m = 5$ and $n = -14$.   ■

Theorem 12.2.   *The greatest common divisor of integers a and b, not both zero, can be expressed as a linear combination of a and b:*

$$(a, b) = am + bn \quad \text{for some integers } m \text{ and } n.$$

PROOF.   We work backward through the equations indicated in (12.1), starting with $r_{k-2} = r_{k-1}q_k + r_k$. Solve this for r_k, to get r_k as a linear combination of r_{k-1} and r_{k-2}.

$$r_k = r_{k-2} - r_{k-1}q_k. \tag{12.4}$$

The next equation up the line in (12.1) would be $r_{k-3} = r_{k-2}q_{k-1} + r_{k-1}$. Solve this for r_{k-1} and substitute in (12.4). The result gives r_k as a linear combination of r_{k-2} and r_{k-3}.

$$r_k = r_{k-2} - (r_{k-3} - r_{k-2}q_{k-1})q_k$$
$$r_k = r_{k-2}(1 + q_{k-1}q_k) - r_{k-3}q_k. \tag{12.5}$$

The next equation up the line in (12.1) would allow us to eliminate r_{k-2} from (12.5) and express r_k as a linear combination of r_{k-3} and r_{k-4}. If we continue in this way, we eventually get r_k as a linear combination of r_2 and r_1, then as a linear combination of r_1 and b, and finally as a linear combination of b and a.   ■

Corollary.   *If a and b are integers, then $(a, b) = 1$ iff there are integers m and n such that $am + bn = 1$.*

PROOF.   If $(a, b) = 1$, then Theorem 12.2 guarantees the existence of m and n such that $am + bn = 1$.

To prove the converse, assume the existence of integers m and n such that $am + bn = 1$. If $d = (a, b)$, then $d \mid a$ and $d \mid b$ so $d \mid (am + bn) = 1$. Therefore, since $d > 0$, we must have $d = 1$.   ∎

Integers having greatest common divisor 1, such as a and b in the preceding corollary, are said to be *relatively prime*.

Although the Euclidean Algorithm can be used to calculate integers m and n such that $(a, b) = am + bn$, it is the mere existence of such m and n, not their actual calculation, which is most often important. In trying to prove statements involving greatest common divisors, it is frequently helpful to begin just by trying to make use of Theorem 12.2. Lemma 13.1 illustrates this point.

Theorem 12.1 can be extended to show that every finite set of integers, not all zero, has a *greatest common divisor*. The greatest common divisor of $a_1, a_2, \ldots, a_n$ is denoted $(a_1, a_2, \ldots, a_n)$. For example, $(-6, 15, 33) = 3$.

Theorem 12.3.   *If a and b are nonzero integers, then there is a unique positive integer m such that*

(a) $a \mid m$ and $b \mid m$, and

(b)   if c is an integer such that $a \mid c$ and $b \mid c$, then $m \mid c$.

Property (a) states that m is a *common multiple* of a and b; property (b) ensures that m is the least positive such multiple. Therefore, the integer m in the theorem is called the *least common multiple* of a and b. It is denoted $[a, b]$. Examples are $[4, -6] = 12$, $[-7, 7] = 7$, and $[25, 33] = 825$.

PROOF.   Let $S = \{x : x \in \mathbb{N}, a \mid x, \text{ and } b \mid x\}$. Then $S \neq \varnothing$ since, for example, $|ab| \in S$. By the Least Integer Principle S has a least element, which we denote by m. By the definition of S, $a \mid m$ and $b \mid m$. Thus to complete the proof it remains only to prove part (b).

Assume $c \in \mathbb{N}$, $a \mid c$, and $b \mid c$. By the Division Algorithm there exists a unique pair of integers q, r such that

$$c = mq + r, \quad 0 \leq r < m. \tag{12.6}$$

Since $a \mid c$, $b \mid c$, $a \mid m$, and $b \mid m$, Equation (12.6) implies that $a \mid r$ and $b \mid r$. Thus $r \in S$, but $0 \leq r < m$, so $r = 0$ by the choice of m as the least element in S. Therefore, $m \mid c$.   ∎

Theorem 12.3 can be extended to show that every finite set of non-zero integers has a *least common multiple*. The least common multiple of $a_1, a_2, \ldots, a_n$ is denoted $[a_1, a_2, \ldots, a_n]$. For example, $[-6, 15, 25] = 150$.

PROBLEMS

In Problems 12.1–12.4, find the greatest common divisor and the least common multiple of the given set of integers.

12.1. $\{6, 14\}$ **12.2.** $\{-75, 105\}$

12.3. $\{36, -60, 90\}$ **12.4.** $\{14, 21, 35\}$

In Problems 12.5–2.8, find the greatest common divisor and write it as a linear combination of the two given integers.

12.5. $(1001, 33)$ **12.6.** $(56, 126)$

12.7. $(-90, 1386)$ **12.8.** $(-2860, -2310)$

12.9. List all of the positive integers that are less than 12 and relatively prime to 12.

12.10. For each integer $n > 1$, let $\phi(n)$ denote the number of positive integers that are less than n and relatively prime to n. For example, $\phi(3) = 2$ and $\phi(4) = 2$. Also, let $\phi(1) = 1$.
(a) Find $\phi(k)$ for $1 \leq k \leq 10$.   (b) If p is a prime, what is $\phi(p)$?

———————

12.11. Verify that $(2, 3)$ can be expressed in at least two different ways as a linear combination of 2 and 3. (This shows that the integers m and n in Theorem 12.2 are not uniquely determined by a and b. Problem 12.12 gives a much stronger statement.)

12.12. Prove that if a and b are integers, not both zero, then there are infinitely many pairs of integers m, n such that $(a, b) = am + bn$.

12.13. Prove that if c is a positive integer, then $(ac, bc) = (a, b)c$.

12.14. Prove that if d is a positive integer, $d \mid a$, and $d \mid b$, then $(a, b) = d$ iff $(a/d, b/d) = 1$.

12.15. Prove that if p is a prime and a is an integer, and $p \nmid a$, then $(a, p) = 1$.

12.16. Prove that if $(a, c) = 1$ and $(b, c) = 1$, then $(ab, c) = 1$.

12.17. Prove that if $c \mid ab$ and $(a, c) = d$, then $c \mid bd$.

12.18. Prove that if $a = bq + r$, then $(a, b) = (b, r)$.

12.19. Prove that if $(a, b) = 1$ and $c \mid a$, then $(c, b) = 1$.

12.20. Prove that if $(a, m) = 1$, then there is a solution (for x) to the congruence $ax \equiv b \pmod{m}$.

12.21. Let n be a positive integer, and let $\mathbb{Z}_{(n)}$ denote the subset of $\mathbb{Z}_n$ defined by

$$\mathbb{Z}_{(n)} = \{[k] : 1 \leq k < n \quad \text{and} \quad (k, n) = 1\}.$$

Thus, for example,

$$\mathbb{Z}_{(12)} = \{[1], [5], [7], [11]\}.$$

Prove that $\mathbb{Z}_{(n)}$ is a group with respect to the operation $\odot$ defined in Section 11. Construct a Cayley table for $\mathbb{Z}_{(12)}$ with respect to $\odot$. [The order of $\mathbb{Z}_{(n)}$ is $\phi(n)$, defined in Problem 12.10.] [*Suggestion*: Use the result in Problem 12.20.]

12.22. Prove that if a, b, c are integers, not all zero, then they have a greatest common divisor, which can be written as a linear combination of $a, b,$ and c.

12.23. Give an alternate proof of Theorem 12.1 by justifying each of the following steps.
(a) Let $S = \{ax + by : x, y \in \mathbb{Z}\}$. The set S contains at least one positive integer.
(b) There is a least positive integer in S; let it be $d = am + bn (m, n \in \mathbb{Z})$. (This will be the d of Theorem 12.1.)
(c) Because $a \in S$ and $b \in S$, to prove Theorem 12.1 (a) it suffices to prove that d divides each positive integer in S.
(d) Let $k = ua + vb$ denote a positive integer in S. There are integers q and r such that

$$k = dq + r, \quad 0 \leq r < d.$$

Therefore,

$$(au + bv) = (am + bn)q + r,$$

and

$$r = (u - mq)a + (v - nq)b,$$

which implies $r \in S$.

   (e) Therefore $r = 0$, and d does divide each positive integer in S. (As stated, this shows that $d \mid a$ and $d \mid b$.)

   (f) If $c \mid a$ and $c \mid b$, then $c \mid d$.

SECTION 13   THE FUNDAMENTAL THEOREM OF ARITHMETIC

Fundamental Theorem of Arithmetic.   *Each integer greater than 1 can be written as a product of primes, and, except for the order in which these primes are written, this can be done in only one way.*

Thus $15 = 3 \cdot 5 = 5 \cdot 3$, $16 = 2^4$, and $17 = 17$, the last example showing that "product" is taken to include the possibility that there is only one factor present. We shall prove two lemmas before proving the theorem.

Lemma 13.1.   If a, b, and c are integers, with $a \mid bc$ and $(a, b) = 1$, then $a \mid c$.

   PROOF.   Since $(a, b) = 1$, by Theorem 12.2 there are integers m and n such that $1 = am + bn$. On multiplying both sides of this equation by c, we get $c = amc + bnc$. Certainly $a \mid amc$. And we are assuming that $a \mid bc$, so that $a \mid bnc$. Therefore, a divides $amc + bnc$, which is equal to c.   ■

Lemma 13.2.   If p is a prime, $a_1, a_2, \ldots, a_n$ are integers, and $p \mid a_1 a_2 \cdots a_n$, then $p \mid a_i$ for some i $(1 \le i \le n)$.

   PROOF.   We use induction on n. The case $n = 1$ is obvious. Assume that $n > 1$. If $p \mid a_1 a_2 \cdots a_{n-1}$, then $p \mid a_i$ for some i $(1 \le i \le n - 1)$ by the induction hypothesis. If $p \nmid a_1 a_2 \cdots a_{n-1}$, then $(p, a_1 a_2 \cdots a_{n-1}) = 1$ because p is a prime. In this case, by Lemma 13.1 (with $a = p, b = a_1 a_2 \cdots a_{n-1}, c = a_n$), $p \mid a_n$.   ■

   PROOF OF THE THEOREM.   Let S denote the set of those integers greater than 1 that *cannot* be written as a product of primes. To prove the first part of the theorem we must show that S is empty. Assume otherwise. Then, by the Least Integer Principle (Section 10), S contains a least element, which we denote by n. The set S contains no primes, so n can be factored as $n = n_1 n_2$, where both n_1 and n_2 are integers and $1 < n_1 < n$ and $1 < n_2 < n$. Because $n_1 < n$ and $n_2 < n$, and n is the least element of S, it follows that $n_1 \notin S$ and $n_2 \notin S$. Thus n_1 and n_2 both can be written as products of primes, so the same is true of $n = n_1 n_2$. This contradicts the fact that $n \in S$. This contradiction proves that S must be empty, as required.

   To prove the last part of the theorem, assume m to be an integer greater than 1, and assume $m = p_1 p_2 \cdots p_s = q_1 q_2 \cdots q_t$, where the p_i and q_j are primes. Then $p_1 \mid q_1 q_2 \cdots q_t$

because $p_1 \mid p_1 p_2 \cdots p_s$. Thus $p_1 \mid q_j$ for some j $(1 \le j \le t)$, by Lemma 13.2. But then $p_1 = q_j$ because both p_1 and q_j are primes. It follows that

$$p_2 p_3 \cdots p_s = q_1 q_2 \cdots q_{j-1} q_{j+1} \cdots q_t,$$

where we have canceled p_1 on the left and q_j on the right. Repeating the argument just used, we see that p_2 must equal one of the remaining prime factors on the right. Continuing in this way, we can pair each prime on the left with a prime on the right. The primes on one side cannot all be canceled before those on the other side because that would imply that 1 is equal to a product of primes, an impossibility. Thus $s = t$ and the lists $p_1, p_2, \ldots, p_s$ and $q_1, q_2, \ldots, q_t$ must be the same except possibly for their arrangement. ∎

By arranging the prime factors in increasing order, we see that each integer $n > 1$ can be written in the form

$$n = p_1^{e_1} p_2^{e_2} \cdots p_k^{e_k} \quad (p_1 < p_2 < \cdots < p_k) \tag{13.1}$$

where the primes $p_1, p_2, \ldots, p_k$ and the positive integers $e_1, e_2, \ldots, e_k$ are uniquely determined by n. We shall call this the *standard form* for n. For example, the standard form for 300 is $2^2 \cdot 3 \cdot 5^2$.

PROBLEMS

Determine the standard form (13.1) for each of the following integers.

13.1. 105   **13.2.** 684   **13.3.** 1375   **13.4.** 139

13.5. Let

$$m = p_1^{s_1} p_2^{s_2} \cdots p_k^{s_k}$$

and

$$n = p_1^{t_1} p_2^{t_2} \cdots p_k^{t_k}$$

where $p_1, p_2, \ldots, p_k$ are distinct prime numbers, $s_i \ge 0$ for $1 \le i \le k$, and $t_i \ge 0$ for $1 \le i \le k$. Prove that $m \mid n$ iff $s_i \le t_i$ for $1 \le i \le k$.

13.6. Let m and n be as in Problem 13.5, and let

$$u_i = \text{the minimum of } s_i \text{ and } t_i \text{ for each } i$$

and

$$v_i = \text{the maximum of } s_i \text{ and } t_i \text{ for each } i.$$

(a) Prove that $(m, n) = p_1^{u_1} p_2^{u_2} \cdots p_k^{u_k}$ (greatest common divisor).

(b) Prove that $[m, n] = p_1^{v_1} p_2^{v_2} \cdots p_k^{v_k}$ (least common multiple).

Use the results of Problem 13.6 to compute the greatest common divisor and least common multiple of each of the following pairs of integers.

13.7. 10, 105   **13.8.** $-39, 54$

13.9. 56, 126   **13.10.** $-2860, -2310$

13.11. Determine all positive integral divisors of each of the following integers.

    (a) 16   (b) 27   (c) $2^3 3^2$ (d) $2^r 3^s$

13.12. Determine the number of positive integral divisors of an integer n having the standard form $p_1^{e_1} p_2^{e_2} \cdots p_k^{e_k}$. (Compare Problem 13.11.)

13.13. Determine $\phi(p^k)$ if p is a prime and k is a positive integer, where ϕ is the function defined in Problem 12.10.

13.14. Prove that if $(a, b) = 1$, $a \mid m$, and $b \mid m$, then $ab \mid m$. (*Suggestion*: If $m = ak$, then $b \mid k$ by Lemma 13.1.)

13.15. An integer is *square-free* if it is not divisible by the square of any integer greater than 1. An integer n is a *perfect square* if $n = k^2$ for some integer k.
   (a) Prove that n is square-free iff in the standard form (13.1) each $e_i = 1$.
   (b) Prove that n is a perfect square iff in the standard form (13.1) each e_i is even.
   (c) Prove that every integer greater than 1 is the product of a square-free integer and a perfect square.

13.16. Prove that if n is an integer, then $\sqrt{n}$ is rational iff n is a perfect square (see Problem 13.15). (*Suggestion*: Apply the Fundamental Theorem of Arithmetic to $a^2 = nb^2$. Compare Theorem 28.1.)

13.17. Prove that $\sqrt[3]{2}$ is irrational. (*Suggestion*: Apply the Fundamental Theorem of Arithmetic to $a^3 = 2b^3$.)

13.18. State and prove a theorem characterizing those integers n for which $\sqrt[3]{n}$ is rational. (Compare Problems 13.16 and 13.17.)

13.19. Prove that if a and b are positive integers, then

$$(a, b)[a, b] = ab.$$

(See Problem 13.6.)

CHAPTER IV

GROUPS

In this chapter the emphasis is less on giving examples of groups and more on proving general theorems about them. The focal points are Lagrange's Theorem, which puts a strong restriction on which subsets of a group can be subgroups, and isomorphism, which makes precise the notion of what it means for groups to be "essentially" alike.

SECTION 14   ELEMENTARY PROPERTIES

Hereafter, whenever a group has no other specified operation, we refer to the operation as multiplication and indicate the product of a and b by ab (juxtaposition). Also, if there is no other established notation, the identity element is denoted by e and the inverse of an element a by a^{-1}. When interpreting general statements in special cases, this notation must be changed accordingly. For instance, ab, e, and a^{-1} become, in additive notation, $a + b, 0$, and $-a$. The reasons for using ab, e, and a^{-1} consistently in general statements are economy and uniformity.

We begin with some remarks about associativity. There are two possible results from inserting parentheses in abc, and the associative law demands that these be equal: $a(bc) = (ab)c$. But what about $abcd$? For example, two of the possibilities here are $a(b(cd))$ and $(ab)(cd)$. One application of the associative law (for three elements) shows these possibilities to be equal: Substitute a for x, b for y, and cd for z in $x(yz) = (xy)z$. Then

$$x(yz) = (xy)z$$

becomes

$$a(b(cd)) = (ab)(cd),$$

as claimed. The other possibilities for $abcd$ are

$$a((bc)d), \quad ((ab)c)d, \quad \text{and} \quad (a(bc))d,$$

and they all give the same result (Problem 3.29). In fact, this is true for any number of elements, by what is known as the *generalized associative law*: If $a_1, a_2, \ldots, a_n$ $(n \geq 2)$ are elements of a set with an associative operation, then the product $a_1 a_2 \ldots a_n$ is unambiguous; that is, the same element will be obtained regardless of how parentheses are inserted in the product, as long as the elements $a_1, a_2, \ldots, a_n$ and their order of appearance are unchanged. (For a proof of this law see either of the last references listed at the end of this chapter.)

Theorem 14.1. *Let G be a group.*

(a) *If $a, b, c \in G$ and $ab = ac$, then $b = c$ (left cancellation law).*

(b) *If $a, b, c \in G$ and $ba = ca$, then $b = c$ (right cancellation law).*

(c) *If $a, b \in G$, then each of the equations $ax = b$ and $xa = b$ has a unique solution in G. In the first, $x = a^{-1}b$; in the second, $x = ba^{-1}$.*

(d) *If $a \in G$, then $(a^{-1})^{-1} = a$.*

(e) *If $a, b \in G$, then $(ab)^{-1} = b^{-1}a^{-1}$.*

PROOF. (a) Assume that $ab = ac$. On multiplying both sides on the left by a^{-1}, we are led to $a^{-1}(ab) = a^{-1}(ac)$, $(a^{-1}a)b = (a^{-1}a)c$, $eb = ec$, and $b = c$.

(b) Similar to part (a) (Problem 14.11).

(c) To see that $x = a^{-1}b$ is a solution of $ax = b$, simply substitute: $a(a^{-1}b) = (aa^{-1})b = eb = b$. To see that there is no other solution, assume that $ax = b$. Then multiplication on the left by a^{-1} leads to $a^{-1}(ax) = a^{-1}b$, $(a^{-1}a)x = a^{-1}b$, $ex = a^{-1}b$, $x = a^{-1}b$; thus $x = a^{-1}b$ is indeed the only solution. The proof for the equation $xa = b$ is similar (Problem 14.12).

(d) The inverse of a^{-1} is the unique element x such that $a^{-1}x = e$. But $a^{-1}a = e$. Therefore, the inverse of a^{-1} must be a.

(e) The inverse of ab is the unique element x such that $(ab)x = e$. But $(ab)(b^{-1}a^{-1}) = a(bb^{-1})a^{-1} = aea^{-1} = aa^{-1} = e$; thus the inverse of ab must be $b^{-1}a^{-1}$. ∎

Here are some observations about the theorem. If a and x are elements of a finite group, then in the Cayley table for the group ax will be in the row labeled by a. If b is also an element of the group, then the existence of a unique solution of $ax = b$ [Theorem 14.1(c)] implies that b appears exactly once in the row labeled by a. Thus

each element of a finite group appears exactly once

in each row of the Cayley table for the group.

(This ignores the row labels at the outside of the table.) Similarly, because there is a unique solution of $xa = b$,

each element of a finite group appears exactly once

in each column of the Cayley table for the group.

Part (e) of Theorem 14.1 shows that the inverse of a product is the product of the inverses, *in reverse order.*

Integral powers of group elements are defined as follows:

$$a^0 = e, a^1 = a, a^2 = aa, \ldots, a^{n+1} = a^n a,$$

so that a^n is equal to the product of n a's for each positive integer n. Also,

$$a^{-n} = (a^{-1})^n \qquad \text{for each positive integer } n.$$

The following laws of exponents can be proved by mathematical induction:

$$a^m a^n = a^{m+n}$$
$$(a^m)^n = a^{mn}$$

for all integers m and n. (See Appendix C.) In additive notation, for n a positive integer a^n becomes $na = a + a + \cdots + a$ (n terms), and $a^{-n} = (a^{-1})^n$ becomes $(-n)a = n(-a)$. In this case the laws above become

$$(ma) + (na) = (m + n)a$$
$$n(ma) = (mn)a$$

for all integers m and n.

Now assume that a is an element of a group G. Consider the set of all integral powers of a, that is, $\{\ldots, a^{-2}, a^{-1}, e, a, a^2, \ldots\}$. Some of these powers may be equal, as illustrated in the following example.

Example 14.1.   In S_3,

$$(1 \quad 2 \quad 3)^3 = (1) \quad \text{and} \quad (1 \quad 2 \quad 3)^0 = (1),$$

so that $(1 \quad 2 \quad 3)^3 = (1 \quad 2 \quad 3)^0$. Also,

$$(1 \quad 2 \quad 3)^4 = (1 \quad 2 \quad 3)^3(1 \quad 2 \quad 3) = (1)(1 \quad 2 \quad 3) = (1 \quad 2 \quad 3),$$

so that $(1 \quad 2 \quad 3)^4 = (1 \quad 2 \quad 3)^1$. Next,

$$(1 \quad 2 \quad 3)^5 = (1 \quad 2 \quad 3)^3(1 \quad 2 \quad 3)^2 = (1)(1 \quad 2 \quad 3)^2 = (1 \quad 2 \quad 3)^2,$$

so that $(1 \quad 2 \quad 3)^5 = (1 \quad 2 \quad 3)^2 = (1 \quad 3 \quad 2)$. Finally,

$$(1 \quad 2 \quad 3)^{-1} = (1 \quad 3 \quad 2),$$

so that $(1 \quad 2 \quad 3)^{-1} = (1 \quad 2 \quad 3)^2$. If we continue in this way, we will soon realize that the set of all integral powers of $(1 \quad 2 \quad 3)$ is just

$$\left\{(1), (1 \quad 2 \quad 3), (1 \quad 2 \quad 3)^2\right\} = \{(1), (1 \quad 2 \quad 3), (1 \quad 3 \quad 2)\}.$$

This set of all powers of $(1 \quad 2 \quad 3)$ is a subgroup of S_3. The next theorem will generalize this example.   ∎

If G is a group and $a \in G$, then $\langle a \rangle$ will denote the set of all integral powers of a. Thus

$$\langle a \rangle = \left\{a^n : n \in \mathbb{Z}\right\}.$$

By Example 14.1,

$$\langle (1 \quad 2 \quad 3) \rangle = \{(1), (1 \quad 2 \quad 3), (1 \quad 3 \quad 2)\}.$$

Theorem 14.2.   *If G is a group and $a \in G$, then $\langle a \rangle$, the set of all integral powers of a, is a subgroup of G.*

PROOF.   It suffices to check the three conditions in Theorem 7.1. First, $\langle a \rangle$ is nonempty since $a \in \langle a \rangle$. Next, the set $\langle a \rangle$ is closed: if $a^m \in \langle a \rangle$ and $a^n \in \langle a \rangle$, then $a^m \cdot a^n = a^{m+n} \in \langle a \rangle$, since $m \in \mathbb{Z}$ and $n \in \mathbb{Z}$ imply $m + n \in \mathbb{Z}$. Finally, $\langle a \rangle$ contains the inverse of each of its elements, because if $a^m \in \langle a \rangle$, then $(a^m)^{-1} = a^{-m} \in \langle a \rangle$.   ∎

The subgroup $\langle a \rangle$ is called the *subgroup generated by* a. If H is a subgroup and $H = \langle a \rangle$ for some $a \in H$, then H is said to be a *cyclic subgroup*. The group of integers is cyclic: $\mathbb{Z} = \langle 1 \rangle$ or $\mathbb{Z} = \langle -1 \rangle$.

The following theorem analyzes what happens when different powers of an element are equal, as in Example 14.1.

Theorem 14.3.   *Assume that G is a group, that $a \in G$, and that there exist unequal integers r and s such that $a^r = a^s$,*
   (a) There is a smallest positive integer n such that $a^n = e$.
   (b) If t is an integer, then $a^t = e$ iff n is a divisor of t.
   (c) The elements $e = a^0, a, a^2, \ldots, a^{n-1}$ are distinct, and $\langle a \rangle = \left\{e, a, a^2, \ldots, a^{n-1}\right\}$.

PROOF.   (a) To prove part (a), it suffices to show that $a^t = e$ for *some* positive integer t; for the Least Integer Principle will then tell us that there is a smallest such integer, which we can call n. Assume $r > s$. (If $s > r$, just interchange r and s in the next sentence.) After multiplying both sides of $a^r = a^s$ by a^{-s}, we obtain $a^{r-s} = e$ with $r - s > 0$. And, as already stated, that is sufficient.

(b) If n is a divisor of t, say $t = nv$, then $a^t = a^{nv} = (a^n)^v = e^v = e$. To prove the other half of part (b), suppose that $a^t = e$. By the Division Algorithm there are integers q and r such that $t = nq + r$, $0 \le r < n$. Thus $a^t = a^{nq+r} = (a^n)^q a^r = e^q a^r = a^r$. But $a^t = e$, so $a^r = e$. This implies that $r = 0$, since $0 \le r < n$ and n is the smallest positive integer such that $a^n = e$. Therefore $t = nq$ so that n is a divisor of t.

(c) To prove that $a^0, a^1, a^2, \ldots, a^{n-1}$ are distinct, suppose that $a^u = a^v$ with $0 \le u < n$ and $0 \le v < n$. We shall prove that u must equal v. Interchanging u and v if necessary, we can assume that $u \ge v$. Then $a^u = a^v$ implies $a^u a^{-v} = a^v a^{-v}$, which implies $a^{u-v} = e$ with $u - v \ge 0$. Therefore, by part (b), n must be a divisor of $u - v$. But $u - v < n$, since $0 \le u < n$ and $0 \le v < n$. Thus n is a divisor of $u - v$ and $0 \le u - v < n$, which can happen only if $u - v = 0$, that is, $u = v$.

Certainly any power of a is in $\langle a \rangle$; hence the proof will be complete if we show that each power of a is in the set $\{e, a, a^2, \ldots, a^{n-1}\}$. Consider a power a^m. By the Division Algorithm there are integers q and r such that $m = nq + r$. This leads to $a^m = a^{nq+r} = (a^n)^q a^r = e^q a^r = a^r$, with $0 \le r < n$, which is just what we need.   ∎

If a is an element of a group, then the smallest positive integer n such that $a^n = e$, if it exists, is called the *order* of a. If there is no such integer, then a is said to have *infinite order*. The order of an element a will be denoted by $o(a)$.

Example 14.2.
(a)   In S_3, $o((1 \quad 2 \quad 3)) = 3$ (Example 14.1).
(b)   In the group of nonzero rational numbers (operation multiplication), 2 has infinite order, because $2^n \ne 1$ for every positive integer n.   ∎

In additive notation, the condition $a^n = e$ becomes $na = 0$. In $\mathbb{Z}_n$, the condition $a^n = e$ becomes $n[a] = [0]$.

Example 14.3.   In $\mathbb{Z}_6$, $o([2]) = 3$, because $[2] \ne [0]$ and

$$2[2] = [2] \oplus [2] = [4] \ne [0]$$

but

$$3[2] = [2] \oplus [2] \oplus [2] = [6] = [0].$$

(We also see that $\langle [2] \rangle = \{[0], [2], [4]\}$.)   ∎

The word *order* has been used in two senses: the order of a *group* (Section 5) and the order of an *element*. The next corollary shows how the two are related.

Corollary.   *If a is an element of a group, then $o(a) = |\langle a \rangle|$.*

PROOF.   If $o(a) = n$ is finite, then $\langle a \rangle = \{e, a, a^2, \ldots, a^{n-1}\}$ by Theorem 14.3(c). Thus $|\langle a \rangle| = n$ and $o(a) = |\langle a \rangle|$. If a is of infinite order, then all of the integral powers of a must be distinct by Theorem 14.3(a), and thus $|\langle a \rangle|$ is infinite.   ∎

PROBLEMS

14.1. Solve the equation $(1\ \ 2)x = (1\ \ 2\ \ 3)$ in S_3.

14.2. Solve the equation $x(1\ \ 3\ \ 2) = (1\ \ 3)$ in S_3.

14.3. Determine the elements in each of the cyclic subgroups of S_3. Also give the order of each element of S_3.

14.4. Determine the elements in each of the cyclic subgroups of $\mathbb{Z}_6$. Also give the order of each element of $\mathbb{Z}_6$.

14.5. Find the order of the element $(1\ \ 2)(3\ \ 4)$ in S_4.

14.6. Find the order of the element $(1\ \ 2)(3\ \ 4\ \ 5)$ in S_5.

14.7. (a) Determine the elements in the subgroup $\langle(1\ \ 2\ \ 3\ \ 4)\rangle$ of S_4.
  (b) Determine the elements in the subgroup $\langle(1\ \ 2\ \ 3\ \ 4\ \ 5)\rangle$ of S_5.
  (c) What is the order of the subgroup $\langle(1\ \ 2\ \ \dots\ \ n)\rangle$ of S_n?

14.8. Determine the elements in each of the following subgroups of the group of symmetries of a square (Table 8.1).
  (a) $\langle\mu_2\rangle$
  (b) $\langle\mu_3\rangle$
  (c) $\langle\mu_4\rangle$

14.9. Let α denote the clockwise rotation of the plane through $90°$ about a fixed point p ($\alpha \in G$ in Example 5.7). What is the order of $\langle\alpha\rangle$?

14.10. (a) Repeat Problem 14.9 with $40°$ in place of $90°$.
  (b) What is the order of $\langle\alpha\rangle$ if α denotes rotation through $(360/k)°$ ($k \in \mathbb{N}$)?

=======

14.11. Prove Theorem 14.1 (b).

14.12. Prove that $xa = b$ has a unique solution in a group. This is the omitted part of the proof of Theorem 14.1(c).

14.13. Prove that $axb = c$ has a unique solution in a group (given a, b, c).

14.14. (a) Prove that if a and b are elements of an Abelian group G, with $o(a) = m$ and $o(b) = n$, then $(ab)^{mn} = e$. Indicate where you use that G is Abelian.
  (b) With G, a, and b as in part (a), prove that $o(ab)$ divides $o(a)o(b)$.
  (c) Give an example of an Abelian group G and elements a and b in G such that $o(ab) \neq o(a)o(b)$. Compare part (b).

14.15. Show with an example that if G is not Abelian, then the statement in Problem 14.14(a) may be false. (There is an example with $G = S_3$.)

14.16. (a) Use Problem 14.14 to prove that in an Abelian group the elements of finite order form a subgroup.
  (b) What are the elements of finite order in the group of positive rationals (operation multiplication)?

14.17. Verify Theorem 14.1(e) for $a = (1\ \ 2\ \ 5)$ and $b = (2\ \ 3\ \ 4)$ in S_5. Is $(ab)^{-1} = a^{-1}b^{-1}$ true for this a and b?

14.18. Assume that a and b are elements of a group G. Prove that $ab = ba$ iff $a^{-1}b^{-1} = b^{-1}a^{-1}$.

14.19. Assume $m, n \in \mathbb{Z}$. Find necessary and sufficient conditions for $\langle m\rangle \subseteq \langle n\rangle$.

14.20. Construct a Cayley table for a group G given that $G = \langle a\rangle$, $a \neq e$, and $a^5 = e$.

14.21. Rewrite Theorem 14.1 (not its proof) for a group written additively, that is, with operation $+$, identity 0, and $-a$ for the inverse of a.

14.22. (a) Prove that if a, b, and c are elements of a group, then any one of the following three equations implies the other two:

$$ab = c, \quad a = cb^{-1}, \quad b = a^{-1}c.$$

(b) Show with an example that $ab = c$ does not always imply $a = b^{-1}c$. (Look in S_3.)

14.23. Prove that a nonidentity element of a group has order 2 iff it is its own inverse.

14.24. Prove that every group of even order has an element of order 2. (Problem 14.23 may help.)

14.25. Prove that a group G is Abelian iff $(ab)^{-1} = a^{-1}b^{-1}$ for all $a, b \in G$.

14.26. There is only one way to complete the following Cayley table so as to get a group. Find it. Why is it unique? (Problem 5.22 may help.)

$*$	a	b	c
a		b	
b			
c			

14.27. Assume that $\{x, y, z, w\}$ is to be a group, with identity x (operation juxtaposition). With any one of the following additional assumptions (a), (b), (c), or (d), there is only one Cayley table yielding a group. Determine that Cayley table in each case.
(a) $y^2 = z$
(b) $y^2 = w$
(c) $y^2 = x$ and $z^2 = x$
(d) $y^2 = x$ and $z^2 = y$

14.28. Prove that if a is a fixed element of a group G, and $\lambda\colon G \to G$ is defined by $\lambda(x) = ax$ for each $x \in G$, then λ is one-to-one and onto.

14.29. Prove that a group is Abelian if each of its nonidentity elements has order 2.

14.30. Prove that if G is a group and $a \in G$, then $o(a^{-1}) = o(a)$.

14.31. Prove that if G is a group and $a, b \in G$, then $o(a^{-1}ba) = o(b)$.

14.32. Prove that if G is a group and a, $b \in G$, then $o(ab) = o(ba)$. (Suggestion: Problem 14.31 may help.)

14.33. Prove or give a counterexample: If a group has a subgroup of order n, then it has an element of order n.

14.34. Prove that if a group G has no subgroup other than G and $\{e\}$, then G is cyclic.

14.35. Prove that if G is a finite group, then Theorem 7.1 is true with condition (c) omitted. Also, give an example to show that (c) cannot be omitted if Theorem 7.1 is to be true for all groups.

14.36. Prove that if α represents an r-cycle in S_n and β represents an s-cycle, and α and β are disjoint, then $o(\alpha\beta) = [r, s]$ (the least common multiple of r and s).

14.37. Determine the largest order of an element of S_n for each n such that $1 \leq n \leq 10$. [*Suggestion*: In $S_5, o((1 \quad 2 \quad 3)(4 \quad 5)) = 6$. Consider Problem 14.36.] Formulate a general statement about how to find the largest order for an element of S_n.

14.38. Prove that if A and B are subgroups of a group G, and $A \cup B$ is also a subgroup, then $A \subseteq B$ or $A \supseteq B$. (Compare Problem 7.13 and Theorem 15.1.)

SECTION 15   GENERATORS. DIRECT PRODUCTS

If G is a group and $a \in G$, then the cyclic subgroup $\langle a \rangle$ is the smallest subgroup containing a. The next two theorems generalize this idea by associating with each subset of a group a unique smallest subgroup containing that subset.

Theorem 15.1.   *If C denotes any collection of subgroups of a group G, then the intersection of all of the groups in C is also a subgroup of G.*

PROOF.   Let H denote the intersection in question; we shall verify that H satisfies the three conditions in Theorem 7.1. Each subgroup in C contains e, the identity of G, so $e \in H$ and H is nonempty. If $a, b \in H$, then a and b belong to each subgroup in C and therefore ab belongs to each subgroup in C; thus $ab \in H$. Finally, if $a \in H$, then a^{-1} belongs to each subgroup in C because a does, and thus $a^{-1} \in H$.   ∎

Example 15.1.   In $\mathbb{Z}$ (operation addition), $\langle 3 \rangle$ consists of all the multiples of 3 and $\langle 4 \rangle$ consists of all the multiples of 4. Because a number is a multiple of both 3 and 4 iff it is a multiple of 12, we have

$$\langle 3 \rangle \cap \langle 4 \rangle = \langle 12 \rangle.$$

Also,

$$\langle 6 \rangle \cap \langle 8 \rangle = \langle 24 \rangle.$$ ∎

Theorem 15.2.   *Let S be any subset of a group G, and let $\langle S \rangle$ denote the intersection of all of the subgroups of G that contain S. Then $\langle S \rangle$ is the unique smallest subgroup of G that contains S, in the sense that*

(a) $\langle S \rangle$ contains S,

(b) $\langle S \rangle$ is a subgroup, and

(c)   if H is any subgroup of G that contains S, then H contains $\langle S \rangle$.

PROOF.   First, notice that there is always at least one subgroup of G containing S, namely, G itself. With $\langle S \rangle$ as defined, $\langle S \rangle$ certainly contains S, since the intersection of any collection of subsets each containing S will contain S, whether the subsets are subgroups or not. Next, $\langle S \rangle$ is a subgroup by Theorem 15.1. Finally, condition (c) is simply a property of the intersection of sets: if H is a subgroup containing S, then H is a member of the collection of subgroups whose intersection is $\langle S \rangle$, and thus H contains $\langle S \rangle$.

To justify use of the term *unique* in the theorem, assume that $[S]$ is a subgroup of G satisfying conditions (a), (b), and (c) with $[S]$ in place of $\langle S \rangle$. Condition (c), with $[S]$ in place of H, implies that $[S] \supseteq \langle S \rangle$. On the other hand, condition (c) with $\langle S \rangle$ in place of H and $[S]$ in place of $\langle S \rangle$ implies that $[S] \subseteq \langle S \rangle$. Thus $[S] = \langle S \rangle$.   ∎

We say that S *generates* $\langle S \rangle$ and that $\langle S \rangle$ is *generated by* S. If $S = \{a\}$ then $\langle S \rangle$ is the cyclic subgroup generated by a, which we denote by $\langle a \rangle$ (Section 14). More generally, if $S = \{a_1, \ldots, a_n\}$, then we denote $\langle S \rangle$ by $\langle a_1, \ldots, a_n \rangle$ rather than $\langle \{a_1, \ldots, a_n\} \rangle$.

Example 15.2.   The subgroup $\langle 9, 12 \rangle$ of the group of integers must contain $12 + (-9) = 3$. Therefore $\langle 9, 12 \rangle$ must contain all multiples of 3. That is, $\langle 9, 12 \rangle \supseteq \langle 3 \rangle$. But also $\langle 9, 12 \rangle \subseteq \langle 3 \rangle$, since both 9 and 12 are multiples of 3. Therefore $\langle 9, 12 \rangle = \langle 3 \rangle$. The next theorem generalizes this example.   ∎

Theorem 15.3.   *Let T_1 and T_2 be subsets of a group G. Then*

$$\langle T_1 \rangle = \langle T_2 \rangle \quad iff \quad T_1 \subseteq \langle T_2 \rangle \quad and \quad \langle T_1 \rangle \supseteq T_2.$$

PROOF.   By Theorem 15.2(a), $T_1 \subseteq \langle T_1 \rangle$. Therefore,

$$\text{if} \quad \langle T_1 \rangle = \langle T_2 \rangle, \quad \text{then} \quad T_1 \subseteq \langle T_2 \rangle.$$

Similarly,

$$\text{if} \quad \langle T_1 \rangle = \langle T_2 \rangle, \quad \text{then} \quad \langle T_1 \rangle \supseteq T_2.$$

Since $\langle T_2 \rangle$ is a subgroup of G, Theorem 15.2(c) implies that

$$\text{if} \quad T_1 \subseteq \langle T_2 \rangle, \quad \text{then} \quad \langle T_1 \rangle \subseteq \langle T_2 \rangle$$

(use $\langle T_2 \rangle$ in place of H and T_1 in place of S). Similarly, since $\langle T_1 \rangle$ is a subgroup of G,

$$\text{if} \quad \langle T_1 \rangle \supseteq T_2, \quad \text{then} \quad \langle T_1 \rangle \supseteq \langle T_2 \rangle.$$

Therefore,

$$\text{if} \quad T_1 \subseteq \langle T_2 \rangle \quad \text{and} \quad \langle T_1 \rangle \supseteq T_2, \quad \text{then} \quad \langle T_1 \rangle = \langle T_2 \rangle. \qquad ∎$$

To determine just which elements are in a subgroup $\langle S \rangle$, we must in general make repeated use of Theorem 7.1, beginning with S and obtaining larger and larger sets until we arrive at a subgroup. At the first step we adjoin to S all elements ab for $a, b \in S$, and also all elements a^{-1} for $a \in S$. This is then repeated with S replaced by the (possibly larger) set consisting of S together with the elements adjoined at the first step. And so on. In this way it can be seen that $\langle S \rangle$ must contain all elements $a_1 a_2 \cdots a_k$, where k is a positive integer and each of $a_1, a_2, \ldots, a_k$ is either an element of S or the inverse of an element of S. In fact, if S is nonempty then $\langle S \rangle$ will consist precisely of the set of all such elements $a_1 a_2 \cdots a_k$ (Problem 15.30). As Example 15.2 shows, however, in special cases $\langle S \rangle$ can be determined more directly than this.

Example 15.3.   Theorem 15.3 and the following calculations show that, in S_4,

$$\langle (1 \quad 2 \quad 4), (2 \quad 3 \quad 4) \rangle = \langle (1 \quad 2 \quad 3), (1 \quad 2)(3 \quad 4) \rangle.$$

First, $\{(1 \quad 2 \quad 4), (2 \quad 3 \quad 4)\} \subseteq \langle (1 \quad 2 \quad 3), (1 \quad 2)(3 \quad 4) \rangle$ because

$$(1 \quad 2 \quad 4) = (1 \quad 2 \quad 3)(1 \quad 2)(3 \quad 4)(1 \quad 2 \quad 3)$$

and

$$(2 \quad 3 \quad 4) = (1 \quad 3 \quad 2)(1 \quad 2)(3 \quad 4) = (1 \quad 2 \quad 3)^{-1}(1 \quad 2)(3 \quad 4).$$

Second, $\langle (1 \quad 2 \quad 4), (2 \quad 3 \quad 4) \rangle \supseteq \{(1 \quad 2 \quad 3), (1 \quad 2)(3 \quad 4)\}$ because

$$(1 \quad 2 \quad 3) = (1 \quad 2 \quad 4)(2 \quad 3 \quad 4)$$

and

$$(1 \quad 2)(3 \quad 4) = (2 \quad 3 \quad 4)(1 \quad 4 \quad 2) = (2 \quad 3 \quad 4)(1 \quad 2 \quad 4)^{-1}.$$ ∎

We now look at *direct products*, which provide a way to construct examples of groups and a way to describe certain groups in terms of less complicated component subgroups. If A and B are groups, then $A \times B$ is the Cartesian product of A and B:

$$A \times B = \{(a, b): a \in A \quad \text{and} \quad b \in B\}.$$

The following theorem shows how to make this set into a group.

Theorem 15.4.   *If A and B are groups, then $A \times B$ is a group with respect to the operation defined by*

$$(a_1, b_1)(a_2, b_2) = (a_1 a_2, b_1 b_2)$$

for all $a_1, a_2 \in A$ and $b_1, b_2 \in B$. The group $A \times B$ (with this operation) is called the direct product *of A and B.*

Remark.   In defining the direct product we have followed the convention of writing groups multiplicatively. If the operation on either A or B is something other than multiplication, then that is taken into account in working with $A \times B$. In $\mathbb{Z} \times \mathbb{Z}$, for example,

$$(a, b)(c, d) = (a + c, b + d)$$

for all $a, b, c, d \in \mathbb{Z}$. Also see Example 15.4.

PROOF.   The identity element of $A \times B$ is (e_A, e_B), where e_A and e_B denote the identity elements of A and B respectively. The inverse of (a, b) is (a^{-1}, b^{-1}). Verification of the associative law is left as an exercise (Problem 15.16).   ∎

Notice that if A and B are finite, then so is $A \times B$, with $|A \times B| = |A| \cdot |B|$.

Example 15.4.   Since $\mathbb{Z}_3 = \{[0], [1], [2]\}$ and $S_2 = \{(1), (1 \quad 2)\}$,

$$\mathbb{Z}_3 \times S_2 = \{([0], (1)), ([0], (1 \quad 2)), ([1], (1)),$$
$$([1], (1 \quad 2)), ([2], (1)), ([2], (1 \quad 2))\}.$$

And, for example,

$$([1], (1 \quad 2))([2], (1)) = ([1] \oplus [2], (1 \quad 2)(1)) = ([0], (1 \quad 2)).$$ ∎

If A and B are groups, then both

$$A \times \{e\} = \{(a, e): a \in A\}$$

and

$$\{e\} \times B = \{(e, b): b \in B\}$$

are subgroups of $A \times B$ (Problem 15.17) .

PROBLEMS

Verify each of the following equalities for subgroups of $\mathbb{Z}$.

15.1. $\langle -20, 8 \rangle = \langle 4 \rangle$

15.2. $\langle 24, -36, 54 \rangle = \langle 6 \rangle$

Verify each of the following equalities for subgroups of S_6.

15.3. $\langle (1 \quad 4 \quad 6 \quad 2 \quad 3 \quad 5) \rangle = \langle (1 \quad 2)(3 \quad 4)(5 \quad 6), (1 \quad 3 \quad 6)(2 \quad 4 \quad 5) \rangle$

15.4. $\langle (1 \quad 2 \quad 3), (4 \quad 5 \quad 6) \rangle = \langle (1 \quad 2 \quad 3)(4 \quad 5 \quad 6), (4 \quad 6 \quad 5) \rangle$

Determine the elements in each of the following subgroups of the group of symmetries of a square (Table 8.1).

15.5. $\langle \mu_2, \mu_5 \rangle$

15.6. $\langle \mu_3, \mu_6 \rangle$

15.7. What is the order of $\mathbb{Z}_4 \times \mathbb{Z}_7$?

15.8. What is the order of $S_4 \times S_4$?

15.9. Simplify the following expression in $\mathbb{Z}_4 \times S_4$.

$$([2], (1 \quad 2 \quad 3))^{-1}([1], (2 \quad 4))([2], (1 \quad 2 \quad 3)).$$

15.10. Simplify the following expression in $A \times B$, where A is the group in Example 5.8 and B is the group of symmetries of a square (Table 8.1).

$$(\alpha_{2,1}, \mu_2)^{-1}(\alpha_{3,2}, \mu_6)(\alpha_{2,1}, \mu_2).$$

15.11. Construct a Cayley table for the group in Example 15.4. Show that the group is cyclic.

15.12. Construct a Cayley table for $\mathbb{Z}_2 \times \mathbb{Z}_3$. Show that the group is cyclic.

15.13. The subgroup $\langle (1 \quad 4 \quad 3 \quad 2), (2 \quad 4) \rangle$ of S_4 has order 8. Determine its elements and write each one as a product of disjoint cycles.

15.14. If $\varnothing$ denotes the empty set, what is $\langle \varnothing \rangle$ (in any group G)?

15.15. Find necessary and sufficient conditions on a subset S of a group G for $S = \langle S \rangle$.

15.16. Prove the associative law for the direct product $A \times B$ of groups A and B.

15.17. Prove that $A \times \{e\}$ is a subgroup of $A \times B$.

15.18. Prove that $A \times B$ is Abelian iff both A and B are Abelian.

15.19. Give an example to show that a direct product of two cyclic groups is not necessarily cyclic. (Compare Problem 15.18.)

15.20. Prove that if A is a subgroup of G and B is a subgroup of H, then $A \times B$ is a subgroup of $G \times H$.

15.21. (a) List the elements of $S_3 \times \mathbb{Z}_2$.
(b) List the elements of the cyclic subgroup $\langle ((1 \quad 2), [1]) \rangle$ of $S_3 \times \mathbb{Z}_2$.
(c) List the elements of the cyclic subgroup $\langle ((1 \quad 2 \quad 3), [1]) \rangle$ of $S_3 \times \mathbb{Z}_2$.

15.22. (a) List the elements in the subgroup $\langle ([2], [2]) \rangle$ of $\mathbb{Z}_4 \times \mathbb{Z}_8$. (The first [2] is in $\mathbb{Z}_4$; the second is in $\mathbb{Z}_8$.)
(b) List the elements in the subgroup $\langle [2] \rangle \times \langle [2] \rangle$ of $\mathbb{Z}_4 \times \mathbb{Z}_8$. (Again, the first [2] is in $\mathbb{Z}_4$, and the second [2] is in $\mathbb{Z}_8$.)

15.23. Prove that if $a, b \in \mathbb{Z}$, then $\langle a, b \rangle = \langle d \rangle$, where d is the greatest common divisor of a and b. Formulate a generalization involving $\langle a_1, a_2, \ldots, a_n \rangle$ for $a_1, a_2, \ldots, a_n \in \mathbb{Z}$.

15.24. Prove that if $a, b \in \mathbb{Z}$, then $\langle a \rangle \cap \langle b \rangle = \langle m \rangle$, where m is the least common multiple of a and b. Formulate a generalization involving $\langle a_1 \rangle \cap \langle a_2 \rangle \cap \cdots \cap \langle a_n \rangle$ for $a_1, a_2, \ldots, a_n \in \mathbb{Z}$.

15.25. Prove that $\langle [a] \rangle = \mathbb{Z}_n$ iff $(a, n) = 1$, where (a, n) denotes the greatest common divisor of a and n.

15.26. Prove that if $(a, n) = d$, then $\langle [a] \rangle = \langle [d] \rangle$ in $\mathbb{Z}_n$. [Here (a, n) denotes the greatest common divisor of a and n.]

15.27. Prove that $\langle [a] \rangle = \langle [b] \rangle$ in $\mathbb{Z}_n$ iff $(a, n) = (b, n)$. (See Problem 15.26.)

15.28. Prove that if A is a group, then $\{(a, a): a \in A\}$ is a subgroup of $A \times A$. This is called the *diagonal subgroup* of $A \times A$. What is it, geometrically, for $A = \mathbb{R}$, with addition as the operation?

15.29. Each subgroup of $\mathbb{Z}$ (operation $+$) is cyclic. Prove this by assuming that H is a subgroup of $\mathbb{Z}$ and giving a reason for each of the following statements.
  (a) If $|H| = 1$, then H is cyclic.
  (b) If $|H| > 1$, then H contains a least positive element; call it b.
  (c) If $a \in H$ and $a = bq + r$, then $r \in H$.
  (d) If $a \in H$ and $a = bq + r$, with $0 \leq r < b$, then $r = 0$.
  (e) $H = \langle b \rangle$.

15.30. Prove that if G is a group with operation $*$, and S is a nonempty subset of G, then $\langle S \rangle$ is the set of all $a_1 * a_2 * \cdots * a_k$, where k is a positive integer and each of $a_1, a_2, \ldots, a_k$ is either an element of S or the inverse of an element of S. [*Question*: Show that the set described satisfies the conditions (a), (b), and (c) in Theorem 15.2, which characterize $\langle S \rangle$.]

SECTION 16   COSETS

We know that congruence modulo n is an equivalence relation on the group of integers. By viewing this in an appropriate way we are led to an idea that is important in the study of all groups. To do this, we first recall that if $n \in \mathbb{Z}$, then $\langle n \rangle$ is the subgroup consisting of all multiples of n. Because

$$a \equiv b (\bmod\, n) \quad \text{iff} \quad a - b \quad \text{is a multiple of } n,$$

we see that

$$a \equiv b (\bmod\, n) \quad \text{iff} \quad a - b \in \langle n \rangle.$$

The next theorem generalizes this by replacing $\mathbb{Z}$ by an arbitrary group G, $\langle n \rangle$ by an arbitrary subgroup H of G, and $a - b$ by the corresponding expression ab^{-1} in our general multiplicative notation.

Theorem 16.1.   *Let H denote a subgroup of a group G, and define a relation $\sim$ on G as follows:*

$$a \sim b \quad \textit{iff} \quad ab^{-1} \in H. \tag{16.1}$$

Then $\sim$ is an equivalence relation on G.

PROOF.   *Reflexive*: If $a \in G$, then $a \sim a$ because $aa^{-1} = e \in H$.
  Symmetric: If $a \sim b$, then $ab^{-1} \in H$, so $ba^{-1} = (ab^{-1})^{-1} \in H$ because H contains the inverse of each of its elements; thus $b \sim a$.

Transitive: If $a \sim b$ and $b \sim c$, then $ab^{-1} \in H$ and $bc^{-1} \in H$, so $ac^{-1} = (ab^{-1})(bc^{-1}) \in H$ because H is closed under products; thus $a \sim c$. ∎

The equivalence classes for this equivalence relation are called the *right cosets* of H in G. (Left cosets will be defined later in the section.) Looking back at the motivating example preceding Theorem 16.1, we see that the right cosets of $\langle n \rangle$ in $\mathbb{Z}$ are simply the congruence classes mod n. Lemma 16.1 will show that the right cosets of H in G have the form described by the following notation: for H a subgroup of a group G and $a \in G$, let

$$Ha = \{ha: h \in H\}.$$

If the group operation is $+$, then $H + a$ is written in place of Ha; similarly for other operations.

Example 16.1.   Let $G = \mathbb{Z}$ and $H = \langle 7 \rangle$. Then

$$H + 3 = \langle 7 \rangle + 3 = \{\ldots, -14, -7, 0, 7, 14, \ldots\} + 3$$
$$= \{\ldots, -11, -4, 3, 10, 17, \ldots\}.$$

This is the congruence class [3] in $\mathbb{Z}_7$. ∎

Example 16.2.   Let $G = S_3$ and $H = \{(1), (1 \quad 2)\}$. Then

$$H(1) = \{(1)(1), (1 \quad 2)(1)\} = \{(1), (1 \quad 2)\}$$
$$H(1 \quad 2 \quad 3) = \{(1)(1 \quad 2 \quad 3), (1 \quad 2)(1 \quad 2 \quad 3)\} = \{(1 \quad 2 \quad 3), (2 \quad 3)\}$$
$$H(1 \quad 3 \quad 2) = \{(1)(1 \quad 3 \quad 2), (1 \quad 2)(1 \quad 3 \quad 2)\} = \{(1 \quad 3 \quad 2), (1 \quad 3)\}.$$

Notice that these three sets form a partition of G. In fact, by the following lemma, they are the right cosets of H in G. ∎

Lemma 16.1.   *If H is a subgroup of a group G, and $a, b \in G$, then the following four conditions are equivalent.*

(a) $ab^{-1} \in H$.

(b) $a = hb$ *for some* $h \in H$.

(c) $a \in Hb$.

(d) $Ha = Hb$.

As a consequence, the right coset of H to which a belongs is Ha.

PROOF.   Problem 16.9. ∎

One right coset of H in G will be $H = He$. To compute all of the right cosets of a subgroup H in a finite group G, first write H, and then choose any element $a \in G$ such that $a \notin H$, and compute Ha. Next, choose any element $b \in G$ such that $b \notin H \cup Ha$, and compute Hb. Continue in this way until the elements of G have been exhausted.

Example 16.3.   Let $G = \mathbb{Z}_{12}$ and $H = \langle [4] \rangle$. The right cosets of H in G are

$$H = \{[0], [4], [8]\}$$
$$H \oplus [1] = \{[1], [5], [9]\}$$
$$H \oplus [2] = \{[2], [6], [10]\}$$
$$H \oplus [3] = \{[3], [7], [11]\}$$

∎

In Example 16.3, H has order three, and each of the right cosets of H also has three elements. The following lemma, which generalizes this observation, will play a crucial role in the next section. In stating the lemma we extend the notation $|H|$ to include subsets, so that for S a finite set, $|S|$ will denote the number of elements in S. Then $|S| = |T|$ if there exists a one-to-one mapping of S onto T. (This is, in fact, the definition of $|S| = |T|$.)

Lemma 16.2.   *If H is a finite subgroup of a group G, and $a \in G$, then $|H| = |Ha|$.*

PROOF.   By the remark preceding the lemma, it suffices to find a one-to-one mapping of H onto Ha. Define $\alpha: H \to Ha$ by $\alpha(h) = ha$ for each $h \in H$. This is a mapping because ha is uniquely determined by h and a. It is onto because Ha consists precisely of the elements of the form ha for $h \in H$. To show that α is one-to-one, assume that $h_1, h_2 \in H$ and $\alpha(h_1) = \alpha(h_2)$. Then $h_1 a = h_2 a$, and therefore, by right cancellation, $h_1 = h_2$. Thus α is one-to-one.   ∎

Left cosets result from replacing $ab^{-1} \in H$ by $a^{-1}b \in H$ in (16.1). Specifically, let H be a subgroup of a group G and define $\sim$ on G by

$$a \sim b \quad \text{iff} \quad a^{-1}b \in H. \tag{16.2}$$

Then $\sim$ is an equivalence relation on G, and the equivalence classes are called the *left cosets* of H in G. These cosets have the form

$$aH = \{ah: h \in H\}.$$

Problems 16.10 and 16.11 ask you to state and prove the analogues of Theorem 16.1 and Lemma 16.1 for left cosets. One application of cosets will appear in the next section, and others in Chapter V and XV.

PROBLEMS

16.1. Determine the right cosets of $\langle [4] \rangle$ in $\mathbb{Z}_8$.

16.2. Determine the right cosets of $\langle [3] \rangle$ in $\mathbb{Z}_{12}$.

16.3. For Example 8.1 (the group of symmetries of a square), determine the right cosets of $\langle \mu_5 \rangle$.

16.4. For Example 8.1, determine the right cosets of $\langle \mu_7 \rangle$.

16.5. Determine the right cosets of $\langle (1 \quad 2 \quad 3) \rangle$ in S_3.

16.6. Determine the right cosets of $\langle (1 \quad 3) \rangle$ in S_3.

16.7. Verify that if $H = \{(x, x): x \in \mathbb{R}\}$, then H is a subgroup of $\mathbb{R} \times \mathbb{R}$ (with $+$ as the operation in each component). Describe the right cosets of H in $\mathbb{R} \times \mathbb{R}$ geometrically. (H is called the *diagonal subgroup* of $\mathbb{R} \times \mathbb{R}$. See Problem 15.28.)

16.8. Consider $\mathbb{Z}$ as a subgroup of $\mathbb{R}$ with addition as the operation. Think of the elements of $\mathbb{R}$ as the points on a directed line in the usual way, and then describe the right cosets of $\mathbb{Z}$ in $\mathbb{R}$ geometrically. (*Suggestion*: Start with a specific example, such as $\mathbb{Z} + \frac{1}{2}$.)

16.9. Prove Lemma 16.1. [*Suggestion*: Prove that (a) implies (b) implies (c) implies (d) implies (a).]

16.10. State and prove Theorem 16.1 with the condition $ab^{-1} \in H$ replaced by $a^{-1}b \in H$.

16.11. State and prove the analogue of Lemma 16.1 for left cosets in place of right cosets.

16.12. Verify that if H is a subgroup of an Abelian group G, and $a \in G$, then the right coset of H to which a belongs is the same as the left coset of H to which a belongs.

16.13. Compute the left cosets of H in G for H and G as in Example 16.2. Verify that in this case the collection of left cosets is different from the collection of right cosets.

16.14. Let $G = S_3$ and $H = \langle (1 \quad 3) \rangle$.
  (a) Determine the right cosets of H in G.
  (b) Determine the left cosets of H in G.
  (c) Verify that the collection of right cosets is different from the collection of left cosets.

16.15. (a) Compute the right and left cosets of $\langle (1 \quad 2 \quad 3) \rangle$ in S_3.
  (b) Verify that for each element π of S_3 the right coset of $\langle (1 \quad 2 \quad 3) \rangle$ to which π belongs is the same as the left coset of $\langle (1 \quad 2 \quad 3) \rangle$ to which π belongs.

16.16. Prove that each right coset of $A \times \{e\}$ in $A \times B$ contains precisely one element from $\{e\} \times B$. (See Problem 15.17.)

16.17. Compute the right cosets of $\langle ((1 \quad 2), [1]) \rangle$ in $S_3 \times \mathbb{Z}_2$.

16.18. Compute the left cosets of $\langle (1 \quad 2) \rangle \times \langle [1] \rangle$ in $S_3 \times \mathbb{Z}_2$.

16.19. Prove that a subset S of a group G cannot be a right coset of two different subgroups of G.

16.20. Assume that H and K are subgroups of a group G and that $a \in G$. The subset HaK of G defined by

$$HaK = \{hak \colon h \in H \quad \text{and} \quad k \in K\}$$

is called a *double coset* of H and K in G. Prove that if HaK and HbK are double cosets of H and K in G, then they are either equal or disjoint; that is, either $HaK = HbK$ or $HaK \cap HbK = \varnothing$.

SECTION 17   LAGRANGE'S THEOREM

Of all the subsets of a finite group, only some will be subgroups. Lagrange's Theorem narrows the field.

Lagrange's Theorem. *If H is a subgroup of a finite group G, then the order of H is a divisor of the order of G.*

Thus, since S_3 has order $3! = 6$, any subgroup of S_3 must have order 1, 2, 3, or 6; S_3 cannot have subgroups of order 4 or 5. A group of order 7 can have only the two obvious subgroups: $\{e\}$, of order 1; and the group itself, of order 7.

PROOF OF LAGRANGE'S THEOREM. The right cosets of H, being equivalence classes, form a partition of G (Theorem 9.1). Thus two right cosets of H are either equal or disjoint. Moreover, since G is finite, there can be only finitely many of these cosets. Choose one element from each coset and let the elements chosen be $a_1, a_2, \ldots, a_k$. Then

$$G = Ha_1 \cup Ha_2 \cup \cdots \cup Ha_k.$$

Each coset Ha_i contains $|H|$ elements (Lemma 16.2), and no element is in more than one coset. It follows that $|G| = |H| \cdot k$. Therefore $|H|$ is a divisor of $|G|$. ∎

The integer k appearing in the proof of Lagrange's Theorem is called the *index* of H in G. This index will be denoted $[G : H]$. Thus $[G : H]$ is the number of right cosets of H in G, and

$$|G| = |H| \cdot [G : H].$$

Notice that this equation shows that $[G : H]$, as well as $|H|$, is a divisor of $|G|$.

Corollary. *If G is a finite group and $a \in G$, then the order of a is a divisor of the order of G.*

PROOF.   By the corollary of Theorem 14.3, $o(a) = |\langle a \rangle|$. But $\langle a \rangle$ is a subgroup of G, and thus $|\langle a \rangle|$ is a divisor of $|G|$ by Lagrange's Theorem. Therefore $o(a)$ is a divisor of $|G|$.   ∎

Corollary. *A group G of prime order contains no subgroup other than $\{e\}$ and G.*

PROOF.   This is a direct consequence of Lagrange's Theorem, since a prime has no positive divisor other than 1 and itself.   ∎

Corollary. *Each group G of prime order is cyclic, generated by any one of its nonidentity elements.*

PROOF.   If $a \in G$ and $a \neq e$, then $\langle a \rangle \neq \{e\}$, so $\langle a \rangle = G$ by the preceding corollary.   ∎

Example 17.1.   In contrast to groups of prime order (the preceding corollary), groups of prime-squared order need not be cyclic. For example, a direct product $\mathbb{Z}_p \times \mathbb{Z}_p$ has order p^2 but is not cyclic because it has no element of order p^2; each of its nonidentity elements has order p. (See Problem 17.24.)   ∎

Example 17.2.   Lagrange's Theorem greatly simplifies the problem of determining all the subgroups of a finite group. Let G denote the group of symmetries of a square (Example 8.1). Aside from $\{\mu_1\}$ and the whole group, any subgroup must have order 2 or 4. Subgroups of order 2 are easy to determine — each contains the identity together with an element of order 2 (these correspond to appearances of μ_1 on the diagonal of the Cayley table). In this case $\langle \mu_3 \rangle$, $\langle \mu_5 \rangle$, $\langle \mu_6 \rangle$, $\langle \mu_7 \rangle$, and $\langle \mu_8 \rangle$ all have order 2. This leaves only order 4, and careful inspection will reveal three subgroups of this order: $\langle \mu_2 \rangle$, $\langle \mu_5, \mu_6 \rangle$, and $\langle \mu_7, \mu_8 \rangle$.   ∎

Figure 17.1 shows the subgroups of G and the inclusion relations between them. This is an example of a *subgroup lattice*. Such a figure is constructed as follows. If A and B are subgroups of G with $A \subsetneq B$, and there is no subgroup C such that $A \subsetneq C \subsetneq B$, then B appears above A and a segment is drawn connecting A and B.

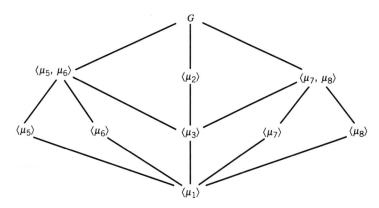

Figure 17.1

Notice that Lagrange's Theorem does *not* say that if n is a divisor of the order of G, then G has a subgroup of order n. That would be false. For example, there is a group of order 12 having no subgroup of order 6 (Problem 17.28). On the other hand, the following theorem shows that every *cyclic* group of finite order n does have a subgroup of every order dividing n. (This theorem should be considered optional.)

Theorem 17.1.   *Let G be a cyclic group of (finite) order n, with $G = \langle a \rangle = \{e, a, a^2, \ldots, a^{n-1}\}$.*

(a) *Every subgroup of G is cyclic.*

(b) *If $1 \le k < n$, then a^k generates a subgroup of order $n/(k, n)$, where (k, n) is the greatest common divisor of k and n.*

(c) *For each positive divisor d of n, G has exactly one subgroup of order d.*

PROOF.   (a) Let H denote a subgroup of G. If $H = \langle e \rangle$, then H is cyclic. Suppose $H \ne \langle e \rangle$, and let m denote the least integer such that $1 \le m < n$ and $a^m \in H$. We shall prove that $H = \langle a^m \rangle$.

Suppose $a^t \in H$. By the Division Algorithm, there are integers q and r such that

$$t = mq + r, \quad 0 \le r < m.$$

Since

$$a^t = a^{mq+r} = (a^m)^q a^r,$$

we have

$$a^r = a^t (a^m)^{-q}.$$

But $a^t \in H$ and $a^m \in H$, so $a^r \in H$. This implies that $r = 0$, since $0 \le r < m$ and m is the least positive integer such that $a^m \in H$. Therefore $a^t = \langle a^m \rangle^q$ and $a^t \in \langle a^m \rangle$. Thus $H = \langle a^m \rangle$ and H is cyclic.

(b) Let $g = (k, n)$. Notice that $|\langle a^k \rangle|$ is the least positive integer s such that $a^{ks} = e$. By Theorem 14.3, $a^{ks} = e$ iff $n \mid ks$. But $n \mid ks$ iff $(n/g) \mid (k/g)s$. Since $(n/g, k/g) = 1$, we have $(n/g) \mid (k/g)s$ iff $(n/g) \mid s$ (Lemma 13.1). The least positive integer s such that $(n/g) \mid s$ is n/g. Therefore, the order of a^k is n/g, that is, $n/(k, n)$.

(c) Suppose $d > 0$ and $d \mid n$. Then $n = du$ for some positive integer u, and $(n, u) = u$ since $u \mid n$. Therefore, by part (b), a^u generates a subgroup of order $n/u = d$. Thus G has at least one subgroup of order d.

To prove that G has a unique subgroup of each possible order, assume that H_1 and H_2 are subgroups of G with $|H_1| = |H_2|$. By the proof of part (a), $H_1 = \langle a^{m_1} \rangle$ and $H_2 = \langle a^{m_2} \rangle$, where m_i is the least positive integer such that $a^{m_i} \in H_i$ for $i = 1, 2$. By part (b), $|\langle a^{m_i} \rangle| = n/(m_i, n)$ for $i = 1, 2$. But $m_i \mid n$ for $i = 1, 2$ because $a^n = e$ so that $a^n \in \langle a^{m_i} \rangle$. Therefore, $(m_i, n) = m_i$ so $n/(m_i, n) = n/m_i$ for $i = 1, 2$. Thus $n/m_1 = n/m_2$, $m_1 = m_2$, $\langle a^{m_1} \rangle = \langle a^{m_2} \rangle$, and $H_1 = H_2$. ∎

Example 17.3. The positive divisors of 12 are 1, 2, 3, 4, 6, and 12. Thus every cyclic group of order 12 has exactly one (cyclic) subgroup of each of these orders. Figure 17.2 illustrates this by showing the subgroup lattice of $\mathbb{Z}_{12}$. The notation a^k in Theorem 17.1 becomes $k[a]$ in this example.

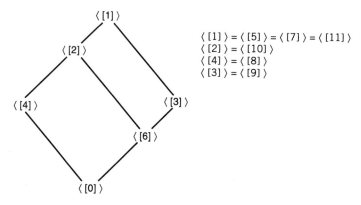

$$\langle [1] \rangle = \langle [5] \rangle = \langle [7] \rangle = \langle [11] \rangle$$
$$\langle [2] \rangle = \langle [10] \rangle$$
$$\langle [4] \rangle = \langle [8] \rangle$$
$$\langle [3] \rangle = \langle [9] \rangle$$

Figure 17.2 ∎

The Norwegian mathematician Ludwig Sylow (1832–1918) proved in 1872 that if p^k is any power of a prime and p^k is a divisor of $|G|$, then G must have a subgroup of order p^k. Thus, for example, any group of order 12 must have subgroups of orders 2, 3, and 4. A proof of Sylow's Theorem, for the case of the highest power of each prime dividing the order of a finite group, will be given in Section 53.

Lagrange's Theorem is named for the French mathematician Joseph-Louis Lagrange (1736–1813), generally regarded as one of the two foremost mathematicians of the eighteenth century, the other being the Swiss-born mathematician Leonhard Euler (1707–1783). Lagrange did not prove this theorem in the form applying to all finite groups; indeed, the general concept of *group* was not studied until after Lagrange. But he did use the theorem in a significant special case, and therefore it is fitting that it be named for him.

PROBLEMS

17.1. Find $[S_3: \langle (1\ \ 2) \rangle]$.

17.2. Find $[S_4: \langle (1\ \ 2\ \ 3) \rangle]$.

17.3. Find $[\mathbb{Z}_{10}: \langle [2] \rangle]$.

17.4. Find $[\mathbb{Z}_{40}: \langle [12], [20] \rangle]$.

17.5. Let G denote the group in Example 8.1 (the group of symmetries of a square). Find $[G: \langle \mu_8 \rangle]$.

17.6. With G as in Problem 17.5, find $[G: \langle \mu_4 \rangle]$.

17.7. A group G has subgroups of orders 4 and 10, and $|G| < 50$. What can you conclude about $|G|$?

17.8. A finite group G has elements of orders p and q, where p and q are distinct primes. What can you conclude about $|G|$?

17.9. Assume that G is a group with a subgroup H such that $|H| = 6$, $[G : H] > 4$, and $|G| < 50$. What are the possibilities for $|G|$?

17.10. Assume that G is a group with a subgroup H such that $|G| < 45$, $|H| > 10$, and $[G : H] > 3$. Find $|G|$, $|H|$, and $[G : H]$.

17.11. Find all of the subgroups of $\mathbb{Z}_6$. Also construct the subgroup lattice.

17.12. Find all of the subgroups of S_3. Also construct the subgroup lattice.

17.13. Find all of the subgroups of $\mathbb{Z}_2 \times \mathbb{Z}_2$.

17.14. Find all of the subgroups of $\mathbb{Z}_3 \times \mathbb{Z}_3$.

17.15. Determine the number of subgroups of $\mathbb{Z}_p \times \mathbb{Z}_p$ if p is a prime. (Compare Problems 17.13 and 17.14.)

17.16. Find all of the subgroups of $\mathbb{Z}_2 \times \mathbb{Z}_4$. (There are eight.)

17.17. Find all of the subgroups of $\mathbb{Z}_{36}$. Also construct the subgroup lattice. Use Theorem 17.1.

17.18. Assume that G is a cyclic group of order n, that $G = \langle a \rangle$, that $k \mid n$, and that $H = \langle a^k \rangle$. Find $[G : H]$.

17.19. Assume that H is a subgroup of a finite group G, and that G contains elements $a_1, a_2, \ldots, a_n$ such that $a_i a_j^{-1} \notin H$ for $1 \leq i \leq n, 1 \leq j \leq n$, and $i \neq j$. What can you conclude about $[G : H]$?

17.20. Assume that A is a subgroup of a finite group G and that B is a subgroup of a finite group H. Then $A \times B$ is a subgroup of $G \times H$ (Problem 15.20). Express $[G \times H: A \times B]$ in terms of $[G : A]$ and $[H : B]$, and prove that your result is correct.

17.21. Assume that A is a finite group and let D denote the diagonal subgroup of $A \times A$ (Problem 15.28). Find $[A \times A : D]$.

17.22. The *exponent* of a group G is the smallest positive integer n such that $a^n = e$ for each $a \in G$, if such an integer n exists. Prove that every finite group has an exponent, and that this exponent divides the order of the group. [*Suggestion*: Consider the least common multiple of $\{o(a): a \in G\}$.]

17.23. Determine the exponent of each of the following groups. (See Problem 17.22.)
(a) S_3 (b) $\mathbb{Z}_n$ (c) $\mathbb{Z}_2 \times \mathbb{Z}_2$
(d) $\mathbb{Z}_2 \times \mathbb{Z}_3$ (e) $\mathbb{Z}_m \times \mathbb{Z}_n$

17.24. Prove that if G is a group of order p^2 (p a prime), and G is not cyclic, then $a^p = e$ for each $a \in G$.

17.25. Prove that if H is a subgroup of G, $[G : H] = 2$, $a, b \in G$, $a \notin H$, and $b \notin H$, then $ab \in H$.

17.26. Verify that S_4 has at least one subgroup of order k for each divisor k of 24.

17.27. Prove that if A and B are finite subgroups of a group G, and $|A|$ and $|B|$ have no common divisor greater than 1, then $A \cap B = \{e\}$. (See Problem 7.13.)

17.28. The subgroup $\langle (1 \quad 2 \quad 3), (1 \quad 2)(3 \quad 4) \rangle$ of S_4 has order 12. Determine its elements. Verify that it has no subgroup of order 6.

17.29. Prove that a finite cyclic group of order n has exactly one subgroup of index m for each positive divisor m of n.

17.30. If H is a subgroup of G and $[G : H] = 2$, then the right cosets of H in G are the same as the left cosets of H in G. Why?

17.31. Write a proof of Lagrange's Theorem using left cosets rather than right cosets.

17.32. Prove that if H is a subgroup of a finite group G, then the number of right cosets of H in G equals the number of left cosets of H in G.

17.33. Prove that a finite cyclic group G of order n has $\phi(n)$ different generators [that is, $\phi(n)$ different elements a such that $G = \langle a \rangle$], where $\phi(n)$ is defined as in Problem 12.10.

SECTION 18   ISOMORPHISM

We speak of the set of integers, but if we were to allow ourselves to be distracted by things that are mathematically irrelevant, we might think that there were many such sets. The integers can appear in Arabic notation $\{\ldots, 1, 2, 3, \ldots\}$, in Roman notation $\{\ldots, \text{I}, \text{II}, \text{III}, \ldots\}$, in German $\{\ldots, \text{ein}, \text{zwei}, \text{drei}, \ldots\}$, and so on; but mathematically we want to think of all these sets as being the same. The idea that filters out such irrelevant differences as names and notation is *isomorphism*. Isomorphism allows us to treat certain groups as being alike just as geometrical congruence allows us to treat certain triangles as being alike. The idea also applies in many cases that are less obvious than that of the integers presented in different language or notation. As a hint of this, consider the subgroup $\langle (1 \quad 2 \quad 3) \rangle$ of S_3 (Table 18.1), and the group $\mathbb{Z}_3$ (Table 18.2). The elements of $\langle (1 \quad 2 \quad 3) \rangle$ are *permutations* and the operation is *composition*; the elements of $\mathbb{Z}_3$ are *congruence classes* and the operation is *addition modulo* 3. Thus the underlying sets and operations arise in totally different ways. Still, these groups are obviously somehow alike: given the correspondence $(1) \leftrightarrow [0]$, $(1 \quad 2 \quad 3) \leftrightarrow [1]$, and $(1 \quad 3 \quad 2) \leftrightarrow [2]$, we could fill in all of one table just by knowing the other. The following definition isolates the idea behind this example.

Table 18.1

	(1)	(1 2 3)	(1 3 2)
(1)	(1)	(1 2 3)	(1 3 2)
(1 2 3)	(1 2 3)	(1 3 2)	(1)
(1 3 2)	(1 3 2)	(1)	(1 2 3)

Table 18.2

$\oplus$	[0]	[1]	[2]
[0]	[0]	[1]	[2]
[1]	[1]	[2]	[0]
[2]	[2]	[0]	[1]

Definition.   Let G be a group with operation $*$, and let H be a group with operation #. An *isomorphism of G* onto H is a mapping $\theta: G \to H$ that is one-to-one and onto and satisfies

$$\theta(a * b) = \theta(a) \,\#\, \theta(b)$$

for all $a, b \in G$. If there is an isomorphism of G onto H, then G and H are said to be *isomorphic* and we write $G \approx H$.

The condition $\theta(a * b) = \theta(a) \,\#\, \theta(b)$ is sometimes described by saying that θ *preserves the operation*. It makes no difference whether we operate in G and then apply θ, or apply θ first and then operate in H—we get the same result either way. See Figure 18.1.

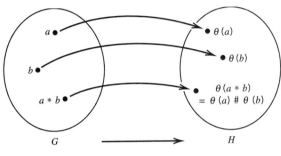

Figure 18.1

Example 18.1.   With the obvious mapping $(\ldots, 1 \mapsto \mathrm{I}, 2 \mapsto \mathrm{II}, 3 \mapsto \mathrm{III}, \ldots)$ from the integers in Arabic notation to the integers in Roman notation, we get the same answer whether we add in Arabic $(1 + 2 = 3)$ and then translate into Roman $(3 \mapsto \mathrm{III})$, or translate first $(1 \mapsto \mathrm{I}, 2 \mapsto \mathrm{II})$ and then add $(\mathrm{I} + \mathrm{II} = \mathrm{III})$. (And this is true for all $m + n$, not just $1 + 2$.)   ∎

Example 18.2.   To illustrate the definition for the case $\langle(1 \quad 2 \quad 3)\rangle$ and $\mathbb{Z}_3$, already considered, use

$$\theta((1)) = [0], \quad \theta((1 \quad 2 \quad 3)) = [1], \quad \theta((1 \quad 3 \quad 2)) = [2].$$

Then, for example,

$$\theta((1 \quad 2 \quad 3)(1 \quad 3 \quad 2)) = \theta((1)) = [0]$$

and

$$\theta((1 \quad 2 \quad 3)) \oplus \theta((1 \quad 3 \quad 2)) = [1] \oplus [2] = [0]$$

so

$$\theta((1 \quad 2 \quad 3)(1 \quad 3 \quad 2)) = \theta((1 \quad 2 \quad 3)) \oplus \theta((1 \quad 3 \quad 2)).$$

There are nine such equations to be checked in this case (one for each entry of the Cayley table). There will be n^2 total equations to check if G and H are finite of order n.   ∎

Example 18.3. Define a mapping θ from the set of all integers to the set of even integers by $\theta(n) = 2n$ for each n. This mapping is one-to-one and onto, and, moreover, it preserves addition:

$$\theta(m + n) = 2(m + n) = 2m + 2n = \theta(m) + \theta(n)$$

for all integers m and n. Thus θ is an isomorphism, and the group of all integers (operation $+$) is isomorphic to the group of even integers (operation $+$). ∎

The preceding example may seem puzzling—isomorphic groups are supposed to be essentially alike, but surely there is a difference between the integers and the even integers. This example shows that, *as groups*, each with addition as operation, there is in fact no essential difference between the integers and the even integers. Remember, however, that we are ignoring multiplication at present; when we take both addition and multiplication into account in the next chapter, we shall be able to detect a difference between the two sets. In Section 29 we shall see precisely what distinguishes the integers mathematically.

Example 18.4. Let $\mathbb{R}^p$ denote the set of positive real numbers, and define $\theta: \mathbb{R}^p \to \mathbb{R}$ by $\theta(x) = \log_{10} x$ for each $x \in \mathbb{R}^p$. Here $\mathbb{R}^p$ is a group with respect to multiplication, $\mathbb{R}$ is a group with respect to addition, and θ is an isomorphism. The mapping θ is one-to-one and onto because it has an inverse, $\phi: \mathbb{R} \to \mathbb{R}^p$ defined by $\phi(x) = 10^x$. The mapping preserves the operation because

$$\theta(xy) = \log_{10}(xy) = \log_{10} x + \log_{10} y = \theta(x) + \theta(y)$$

for all $x, y \in \mathbb{R}^p$.

This isomorphism explains the historical importance of logarithms in simplifying calculations: When used in conjunction with a table of logarithms, it allows one to replace a multiplication problem by an addition problem. Now it is much easier to use a calculator or computer, of course. ∎

The following theorem shows that any group isomorphic to an Abelian group must also be Abelian. This can be taken as an illustration that isomorphic groups share significant properties. On the other hand, it can also be taken as showing that the property of being Abelian (or non-Abelian) is a significant property of groups. For the significant properties of groups, as groups, are those properties that are shared by isomorphic groups—that is what isomorphism is all about.

Theorem 18.1. *If G and H are isomorphic groups and G is Abelian, then H is Abelian.*

PROOF. Let the operations on G and H be $*$ and $\#$, respectively, and let $\theta: G \to H$ be an isomorphism. If $x, y \in H$, then there are elements $a, b \in G$ such that $\theta(a) = x$ and $\theta(b) = y$. Since θ preserves the operation and G is Abelian,

$$x \# y = \theta(a) \# \theta(b) = \theta(a * b) = \theta(b * a) = \theta(b) \# \theta(a) = y \# x.$$

This proves that H is Abelian. ∎

Other examples of properties shared by isomorphic groups will be given in the next section. The following theorem gives some technical facts about isomorphisms. Notice that, except for part (e), the theorem does not assume that θ is one-to-one or onto; only that it preserves the operation. Such a mapping is called a *homomorphism*. That is, if G and H are groups with operations $*$ and #, respectively, then $\theta: G \to H$ is a homomorphism if

$$\theta(a * b) = \theta(a) \, \# \, \theta(b)$$

for all $a, b \in G$. Group homomorphisms are studied in more detail in Chapter V.

Theorem 18.2. *Let G be a group with operation $*$, let H be a group with operation #, and let $\theta: G \to H$ be a mapping such that $\theta(a * b) = \theta(a) \, \# \, \theta(b)$ for all $a, b \in G$. Then:*

(a) $\theta(e_G) = e_H$,

(b) $\theta(a^{-1}) = \theta(a)^{-1}$ *for each $a \in G$,*

(c) $\theta(a^k) = \theta(a)^k$ *for each $a \in G$ and each integer k,*

(d) $\theta(G)$, *the image of θ, is a subgroup of H, and*

(e) *if θ is one-to-one, then $G \approx \theta(G)$.*

PROOF. (a) Because θ preserves the operation and $e_G * e_G = e_G$, we have $\theta(e_G) \, \# \, \theta(e_G) = \theta(e_G * e_G) = \theta(e_G)$. But $\theta(e_G) \in H$, so $\theta(e_G) = \theta(e_G) \, \# \, e_H$. This gives $\theta(e_G) \, \# \, \theta(e_G) = \theta(e_G) \, \# \, e_H$, from which $\theta(e_G) = e_H$ by left cancellation.

(b) Using, in order, the properties of θ and a^{-1}, and part (a), we can write $\theta(a) \, \# \, \theta(a^{-1}) = \theta(a * a^{-1}) = \theta(e_G) = e_H$. Therefore $\theta(a^{-1})$ must equal $\theta(a)^{-1}$, because $\theta(a)^{-1}$ is the unique solution of $\theta(a) \, \# \, x = e_H$ in H.

(c) We use induction for $k > 0$, and leave $k \leq 0$ as an exercise. The case $k = 1$ is obvious. Assume $\theta(a^k) = \theta(a)^k$. Then

$$\theta(a^{k+1}) = \theta(a^k * a) = \theta(a^k) \, \# \, \theta(a) = \theta(a)^k \, \# \, \theta(a) = 0(a)^{k+1}.$$

(d) By parts (a) and (b), $\theta(G)$ contains e_H, and also along with any element the inverse of that element. Thus it now suffices to show that $\theta(G)$ is closed with respect to #. Assume that $x, y \in \theta(G)$. Then $x = \theta(a)$ and $y = \theta(b)$ for some $a, b \in G$; thus $x \, \# \, y = \theta(a) \, \# \, \theta(b) = \theta(a * b) \in \theta(G)$, which establishes closure.

(e) By assumption, θ preserves the operation and is one-to-one. Also, thought of as a mapping from G to $\theta(G)$, θ is onto. Therefore $\theta: G \to \theta(G)$ is an isomorphism.   ∎

PROBLEMS

18.1. Prove that $\mathbb{Z}$ is isomorphic to the multiplicative group of all rational numbers of the form 2^m $(m \in \mathbb{Z})$.

18.2. Prove that $\mathbb{Z} \times \mathbb{Z}$ is isomorphic to the multiplicative group of all rational numbers of the form $2^m 3^n$ $(m, n \in \mathbb{Z})$.

18.3. Fill in the blanks in the following table to obtain a group isomorphic to $\mathbb{Z}_4$. What is the isomorphism?

$*$	a	b	c	d
a				
b				
c				
d				

18.4. Repeat Problem 18.3, with $\mathbb{Z}_4$ replaced by $\mathbb{Z}_2 \times \mathbb{Z}_2$.

18.5. Assume that $H = \{u, v, w, x, y, z\}$ is a group with respect to multiplication and that $\theta \colon S_3 \to H$ is an isomorphism with

$$\theta((1)) = u, \quad \theta((1 \ \ 2 \ \ 3)) = v, \quad \theta((1 \ \ 3 \ \ 2)) = w,$$
$$\theta((1 \ \ 2)) = x, \quad \theta((1 \ \ 3)) = y, \quad \theta((2 \ \ 3)) = z.$$

Replace each of the following by the appropriate letter, either u, v, w, x, y, or z.
(a) xw (b) w^{-1} (c) v^5 (d) $zv^{-1}x$

18.6. Assume that $H = \{u, v, w, x, y, z\}$ is a group with respect to multiplication and that $\theta \colon \mathbb{Z}_6 \to H$ is an isomorphism with

$$\theta([0]) = u, \quad \theta([1]) = v, \quad \theta([2]) = w,$$
$$\theta([3]) = x, \quad \theta([4]) = y, \quad \theta([5]) = z.$$

Replace each of the following by the appropriate letter, either u, v, w, x, y, or z.
(a) xw (b) w^{-1} (c) v^5 (d) $zv^{-1}x$

18.7. One of the conditions in the definition of isomorphism was not used in the proof of Theorem 18.1. Which one?

18.8. Describe an isomorphism between the two groups in Example 5.5.

18.9. Prove that if G, H, and K are groups, and $\theta \colon G \to H$ and $\phi \colon H \to K$ are isomorphisms, then $\phi \circ \theta \colon G \to K$ is an isomorphism. (Use multiplication for the group operations.)

18.10. Prove that if G and H are groups, then $G \times H \approx H \times G$. (Let the operations on G and H be denoted by $*$ and $\#$, respectively.)

18.11. Prove that $\theta(x) = e^x$ defines an isomorphism of the group $\mathbb{R}$ of all real numbers (operation addition) onto the group $\mathbb{R}^p$ of all positive real numbers (operation multiplication). What is the inverse of the mapping θ? Is the inverse an isomorphism?

18.12. Verify that $\mathbb{Z}_4$ (operation $\oplus$) is isomorphic to $\mathbb{Z}_5^{\#}$ (operation $\odot$). (See Example 11.4.)

18.13. Use the mapping $\theta([a]_6) = ([a]_2, [a]_3)$ to show that $\mathbb{Z}_6 \approx \mathbb{Z}_2 \times \mathbb{Z}_3$. First verify that θ is well-defined.

18.14. Prove that if m and n are relatively prime (that is, have greatest common divisor 1), then

$$\mathbb{Z}_{mn} \approx \mathbb{Z}_m \times \mathbb{Z}_n.$$

(Problem 18.13 is a special case.)

18.15. Assume that G, H, and θ are as in Theorem 18.2. Assume also that B is a subgroup of H and that

$$A = \{x \in G \colon \theta(x) \in B\}.$$

Prove that A is a subgroup of G. (A is called the *inverse image* of B with respect to θ. Thus the inverse image of a subgroup with respect to a homomorphism is a subgroup.)

SECTION 19   MORE ON ISOMORPHISM

If two finite groups are isomorphic, then they must have the same order because an isomorphism is, among other things, one-to-one and onto. Turning this around (that is, using the contrapositive), we get the simplest of all tests for showing that two groups are not isomorphic: If G and H are groups and $|G| \neq |H|$, then G and H are not isomorphic. It is useful to have a list of other properties that are shared by isomorphic groups. Such a list will frequently make it much easier to determine quickly if two groups are not isomorphic.

Theorem 19.1.   *Assume that G and H are groups and that $G \approx H$.*

(a) $|G| = |H|$.

(b)   *If G is Abelian, then H is Abelian.*

(c)   *If G is cyclic, then H is cyclic.*

(d)   *If G has a subgroup of order n (for some positive integer n), then H has a subgroup of order n.*

(e)   *If G has an element of order n, then H has an element of order n.*

(f)   *If every element of G is its own inverse, then every element of H is its own inverse.*

(g)   *If every element of G has finite order, then every element of H has finite order.*

PROOF.   Statement (a) is the observation made at the beginning of this section. Statement (b) was proved as Theorem 18.1. The proofs of the remaining statements are left to the problems at the end of this section.   ∎

Many other properties could be added to the list in Theorem 19.1, but the ones given there should help to give better insight into the nature of isomorphism. It is also important to be able to determine if groups are isomorphic, of course. This problem is considered in the following more general discussion.

In Example 18.2 we saw that two particular groups of order 3 are isomorphic. It will follow from Theorem 19.3 that *any* two groups of order 3 are isomorphic. This means that in essence there is only one group of order 3. More precisely, it means that there is only one *isomorphism class* of groups of order 3, where by an isomorphism class we mean an equivalence class for the equivalence relation imposed on groups by isomorphism, as described by the following theorem. We revert to the convention of using multiplicative notation for unspecified group operations.

Theorem 19.2.   *Isomorphism is an equivalence relation on the class of all groups.*

PROOF.   If G is a group, then it is easy to verify that the identity mapping $\iota : G \to G$ is an isomorphism. Thus $G \approx G$, and $\approx$ is reflexive.

Assume that $G \approx H$ and that $\theta : G \to H$ is an isomorphism. Then θ is one-to-one and onto, so there is an inverse mapping $\theta^{-1} : H \to G$ (Theorem 2.2). We shall show that θ^{-1} is an isomorphism; it is necessarily one-to-one and onto. Suppose that $a, b \in H$. We must show that $\theta^{-1}(ab) = \theta^{-1}(a)\theta^{-1}(b)$. Let $\theta^{-1}(a) = x$ and $\theta^{-1}(b) = y$. Then $a = \theta(x)$ and $b = \theta(y)$ so that $ab = \theta(x)\theta(y) = \theta(xy)$. This implies that $\theta^{-1}(ab) = xy = \theta^{-1}(a)\theta^{-1}(b)$, as required. Therefore $H \approx G$, and $\approx$ is symmetric.

Finally, it is easy to show that if G, H, and K are groups, and $\theta: G \to H$ and $\phi: H \to K$ are isomorphisms, then $\phi \circ \theta: G \to K$ is also an isomorphism (Problem 18.9). Thus $\approx$ is transitive.   ∎

Theorem 19.3.   *If p is a prime and G is a group of order p, then G is isomorphic to $\mathbb{Z}_p$.*

PROOF. Let the operation on G be $*$, and let a be a nonidentity element of G. Then $\langle a \rangle$ is a subgroup of G and $\langle a \rangle \neq \{e\}$, so $\langle a \rangle = G$ by the second corollary of Lagrange's Theorem (Section 17). Thus $G = \{e, a, a^2, \ldots, a^{p-1}\}$.

Define $\theta: G \rightarrow \mathbb{Z}_p$ by $\theta(a^k) = [k]$. This mapping is well-defined and one-to-one because

$$a^{k_1} = a^{k_2} \quad \text{iff} \quad a^{k_1 - k_2} = e \quad \text{iff} \quad p \mid (k_1 - k_2) \quad \text{iff} \quad [k_1] = [k_2].$$

Also, θ is onto. Finally, if $a^m, a^n \in G$, then

$$\theta(a^m * a^n) = \theta(a^{m+n}) = [m + n] = [m] \oplus [n] = \theta(a^m) \oplus (a^n).$$

Therefore, θ is an isomorphism and $G \approx \mathbb{Z}_p$.   ∎

With Theorem 19.3 we have completely classified all groups of prime order. The principal problem of finite group theory is to do the same for groups of all finite orders. An immense amount of work has been done on this problem. Although much of this work is well beyond the range of this book, it will still be interesting to look at what is known in some special cases. Proofs will be omitted.

Table 19.1 shows the number of isomorphism classes of groups of order n for each n from 1 to 32. The label "number of groups" is what is conventionally used in place of the more accurate but longer phrase "number of isomorphism classes of groups." Whenever we ask for "all" groups having a property (such as being Abelian and of order n, for example), we are really asking for one group from each isomorphism class of groups with that property.

Notice from Table 19.1 that there is just one group of each prime order—Theorem 19.3 guarantees that. But notice that there is just one group of order 15, and 15 is not a prime. The key is this:

There is just one group of order n

iff

n is a prime or a product of distinct primes $p_1, p_2, \ldots, p_k$
such that $p_j \nmid (p_i - 1)$
for $1 \leq i \leq k, 1 \leq j \leq k$.

(For a proof, see [6, Theorem 9.2.7].) Thus, for instance, there is also only one group of order 33, since $33 = 3 \cdot 11$ and $3 \nmid 10$ and $11 \nmid 2$. For such n, any group of order n will be isomorphic to $\mathbb{Z}_n$.

Another easy-to-describe case is $n = p^2$, the square of a prime. There are two isomorphism classes of groups of order p^2: the group $\mathbb{Z}_{p^2}$ is in one class, and $\mathbb{Z}_p \times \mathbb{Z}_p$ is

Table 19.1

Order	1	2	3	4	5	6	7	8	9	10	11	12	13	14	15	16
Number of groups	1	1	1	2	1	2	1	5	2	2	1	5	1	2	1	14

Order	17	18	19	20	21	22	23	24	25	26	27	28	29	30	31	32
Number of groups	1	5	1	5	2	2	1	15	2	2	5	4	1	4	1	51

in the other. (Compare the entries for $n = 4, 9$, and 25 in Table 19.1.) Notice that all of these groups of order p^2 are Abelian. If $n = p^3$, the cube of a prime, then there are five isomorphism classes: three of these classes consist of Abelian groups and are represented by $\mathbb{Z}_{p^3}$, $\mathbb{Z}_{p^2} \times \mathbb{Z}_p$, and $\mathbb{Z}_p \times \mathbb{Z}_p \times \mathbb{Z}_p$; the other two classes consist of non-Abelian groups. (For the definition of direct products of more than two groups, see Problem 19.24.)

If only finite Abelian groups are considered, then the problem of determining all isomorphism classes is completely settled by the following theorem, which has been known since at least the 1870s. (Each book listed at the end of this chapter gives a proof.)

Fundamental Theorem of Finite Abelian Groups.   *If G is a finite Abelian group, then G is the direct product of cyclic groups of prime power order. Moreover, if*

$$G \approx A_1 \times A_2 \times \cdots \times A_s$$

and

$$G \approx B_1 \times B_2 \times \cdots \times B_t,$$

where each A_i and each B_j is cyclic of prime power order, then $s = t$ and, after suitable relabeling of subscripts, $\left| A_i \right| = \left| B_i \right|$ for $1 \le i \le s$.

Because each cyclic group of prime power order p^k is isomorphic to $\mathbb{Z}_{p^k}$, we can use this theorem to exhibit one group from each isomorphism class of finite Abelian groups.

Example 19.1.   Let $n = 125 = 5^3$. To apply the theorem, first determine all possible ways of factoring 125 as a product of (not necessarily distinct) prime powers: $5^3, 5^2 \cdot 5, 5 \cdot 5 \cdot 5$. Each factorization gives a different isomorphism class, so there are three isomorphism classes of Abelian groups of order 125. Here is one representative from each class:

$$\mathbb{Z}_{5^3}, \quad \mathbb{Z}_{5^2} \times \mathbb{Z}_5, \quad \mathbb{Z}_5 \times \mathbb{Z}_5 \times \mathbb{Z}_5.$$

This same idea accounts for the general statement, made earlier, that there are three isomorphism classes of Abelian groups of order p^3 for each prime p. ∎

Example 19.2.   There are six isomorphism classes of Abelian groups of order $200 = 2^3 \times 5^2$, Here is one representative from each class:

$$\mathbb{Z}_{2^3} \times \mathbb{Z}_{5^2}, \quad \mathbb{Z}_{2^3} \times \mathbb{Z}_5 \times \mathbb{Z}_5, \quad \mathbb{Z}_{2^2} \times \mathbb{Z}_2 \times \mathbb{Z}_{5^2},$$
$$\mathbb{Z}_{2^2} \times \mathbb{Z}_2 \times \mathbb{Z}_5 \times \mathbb{Z}_5, \quad \mathbb{Z}_2 \times \mathbb{Z}_2 \times \mathbb{Z}_2 \times \mathbb{Z}_{5^2},$$
$$\mathbb{Z}_2 \times \mathbb{Z}_2 \times \mathbb{Z}_2 \times \mathbb{Z}_5 \times \mathbb{Z}_5.$$
∎

Further remarks about the problem of classifying finite groups can be found in Section 33. As the case of Abelian groups suggests, the number of groups of order n is influenced largely by the character of the prime factorization of n, and not by the size of n alone. Table 19.1 gives a further hint of this, and here is one more remark to make it even clearer: There are 267 groups of order 64.

PROBLEMS

In each of Problems 19.1–19.10, the two groups are not isomorphic. Give a reason in each case.

19.1. $\mathbb{Z}_5, \mathbb{Z}_6$

19.2. $\mathbb{Z}_5 \times \mathbb{Z}_2, \mathbb{Z}_7$

19.3. $\mathbb{Z}_6, S_3$

19.4. $\mathbb{Z}_4 \times \mathbb{Z}_2$, symmetry group of a square

19.5. $\mathbb{Z}_4 \times \mathbb{Z}_2, \mathbb{Z}_2 \times \mathbb{Z}_2 \times \mathbb{Z}_2$

19.6. $\mathbb{Z}, \mathbb{Q}$ (both with +)

19.7. $\mathbb{Z}, \mathbb{R}$ (both with +)

19.8. $\mathbb{Z}_8 \times \mathbb{Z}_4, \mathbb{Z}_{16} \times \mathbb{Z}_2$

19.9. $S_3 \times \mathbb{Z}_2$, the group in Problem 17.28.

19.10. $\mathbb{Q}$ (operation +), positive rational numbers (operation ·).

19.11. Is there a noncyclic group of order 59? Why?

19.12. Is there a noncyclic group of order 39? Why? Problem 19.17 may help.

19.13. Give an example of a noncyclic group of order 49.

19.14. Give examples of two nonisomorphic groups of order 27.

19.15. If p is a prime, then there are five isomorphism classes of Abelian groups of order p^4. Give one group from each class. (Compare Example 19.1.)

19.16. If p is a prime, then there are seven isomorphism classes of Abelian groups of order p^5. Give one group from each class. (Compare Example 19.1.)

19.17. Prove that if G is a cyclic group of order n, then $G \approx \mathbb{Z}_n$. (*Suggestion*: Write G as $G = \langle a \rangle = \{e, a, \ldots, a^{n-1}\}$.)

19.18. Prove that if G is an infinite cyclic group, then $G \approx \mathbb{Z}$. (Compare Problem 19.17.)

Assume that G and H are groups and that $G \approx H$. Prove each of the statements in Problems 19.19–19.23.

19.19. If G is cyclic, then H is cyclic.

19.20. If G has a subgroup of order n, then H has a subgroup of order n.

19.21. If G has an element of order n, then H has an element of order n.

19.22. If $x^{-1} = x$ for each $x \in G$, then $x^{-1} = x$ for each $x \in H$.

19.23. If each element of G has finite order, then each element of H has finite order.

19.24. If $\{A_1, A_2, \ldots, A_n\}$ is any collection of groups, each with juxtaposition as operation, then the *direct product* $A_1 \times A_2 \times \cdots \times A_n$ is $\{(a_1, a_2, \ldots, a_n) : a_i \in A_i\}$ with operation

$$(a_1, a_2, \ldots, a_n)(b_1, b_2, \ldots, b_n) = (a_1 b_1, a_2 b_2, \ldots, a_n b_n).$$

(a) Verify that this direct product is a group.
(b) If each group A_i is finite, what is the order of the direct product?

19.25. An isomorphism of a group onto itself is called an *automorphism*. Prove that the set of all automorphisms of a group is itself a group with respect to composition. [The group of all automorphisms of a group G is called the *automorphism group* of G, and will be denoted Aut(G). Notice that the elements of Aut(G) are mappings from G onto G. Examples are given in the problems that follow, and also in Problem 51.10.]

19.26. Verify that if G is an Abelian group, then $\theta: G \to G$ defined by $\theta(a) = a^{-1}$ for each $a \in G$ is an automorphism of G. (See Problem 19.25.)

19.27. Prove that if $[a]$ is a generator for $\mathbb{Z}_n$, and $\theta: \mathbb{Z}_n \to \mathbb{Z}_n$ is defined by $\theta([k]) = [ka]$, then $\theta \in$ Aut($\mathbb{Z}_n$). (See Problem 19.25.)

19.28. Prove that $\left| \text{Aut}(\mathbb{Z}_2) \right| = 1$.

19.29. Prove that Aut($\mathbb{Z}_3$) $\approx \mathbb{Z}_2$. (See Problem 19.27.)

19.30. Prove that Aut($\mathbb{Z}_4$) $\approx \mathbb{Z}_2$.

19.31. Prove that if p is a prime, then Aut($\mathbb{Z}_p$) $\approx \mathbb{Z}_{p-1}$.

19.32. Prove that Aut($\mathbb{Z}$) $\approx \mathbb{Z}_2$.

19.33. Assume that n is an integer, $n > 1$, and that the standard form for n, in the sense of (13.1), is

$$n = p_1^{e_1} p_2^{e_2} \cdots p_k^{e_k} \quad (p_1 < p_2 < \cdots < p_k).$$

Prove that

$$\mathbb{Z}_n \approx \mathbb{Z}_{p_1^{e_1}} \times \mathbb{Z}_{p_2^{e_2}} \times \cdots \times \mathbb{Z}_{p_k^{e_k}}.$$

[*Suggestion*: Consider $\theta([a]) = ([a], [a], \ldots, [a])$. Be sure to prove that, among other things, θ is well-defined. Also, Problem 13.14 may help.]

SECTION 20   CAYLEY'S THEOREM

We have seen that the nature of groups can vary widely—from groups of numbers to groups of permutations to groups defined by tables. Cayley's Theorem asserts that in spite of this broad range of possibilities, each group is isomorphic to some group of permutations. This is an example of what is known as a *representation theorem*—it tells us that any group can be represented as (is isomorphic to) something reasonably concrete. In place of studying the given group, we can just as well study the concrete object (permutation group) representing it; and this can be an advantage. However, it can also be a disadvantage, for part of the power of abstraction comes from the fact that abstraction filters out irrelevancies, and in concentrating on any concrete object we run the risk of being distracted by irrelevancies. Still, Cayley's Theorem has proved to be useful, and its proof ties together several of the important ideas that we have studied.

In proving Cayley's Theorem, we associate with each element of a group G a permutation of the set G. The way in which this is done is suggested by looking at the Cayley table for a finite group. As we observed after Theorem 14.1, each element of a finite group appears exactly once in each row of the Cayley table for the group (if we ignore the row labels at the outside of the table). Thus the elements in each row of the table are merely a permutation of the elements in the first row. What we do is simply associate with each element a of G the permutation whose first row (in two-row form) is the first row of the Cayley table and whose second row is the row labeled by a. If the elements in the first row are $a_1, a_2, \ldots, a_n$ (in that order), then the elements in the row labeled by a will be $aa_1, aa_2, \ldots, aa_n$ (in that order).

Example 20.1.   Consider the Cayley table for $\mathbb{Z}_6$, given in Example 11.2. The permutation associated with [3] by the idea just described is

$$\begin{pmatrix} [0] & [1] & [2] & [3] & [4] & [5] \\ [3] & [4] & [5] & [0] & [1] & [2] \end{pmatrix}.$$ ∎

Cayley's Theorem extends this idea to groups that are not necessarily finite, and also establishes that this association of group elements with permutations is an isomorphism.

Cayley's Theorem.   *Every group is isomorphic to a permutation group on its set of elements.*

PROOF.   Our isomorphism will be a mapping $\theta: G \to \text{Sym}(G)$. We begin by describing the permutation that θ will assign to an element $a \in G$.

For $a \in G$, define $\lambda_a: G \to G$ by $\lambda_a(x) = ax$ for each $x \in G$. Each such mapping λ_a is one-to-one and onto because each equation $ax = b$ ($b \in G$) has a unique solution in G [Theorem 14.1(c)]. Thus $\lambda_a \in \text{Sym}(G)$ for each $a \in G$.

Now define $\theta: G \to \text{Sym}(G)$ by $\theta(a) = \lambda_a$ for each $a \in G$. To prove that θ is one-to-one, suppose that $\theta(a) = \theta(b)$; we shall deduce that $a = b$. From $\theta(a) = \theta(b)$ we have $\lambda_a = \lambda_b$, and thus in particular $\lambda_a(e) = \lambda_b(e)$, since the mappings λ_a and λ_b can be equal only if they give the same image for each element in their common domain. But $\lambda_a(e) = ae = a$ and $\lambda_b(e) = be = b$, so $\lambda_a(e) = \lambda_b(e)$ implies $a = b$. Thus θ is one-to-one.

Finally, if $a, b \in G$, then $\theta(ab) = \lambda_{ab}$ and $\theta(a) \circ \theta(b) = \lambda_a \circ \lambda_b$, implying that $\theta(ab) = \theta(a) \circ \theta(b)$ if $\lambda_{ab} = \lambda_a \circ \lambda_b$. To verify the latter, let $x \in G$ and write $\lambda_{ab}(x) = (ab)x = a(bx) = \lambda_a(bx) = \lambda_a(\lambda_b x) = (\lambda_a \circ \lambda_b)(x)$. Thus we have proved that $\theta(ab) = \theta(a) \circ \theta(b)$ for all $a, b \in G$.

It follows from Theorem 18.2(e) that G is isomorphic to $\theta(G)$, a subgroup of $\text{Sym}(G)$. This completes the proof.   ∎

Corollary.   *Every group of finite order n is isomorphic to a subgroup of S_n.*

PROOF.   Label the elements of the group $a_1, a_2, \ldots, a_n$. The construction in the proof of Cayley's Theorem will assign to an element a in the group the permutation

$$\begin{pmatrix} a_1 & a_2 & \cdots & a_n \\ aa_1 & aa_2 & \cdots & aa_n \end{pmatrix}.$$

As remarked preceding Example 20.1, the elements $aa_1, aa_2, \ldots, aa_n$ are just $a_1, a_2, \ldots, a_n$ in some order. If we replace each aa_i by that a_j such that $aa_i = a_j$, and then replace each a_k in the permutation by the number k, we obtain an element of S_n. (See Example 20.2, for instance.) By assigning each a to the element of S_n obtained in this way, we get the desired isomorphism.   ∎

Example 20.2.   Label the elements of $\mathbb{Z}_3$ as follows: $a_1 = [0], a_2 = [1], a_3 = [2]$. Then the construction in the proof of Cayley's Theorem yields

$$\theta(a_3) = \theta([2]) = \begin{pmatrix} [0] & [1] & [2] \\ [2] & [0] & [1] \end{pmatrix} = \begin{pmatrix} a_1 & a_2 & a_3 \\ a_3 & a_1 & a_2 \end{pmatrix}.$$

The idea in the proof of the corollary is simply to delete the a's and keep the subscripts, so that

$$\begin{pmatrix} 1 & 2 & 3 \\ 3 & 1 & 2 \end{pmatrix}$$

is assigned to a_3. Notice that this is an element of S_3 because $|\mathbb{Z}_3| = 3$.   ∎

Corollary.   *For each positive integer n, there are only finitely many isomorphism classes of groups of order n.*

    PROOF.   By the previous corollary every group of order n is isomorphic to a subgroup of S_n. But S_n is finite, and thus has only finitely many subgroups. Hence there can be only finitely many isomorphism classes of groups of order n.   ∎

PROBLEMS

20.1. Write the permutation associated with each element of $\mathbb{Z}_5$ by the isomorphism θ in the proof of Cayley's Theorem.

20.2. Write the permutation associated with each element of the symmetry group of a square (Example 8.1) by the isomorphism θ in the proof of Cayley's Theorem.

20.3. Write the permutation associated with each element of $\langle (1\ \ 2\ \ 3) \rangle$ by the isomorphism θ in the proof of Cayley's Theorem.

20.4. Give an alternative proof of Cayley's Theorem by replacing λ_a (lambda for "left" multiplication) by ρ_a (rho for "right" multiplication) defined as follows: $\rho_a(x) = xa^{-1}$ for each $x \in G$. Why is xa^{-1} used here, rather than xa?

20.5. Assume that G is a group, and that $\lambda_a: G \to G$ and $\rho_a: G \to G$ are defined by $\lambda_a(x) = ax$ and $\rho_a(x) = xa^{-1}$ for each $x \in G$. For each $a \in G$, define $\gamma_a: G \to G$ by $\gamma_a = \rho_a \circ \lambda_a$. Prove that each γ_a is an isomorphism of G onto G. [In the language of Problem 19.25, each γ_a is an automorphism of G. Notice that, in particular, each $\gamma_a \in \mathrm{Sym}(G)$.]

20.6. Let γ_a be defined as in Problem 20.5, and define $\phi: G \to \mathrm{Sym}(G)$ by $\phi(a) = \gamma_a$ for each $a \in G$. Prove that $\phi(a) = \iota_G$ iff $ax = xa$ for all $a \in G$.

NOTES ON CHAPTER IV

Here are six standard references on group theory. The first is a classic and is listed because of its historical importance. The second emphasizes infinite groups.

1.   Burnside, W., *Theory of Groups of Finite Order*, 2nd ed., Cambridge University Press, Cambridge, England, 1911; Dover, New York, 1955.

2.   Kurosh, A., *Theory of Groups*, Vols. I and II, trans. from the Russian by K. A. Hirsch; Chelsea, New York, 1979.

3.   Robinson, D. J. S., *A Course in the Theory of Groups*, 2nd ed., Springer-Verlag, New York, 1996.

4.   Rose, John S., *A Course on Group Theory*, Dover, 1994.

5.   Rotman, J. J., *An Introduction to the Theory of Groups*, 4th ed., Springer-Verlag, New York, 1995.

6.   Scott, W. R., *Group Theory*, Dover, 1985.

CHAPTER V

GROUP HOMOMORPHISMS

One way to study a relatively large and complicated group is to study its smaller and less complicated subgroups. But it would also be useful to be able to study the group as a whole, much as a globe allows us to study the earth's surface—brought down to manageable size in a way that preserves as many essential features as possible. Homomorphisms, which are more general than isomorphisms, can help us do just that. A homomorphism is a mapping from one group to another that preserves the operation but is not necessarily one-to-one. Thus the image of a homomorphism can be smaller than the domain, but it will generally reflect some essential features of the domain. Even more important, subgroups and images of homomorphisms can be used together to show that most groups are built up from smaller component groups, as we shall see at the end of this chapter. The concept of homomorphism also extends to other algebraic structures that we'll study; it is unquestionably one of the most important concepts in algebra.

SECTION 21   HOMOMORPHISMS OF GROUPS. KERNELS

Definition.   If G is a group with operation $*$, and H is a group with operation #, then a mapping $\theta : G \rightarrow H$ is a *homomorphism* if

$$\theta(a * b) = \theta(a) \, \# \, \theta(b)$$

for all $a, b \in G$.

Every isomorphism is a homomorphism. But a homomorphism need not be one-to-one, and it need not be onto.

Example 21.1.   For any positive integer n, define $\theta : \mathbb{Z} \rightarrow \mathbb{Z}_n$ by $\theta(a) = [a]$ for each $a \in \mathbb{Z}$. Then $\theta(a + b) = [a + b] = [a] \oplus [b] = \theta(a) \oplus \theta(b)$ for all $a, b \in \mathbb{Z}$, so that θ is a homomorphism. It is onto but not one-to-one.   ∎

Example 21.2.   Define $\theta : \mathbb{Z} \rightarrow \mathbb{Z}$ by $\theta(a) = 2a$ for each $a \in \mathbb{Z}$. Then $\theta(a + b) = 2(a + b) = 2a + 2b = \theta(a) + \theta(b)$ for all $a, b \in \mathbb{Z}$. Thus θ is a homomorphism. It is one-to-one, but not onto.   ∎

Example 21.3.   For each $r \in \mathbb{R}$, define $\rho_r : \mathbb{R} \rightarrow \mathbb{R}$ by $\rho_r(a) = ar$ for each $a \in \mathbb{R}$. Then ρ_r is a homomorphism from the additive group of R to itself: $\rho_r(a + b) = (a + b)r =$

$ar + br = \rho_r(a) + \rho_r(b)$ for all $a, b \in \mathbb{R}$. Notice that ρ_r is a homomorphism precisely because of the distributive law $(a + b)r = ar + br$. ∎

Example 21.4. Let A and B be groups, and $A \times B$ their direct product (defined in Section 15). Then $\pi_1 : A \times B \to A$ defined by $\pi_1((a, b)) = a$ is a homomorphism from $A \times B$ onto A:

$$\pi_1((a_1, b_1)(a_2, b_2)) = \pi_1((a_1 a_2, b_1 b_2)) = a_1 a_2 = \pi_1((a_1, b_1))\pi_1((a_2, b_2)).$$

(Problem 21.1 asks you to justify each step.) Also, $\pi_2 : A \times B \to B$ defined by $\pi_2((a, b)) = b$ is a homomorphism (Problem 21.2). ∎

We can see now that the basic assumption of Theorem 18.2 is that the mapping θ there is a homomorphism. Thus that theorem tells us that if $\theta : G \to H$ is a homomorphism, then

1. $\theta(e_G) = e_H$.
2. $\theta(a^{-1}) = \theta(a)^{-1}$ for each $a \in G$,
3. $\theta(a^k) = \theta(a)^k$ for each $a \in G$ and each integer k,
4. $\theta(G)$, the image of θ, is a subgroup of H, and
5. if θ is one-to-one, then $G \approx \theta(G)$.

The subgroup $\theta(G)$ is called a *homomorphic image* of G. We shall see that nearly everything about a homomorphic image is determined by the domain of the homomorphism and the following subset of the domain.

Definition. If $\theta : G \to H$ is a homomorphism, then the *kernel* of θ is the set of all elements $a \in G$ such that $\theta(a) = e_H$. This set will be denoted by *Ker* θ.

The kernel of a homomorphism is always a subgroup of the domain. Before proving that, however, let us look at some examples.

Example 21.5. For the homomorphism $\theta : \mathbb{Z} \to \mathbb{Z}_n$ in Example 21.1, $a \in$ Ker θ iff $\theta(a) = [a] = [0]$. Therefore, Ker θ consists of the set of all integral multiples of n. ∎

Example 21.6. In Example 21.2, Ker $\theta = \{0\}$. ∎

Example 21.7. In Example 21.4, Ker $\pi_1 = \{e_A\} \times B$. ∎

Theorem 21.1. *If $\theta : G \to H$ is a homomorphism, then Ker θ is a subgroup of G. Moreover, θ is one-to-one iff Ker $\theta = \{e_G\}$.*

PROOF. We use multiplicative notation for the operations on both G and H. To show that Ker θ is a subgroup of G, we use Theorem 7.1. Theorem 18.2(a) shows that $e_G \in$ Ker θ. If $a, b \in$ Ker θ, then $\theta(ab) = \theta(a)\theta(b) = e_H e_H = e_H$, so that $ab \in$ Ker θ. Theorem 18.2(b) shows that $\theta(a^{-1}) = \theta(a)^{-1}$; therefore, if $a \in$ Ker θ, then $\theta(a^{-1}) = \theta(a)^{-1} = e_H^{-1} = e_H$, and $a^{-1} \in$ Ker θ. Thus Ker θ is a subgroup.

Since $e_G \in \text{Ker }\theta$, it is clear that if θ is one-to-one, then $\text{Ker }\theta = \{e_G\}$. Assume, on the other hand, that $\text{Ker }\theta = \{e_G\}$. If $a, b \in G$ and $\theta(a) = \theta(b)$, then $\theta(a)\theta(b)^{-1} = e_H$, $\theta(a)\theta(b^{-1}) = e_H$, $\theta(ab^{-1}) = e_H$, $ab^{-1} \in \text{Ker }\theta = \{e_G\}$; hence $ab^{-1} = e_G$ and $a = b$. This proves that θ is one-to-one. ∎

Kernels have one more property in common: they are all *normal*, in the sense of the next definition.

Definition. A subgroup N of a group G is a *normal subgroup* of G if $gng^{-1} \in N$ for all $n \in N$ and all $g \in G$. If N is a normal subgroup, we write $N \triangleleft G$.

Example 21.8. If N is a subgroup of an Abelian group G, and $n \in N$ and $g \in G$, then $gng^{-1} = n \in N$. Thus every subgroup of an Abelian group is a normal subgroup. ∎

Example 21.9. The subgroup $\langle (1 \quad 2) \rangle = \{(1), (1 \quad 2)\}$ of S_3 is not normal, since, for example,

$$(1 \quad 2 \quad 3)(1 \quad 2)(1 \quad 2 \quad 3)^{-1} = (2 \quad 3) \notin \langle (1 \quad 2) \rangle.$$ ∎

It often helps to realize that

$$gng^{-1} \in N \quad \text{for all} \quad n \in N \quad \text{and} \quad g \in G$$

iff

$$g^{-1}ng \in N \quad \text{for all} \quad n \in N \quad \text{and} \quad g \in G.$$

Thus $N \triangleleft G$ iff $g^{-1}ng \in N$ for all $n \in N$ and $g \in G$ (Problem 21.20).

A large collection of examples of normal subgroups—in fact, all examples, as we shall see in the next section—is given by the following theorem.

Theorem 21.2. *If G and H are groups and $\theta : G \to H$ is a homomorphism, then $\text{Ker }\theta \triangleleft G$.*

PROOF. We know from Theorem 21.1 that $\text{Ker }\theta$ is a subgroup of G; thus it suffices to show that it is normal. Let $n \in \text{Ker }\theta$ and $g \in G$. Then $\theta(n) = e$ so that $\theta(gng^{-1}) = \theta(g)\theta(n)\theta(g^{-1}) = \theta(g)e\theta(g)^{-1} = e$. Therefore $gng^{-1} \in \text{Ker }\theta$, as required. ∎

A homomorphism that is one-to-one is sometimes called a *monomorphism*; a homomorphism that is onto is sometimes called an *epimorphism*. We shall not use either of these terms.

PROBLEMS

21.1. Justify each step in the proof that π_1 is a homomorphism in Example 21.4.

21.2. Prove that $\pi_2 : A \times B \to B$, as defined in Example 21.4, is a homomorphism.

21.3. Consider $M(2, \mathbb{Z})$ as a group with respect to addition (Example 3.5 and Problem 5.16). Define $\rho_r : M(2, \mathbb{Z}) \rightarrow M(2, \mathbb{Z})$ by $\rho_r(x) = xr$ for each $x \in M(2, \mathbb{Z})$, where

$$r = \begin{bmatrix} 1 & 0 \\ 0 & 0 \end{bmatrix}.$$

Prove that ρ_r is a homomorphism.

21.4. Find $\mathrm{Ker}\, \rho_r$ for ρ_r in Problem 21.3.

21.5. Define $\theta : \mathbb{Z}_6 \rightarrow \mathbb{Z}_3$ by $\theta([a]_6) = [a]_3$ for each $[a]_6 \in \mathbb{Z}_6$.
(a) Prove that θ is well-defined.
(b) Prove that θ is a homomorphism.
(c) What is $\mathrm{Ker}\, \theta$?

21.6. (a) Show that $\alpha : \mathbb{Z}_3 \rightarrow \mathbb{Z}_6$ by $\alpha([a]_3) = [a]_6$ is not well-defined. (Compare Problem 21.5.)
(b) For which pairs m, n is $\beta : \mathbb{Z}_m \rightarrow \mathbb{Z}_n$ by $\beta([a]_m) = [a]_n$ well-defined?

21.7. Prove that every homomorphic image of an Abelian group is Abelian.

21.8. Prove that every homomorphic image of a cyclic group is cyclic.

21.9. Prove that if $\theta : G \rightarrow H$ is a homomorphism and A is a subgroup of G, then $\theta(A)$ is a subgroup of H.

21.10. Prove that if $\theta : G \rightarrow H$ is a homomorphism and B is a subgroup of H, then $\theta^{-1}(B)$ is a subgroup of G, where $\theta^{-1}(B) = \{g \in G : \theta(g) \in B\}$, the inverse image of B under θ.

21.11. Prove that if $\alpha : G \rightarrow H$ is a homomorphism and $\beta : H \rightarrow K$ is a homomorphism, then $\beta \circ \alpha : G \rightarrow K$ is a homomorphism.

21.12. (a) With α and β as in Problem 21.11, prove that $\mathrm{Ker}\, \alpha \subseteq \mathrm{Ker}\, \beta \circ \alpha$.
(b) Give an example of $\alpha : G \rightarrow H$ and $\beta : H \rightarrow K$ for which $\mathrm{Ker}\, \alpha \neq \mathrm{Ker}\, \beta \circ \alpha$.

21.13. Let n and k denote positive integers, and define $\theta : \mathbb{Z} \rightarrow \mathbb{Z}_n$ by

$$\theta(a) = [ka] \quad \text{for each } a \in \mathbb{Z}.$$

Prove that θ is a homomorphism.

21.14. Determine $\mathrm{Ker}\, \theta$ in each of the following cases, for θ, n, and K as in Problem 21.13.
(a) $n = 6, k = 5$.   (b) $n = 6, k = 3$.   (c) General n and k.

21.15. Let G denote the subgroup $\{1, -1, i, -i\}$ of complex numbers (operation multiplication). Define $\theta : \mathbb{Z} \rightarrow G$ by $\theta(n) = i^n$ for each $n \in \mathbb{Z}$. Verify that θ is a homomorphism and determine $\mathrm{Ker}\, \theta$. (Recall that $i^2 = -1$.)

21.16. Define λ_r from $M(2, \mathbb{Z})$ to itself by $\lambda_r(x) = rx$ for each $x \in M(2, \mathbb{Z})$, where $r = \begin{bmatrix} 1 & 1 \\ 0 & 0 \end{bmatrix}$.
Prove that λ_r is a homomorphism and then find $\mathrm{Ker}\, \lambda_r$.

21.17. Prove that if $\theta : G \rightarrow H$ is a homomorphism and $a \in G$, then $o(\theta(a)) \mid o(a)$.

21.18. There is a unique homomorphism $\theta : \mathbb{Z}_6 \rightarrow S_3$ such that $\theta([1]) = (1 \quad 2 \quad 3)$. Determine $\theta([k])$ for each $[k] \in \mathbb{Z}_6$. Which elements are in $\mathrm{Ker}\, \theta$?

21.19. True or false: If $N \triangleleft G$, then $gng^{-1} = n$ for all $n \in N$ and $g \in G$. Justify your answer.

21.20. Prove that $N \triangleleft G$ iff $g^{-1}ng \in N$ for all $n \in N$ and $g \in G$. (In other words, prove that $gng^{-1} \in N$ for all $n \in N$ and $g \in G$ iff $g^{-1}ng \in N$ for all $n \in N$ and $g \in G$.)

21.21. By choosing a rectangular coordinate system, the points of a plane can be identified with the elements of $\mathbb{R} \times \mathbb{R}$. What is the geometric interpretation of $\pi_1 : \mathbb{R} \times \mathbb{R} \rightarrow \mathbb{R}$ defined by $\pi_1((a, b)) = a$ (as in Example 21.4)?

21.22. Determine all of the normal subgroups of the group of symmetries of a square. (See Example 17.2.)

21.23. Prove that if θ is a homomorphism from G onto H, and $N \lhd G$, then $\theta(N) \lhd H$.

21.24. If A and B are groups, then $\{e\} \times B \lhd A \times B$. Give two different proofs of this, one using the definition of normal subgroup, and the other using Theorem 21.2 and Example 21.4.

21.25. Prove that if C denotes any collection of normal subgroups of a group G, then the intersection of all the groups in C is also a normal subgroup of G. (See Theorem 15.1.)

21.26. Prove that if N is a subgroup of G, then $N \lhd G$ iff $Ng = gN$ for each $g \in G$.

21.27. Prove that if N is a subgroup of G and $[G : N] = 2$, then $N \lhd G$. (*Suggestion*: Use Problem 21.26.)

21.28. Determine all of the normal subgroups of S_3. (See Problem 17.12.)

21.29. Prove that if H and N are subgroups of a group G and $N \lhd G$, then $H \cap N \lhd H$.

21.30. For $\mathbb{C}$ the complex numbers, let E, I, J, and K be the elements of $M(2, \mathbb{C})$ defined as follows:

$$E = \begin{bmatrix} 1 & 0 \\ 0 & 1 \end{bmatrix}, \quad I = \begin{bmatrix} 0 & 1 \\ -1 & 0 \end{bmatrix}, \quad J = \begin{bmatrix} 0 & i \\ i & 0 \end{bmatrix}, \quad K = \begin{bmatrix} i & 0 \\ 0 & -i \end{bmatrix}.$$

(If you are unfamiliar with complex numbers, do this problem after reading Section 32.)
  (a) Form a Cayley table to verify that, with matrix multiplication as the operation, $\{\pm E, \pm I, \pm J, \pm K\}$ is a group of order 8. This group is called the *quaternion group*; denote it by Q_8.
  (b) The group Q_8 has one subgroup of order 1, one subgroup of order 2, three subgroups of order 4, and one subgroup of order 8. Find the elements in each subgroup. (Each subgroup of order less than 8 is cyclic.)
  (c) Verify that every subgroup of Q_8 is a normal subgroup of Q_8.
  (d) Verify that Q_8 is not Abelian. With part (c), this shows that Q_8 is a non-Abelian group in which every subgroup is normal. Compare Example 21.8.

21.31. If $G = \langle a \rangle$, and $\theta : G \rightarrow H$ is a homomorphism, then θ is completely determined by $\theta(a)$. Explain.

21.32. There is only one homomorphism from $\mathbb{Z}_2$ to $\mathbb{Z}_3$. Why?

21.33. Verify that if G and H are any groups, and $\theta : G \rightarrow H$ is defined by $\theta(g) = e_H$ for each $g \in G$, then θ is a homomorphism.

21.34. Define $\theta : \mathbb{Z} \times \mathbb{Z} \rightarrow \mathbb{Z}$ by $\theta((a, b)) = a + b$. Verify that θ is a homomorphism and determine Ker θ.

21.35. Determine all of the homomorphisms of $\mathbb{Z}$ onto $\mathbb{Z}$.

SECTION 22   QUOTIENT GROUPS

It may not be obvious at the outset, but quotient groups, which we introduce in this section, are essentially the same as homomorphic images. The proof that they are essentially the same comes with Theorem 22.2 and the Fundamental Homomorphism Theorem (in Section 23).

Each group $\mathbb{Z}_n$ is constructed in a simple way from the group of integers. The set of all multiples of the integer n forms a subgroup, $\langle n \rangle$, of $\mathbb{Z}$, and the elements of $\mathbb{Z}_n$ are the right cosets of that subgroup (Section 16). Moreover, the operation $\oplus$ of $\mathbb{Z}_n$ depends in a natural way on the operation $+$ of the integers: $[a] \oplus [b] = [a + b]$. We shall now see how this idea can be used to construct new groups in much more general circumstances. Indeed, the following theorem shows that $\mathbb{Z}$ can be replaced by any group G, and $\langle n \rangle$ by any normal subgroup N of G. (Notice that $\langle n \rangle \lhd \mathbb{Z}$ because $\mathbb{Z}$ is Abelian.) It may help to review Section 16, especially Theorem 16.1 and Lemma 16.1, before reading this section. We continue to use juxtaposition to denote unspecified group operations.

Theorem 22.1.   *Let N be a normal subgroup of G, and let G/N denote the set of all right cosets of N in G. For*

$$Na \in G/N \quad and \quad Nb \in G/N, \quad let \quad (Na)(Nb) = N(ab).$$

With this operation G/N is a group called the quotient group *(or* factor group*) of G by N.*

Remark.   Figure 22.1 represents the idea behind Theorem 22.1. Each horizontal section represents a coset of N. For example, Na is the coset to which a belongs. The cosets are the elements of G/N. The "product" of the cosets Na and Nb is $N(ab)$, the coset to which ab belongs. The first part of the following proof shows that if $N \lhd G$, then it does not matter which element is chosen from the coset Na and which is chosen from the coset Nb, their "product" will be in the coset $N(ab)$.

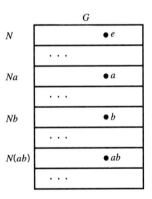

Figure 22.1

PROOF.   We must first prove that the operation on G/N is well-defined, that is,

$$\text{if } Na_1 = Na_2 \quad and \quad Nb_1 = Nb_2, \quad then \quad N(a_1b_1) = N(a_2b_2).$$

From $Na_1 = Na_2$ we have $a_1 = n_1 a_2$ for some $n_1 \in N$; from $Nb_1 = Nb_2$ we have $b_1 = n_2 b_2$ for some $n_2 \in N$. Therefore $a_1 b_1 = n_1 a_2 n_2 b_2$. But $a_2 n_2 a_2^{-1} = n_3$ for some $n_3 \in N$ because $N \lhd G$. This gives $a_2 n_2 = n_3 a_2$, so that $a_1 b_1 = n_1 n_3 a_2 b_2$ with $n_1 n_3 \in N$. This proves that $N(a_1 b_1) = N(a_2 b_2)$, as required.

The operation on G/N is associative because if $a, b, c \in G$, then

$$Na(NbNc) = Na(Nbc) = N(a(bc)) = N((ab)c) = N(ab)Nc = (NaNb)Nc.$$

The element Ne is an identity element because if $a \in G$, then

$$NeNa = N(ea) = Na \quad and \quad NaNe = Nae = Na.$$

Finally, Na^{-1} is an inverse for Na because

$$NaNa^{-1} = N(aa^{-1}) = Ne \quad \text{and} \quad Na^{-1}Na = N(a^{-1}a) = Ne.$$

This proves that G/N is a group.   ∎

To emphasize: the *elements* of G/N are *subsets* of G. If G is finite, then the order of G/N is the number of right cosets of N in G; this is $[G : N]$, the index of N in G (Section 17). From the remarks after Lagrange's Theorem, $[G : N] = |G|/|N|$. Therefore,

$$|G/N| = |G|/|N|.$$

Example 22.1.   Consider G/N for $G = S_3$ and $N = \langle (1 \quad 2 \quad 3) \rangle$. We have

$$N = \{(1), (1 \quad 2 \quad 3), (1 \quad 3 \quad 2)\} \lhd S_3$$

and

$$|S_3/N| = 6/3 = 2.$$

The elements of S_3/N are

$$N = \{(1), (1 \quad 2 \quad 3), (1 \quad 3 \quad 2)\}$$

and

$$N(1 \quad 2) = \{(1 \quad 2), (1 \quad 3), (2 \quad 3)\}.$$

The element (coset) N is the identity, and

$$\begin{aligned} N(1 \quad 2) \cdot N(1 \quad 2) &= N(1 \quad 2)^2 \\ &= N(1) \\ &= N. \end{aligned}$$

Table 22.1 is the Cayley table.

Table 22.1

	N	$N(1 \quad 2)$
N	N	$N(1 \quad 2)$
$N(1 \quad 2)$	$N(1 \quad 2)$	N

∎

Example 22.2.   Let G be the group of symmetries of a square (Example 8.1), and let $N = \langle \mu_3 \rangle$. Then $N = \{\mu_1, \mu_3\}$ and $N \lhd G$. (In fact, if μ is any element of G, then $\mu \circ \mu_1 \circ \mu^{-1} = \mu_1$ and $\mu \circ \mu_3 \circ \mu^{-1} = \mu_3$.) The elements of G/N are $\{\mu_1, \mu_3\}$, $\{\mu_2, \mu_4\}$, $\{\mu_5, \mu_6\}$, and $\{\mu_7, \mu_8\}$. If we denote these cosets by $[\mu_1]$, $[\mu_2]$, $[\mu_5]$, and $[\mu_7]$, respectively, then the Cayley table for G/N is as shown in Table 22.2. For instance, $[\mu_2][\mu_7] = [\mu_5]$ because $[\mu_2][\mu_7] = N\mu_2 N\mu_7 = N(\mu_2 \circ \mu_7) = N\mu_6$, and $N\mu_6 = N\mu_5$.

Table 22.2

	$[\mu_1]$	$[\mu_2]$	$[\mu_5]$	$[\mu_7]$
$[\mu_1]$	$[\mu_1]$	$[\mu_2]$	$[\mu_5]$	$[\mu_7]$
$[\mu_2]$	$[\mu_2]$	$[\mu_1]$	$[\mu_7]$	$[\mu_5]$
$[\mu_5]$	$[\mu_5]$	$[\mu_7]$	$[\mu_1]$	$[\mu_2]$
$[\mu_7]$	$[\mu_7]$	$[\mu_5]$	$[\mu_2]$	$[\mu_1]$

■

Example 22.3.   Let $G = \mathbb{Z}_{12} \times \mathbb{Z}_4$ and $N = \langle [3] \rangle \times \langle [2] \rangle$, with the notation interpreted in the following natural way. The first factor, $\langle [3] \rangle$, is a subgroup of $\mathbb{Z}_{12}$:

$$\langle [3] \rangle = \{[0], [3], [6], [9]\} \subseteq \mathbb{Z}_{12}.$$

The second factor is

$$\langle [2] \rangle = \{[0], [2]\} \subseteq \mathbb{Z}_4.$$

Because G is Abelian, $N \lhd G$. And $|G/N| = 48/8 = 6$. A complete list of right coset representatives of N in G is

$$([0], [0]), ([0], [1]), ([1], [0]), ([1], [1]), ([2], [0]), ([2], [1]).$$

Denote the corresponding cosets of N by

$$N_{0,0}, N_{0,1}, N_{1,0}, N_{1,1}, N_{2,0}, N_{2,1},$$

respectively. Then, for example,

$$\begin{aligned}
N_{2,1}N_{1,0} &= N([2], [1])N([1], [0]) \\
&= N([2] \oplus [1], [1] \oplus [0]) \\
&= N([3], [1]) \\
&= N([0], [1]) \\
&= N_{0,1}.
\end{aligned}$$

Problem 22.4 asks for the Cayley table of this quotient group.   ■

It can be proved that if N is not normal, then the operation on G/N from Theorem 22.1 will not be well-defined (Problem 22.15). Thus the concepts of quotient group and normal subgroup are inseparable. Also, a subgroup N of a group G is normal iff the right coset of N determined by each element is the same as the left coset of N determined by that same element (Problem 21.26). Thus in working with cosets of a normal subgroup N, it is immaterial whether we use Na or aN; we shall always use Na.

By Theorem 21.2, a kernel of a homomorphism is necessarily a normal subgroup. The next theorem shows that, conversely, every normal subgroup is a kernel of some homomorphism.

Theorem 22.2.   *If G is a group and $N \lhd G$, then the mapping $\eta : G \to G/N$ defined by*

$$\eta(a) = Na \quad \text{for each} \quad a \in G$$

is a homomorphism of G onto G/N, and $\operatorname{Ker} \eta = N$.

PROOF. The mapping η is clearly well-defined and onto. Also, if $a, b \in G$, then $\eta(ab) = N(ab) = (Na)(Nb) = \eta(a)\eta(b)$, so η is a homomorphism. Finally, if $a \in G$, then $a \in \text{Ker } \eta$ iff $\eta(a) = Na = Ne$, because Ne is the identity element of G/N. Thus $a \in \text{Ker } \eta$ iff $a \in N$, so $\text{Ker } \eta = N$.   ∎

The mapping η in Theorem 22.2 is called the *natural homomorphism* of G onto G/N.

Example 22.4. If n is a positive integer, then $\mathbb{Z}/\langle n \rangle = \mathbb{Z}_n$. The natural homomorphism $\eta : \mathbb{Z} \to \mathbb{Z}/\langle n \rangle$ is given by $\eta(a) = \langle n \rangle + a = [a]$. And $\text{Ker } \eta = \langle n \rangle$. This η is the same as the mapping θ in Example 21.1, whose kernel we computed in Example 21.5.   ∎

PROBLEMS

Determine the order of each of the following quotient groups.

22.1. (a) $\mathbb{Z}_8/\langle [4] \rangle$ (b) $\langle 2 \rangle/\langle 8 \rangle$, where $\langle 8 \rangle \subseteq \langle 2 \rangle \subseteq \mathbb{Z}$

22.2. (a) $\mathbb{Z}_8/\langle [3] \rangle$ (b) $\langle 3 \rangle/\langle 6 \rangle$, where $\langle 6 \rangle \subseteq \langle 3 \rangle \subseteq \mathbb{Z}$

22.3. Construct the Cayley table for $\mathbb{Z}_{12}/\langle [4] \rangle$. (*Suggestion:* Use $[[k]]$ to denote the coset to which $[k]$ belongs.)

22.4. Construct the Cayley table for the group in Example 22.3.

22.5. Prove that every quotient group of an Abelian group is Abelian.

22.6. Prove that every quotient group of a cyclic group is cyclic.

22.7. If m and n are positive integers and $m \mid n$, then $\langle n \rangle$ is a normal subgroup of $\langle m \rangle$ (in Z). What is $|\langle m \rangle/\langle n \rangle|$? Justify your answer. (Problems 22.1 and 22.2 contain special cases.)

22.8. Give a reason for each step in the proof of Theorem 22.1 for why the operation on G/N is associative.

22.9. Assume $N \triangleleft G$.
(a) Prove that if $[G : N]$ is a prime, then G/N is cyclic.
(b) Prove or disprove the converse of the statement in part (a).

22.10. Determine the order of $(\mathbb{Z}_{12} \times \mathbb{Z}_4)/\langle ([3], [2]) \rangle$. Explain the difference between this quotient group and the one in Example 22.3.

22.11. Prove that if $N \triangleleft G$ and $a \in G$, then $o(Na) \mid o(a)$. [Here $o(Na)$ denotes the order of Na as an element of G/N.] How is this problem related to Problem 21.17?

22.12. Prove that every element of $\mathbb{Q}/\mathbb{Z}$ has finite order. Also show that $\mathbb{Q}/\mathbb{Z}$ has an element of order n for each positive integer n.

22.13. The elements of finite order in an Abelian group form a subgroup (Problem 14.16). Show that the subgroup of elements of finite order in $\mathbb{R}/\mathbb{Z}$ is $\mathbb{Q}/\mathbb{Z}$. (See Problem 22.12).

22.14. Prove that G/N is Abelian iff $aba^{-1}b^{-1} \in N$ for all $a, b \in G$.

22.15. Prove that if N is a subgroup of G, and the operation $(Na)(Nb) = N(ab)$ is well-defined on the set G/N of all right cosets of N in G, then $N \triangleleft G$.

SECTION 23   THE FUNDAMENTAL HOMOMORPHISM THEOREM

The natural homomorphism $\eta : G \to G/N$ shows that each quotient group of a group G is a homomorphic image of G (Theorem 22.2). The next theorem shows that the converse is also true: each homomorphic image of G is (isomorphic to) a quotient group of G. Thus the claim made at the beginning of Section 22 is justified: quotient groups are essentially the same as homomorphic images.

Theorem 23.1 (Fundamental Homomorphism Theorem).   *Let G and H be groups, and let $\theta : G \to H$ be a homomorphism from G onto H with $\mathrm{Ker}\, \theta = K$. Then the mapping $\phi : G/K \to H$ defined by*

$$\phi(Ka) = \theta(a) \qquad \text{for each} \qquad Ka \in G/K$$

is an isomorphism of G/K onto H. Therefore

$$G/K \approx H.$$

PROOF.   We must first verify that ϕ is well-defined. If $Ka_1 = Ka_2$, then $ka_1 = a_2$ for some $k \in K = \mathrm{Ker}\, \theta$, so $\theta(ka_1) = \theta(a_2)$. But $\theta(ka_1) = \theta(k)\theta(a_1) = e\theta(a_1) = \theta(a_1)$, so that $\theta(a_1) = \theta(a_2)$. Therefore, $\theta(a)$ is determined solely by the coset of K to which a belongs, so ϕ is well-defined.

To prove that ϕ preserves the operation, assume that $Ka \in G/K$ and $Kb \in G/K$. Then $\phi((Ka)(Kb)) = \phi(K(ab)) = \theta(ab) = \theta(a)\theta(b) = \phi(Ka)\phi(Kb)$, as required. Clearly ϕ is onto, because θ is onto. It remains only to prove that ϕ is one-to-one; or, equivalently, by Theorem 21.1, that $\mathrm{Ker}\, \phi$ contains only the identity element, Ke, of G/K. This is true because if $Ka \in \mathrm{Ker}\, \phi$, then $\theta(a) = \phi(Ka) = e$, and therefore $a \in \mathrm{Ker}\, \theta = K$, or $Ka = Ke$.   ∎

If a homomorphism $\theta : G \to H$ is not onto, then H should be replaced by $\theta(G)$ in the last two sentences of the theorem. Then the last statement of the theorem becomes $G/K \approx \theta(G)$. In any case, with θ, ϕ, and K as in the theorem, and $\eta : G \to G/K$ the natural homomorphism, it can be verified that $\phi \circ \eta = \theta$. Schematically, the two ways (θ and $\phi \circ \eta$) of getting from G to H in Figure 23.1 give the same result for every element of G (Problem 23.7). This is described by saying the *diagram commutes*.

Figure 23.1

Example 23.1.   Let G denote the group of all rotations of the plane about a fixed point p (Example 5.7). For each real number r, let $\theta(r)$ in G denote clockwise rotation through r radians. Then θ is a homomorphism of $\mathbb{R}$ onto G, and

$$\mathrm{Ker}\, \theta = \langle 2\pi \rangle = \{2k\pi : k \in \mathbb{Z}\}.$$

The Fundamental Homomorphism Theorem shows that $\mathbb{R}/\langle 2\pi \rangle \approx G$.   ∎

Example 23.2.   For $a \in \mathbb{Z}$, let $[a]_{12}$ and $[a]_4$ denote the congruence classes determined by a in $\mathbb{Z}_{12}$ and $\mathbb{Z}_4$, respectively. Define

$$\theta: \mathbb{Z}_{12} \to \mathbb{Z}_4 \text{ by } \theta([a]_{12}) = [a]_4.$$

θ is well-defined because if $[a]_{12} = [b]_{12}$, then $12 \mid (a - b)$, and therefore $4 \mid (a - b)$ and $[a]_4 = [b]_4$. Also, θ is a homomorphism:

$$\theta([a]_{12} \oplus [b]_{12}) = \theta([a + b]_{12}) = [a + b]_4 = [a]_4 \oplus [b]_4$$
$$= \theta([a]_{12}) \oplus \theta([b]_{12}).$$

Clearly θ is onto, and

$$\operatorname{Ker} \theta = \{[0]_{12}, [4]_{12}, [8]_{12}\} = \langle [4]_{12} \rangle$$

Therefore, by the Fundamental Homomorphism Theorem,

$$\frac{\mathbb{Z}_{12}}{\langle [4]_{12} \rangle} \approx \mathbb{Z}_4.$$

Notice that from $|\operatorname{Ker} \theta| = 3$ alone we could deduce that $|\mathbb{Z}_{12}/(\operatorname{Ker} \theta)| = 4$, but this would still leave the possibility that $\mathbb{Z}_{12}/(\operatorname{Ker} \theta) \approx \mathbb{Z}_2 \times \mathbb{Z}_2$.   ∎

Example 23.3.   Let G be the group of symmetries of a square (Example 8.1); we shall determine all of the homomorphic images of G. By the Fundamental Homomorphism Theorem, this is equivalent to determining all of the quotient groups of G. To do that, we must determine all of the normal subgroups of G. There are six in all (Problem 21.22). Any normal subgroup of order 4 will produce a quotient group of order 2, so we can ignore two of the three of order 4, and work with these:

$$G, \quad \langle \mu_2 \rangle, \quad \langle \mu_3 \rangle, \quad \langle \mu_1 \rangle.$$

These are of order 8, 4, 2, 1, respectively. Therefore,

$$\left| \frac{G}{G} \right| = 1, \qquad \left| \frac{G}{\langle \mu_2 \rangle} \right| = 2, \qquad \left| \frac{G}{\langle \mu_3 \rangle} \right| = 4, \qquad \left| \frac{G}{\langle \mu_1 \rangle} \right| = 8.$$

Any group of order 1 is isomorphic to $\mathbb{Z}_1$, and any group of order 2 is isomorphic to $\mathbb{Z}_2$. Therefore,

$$\frac{G}{G} \approx \mathbb{Z}_1, \qquad \frac{G}{\langle \mu_2 \rangle} \approx \mathbb{Z}_2.$$

There are two isomorphism classes of groups of order 4; namely, those determined by $\mathbb{Z}_4$ and $\mathbb{Z}_2 \times \mathbb{Z}_2$ (Section 19). Problem 23.9 suggests how to show that $G/\langle \mu_3 \rangle$ is isomorphic to the latter. Finally, $G/\langle \mu_1 \rangle \approx G$ (see Problem 23.8).

Summarizing, we see that any homomorphic image of this group G is isomorphic to $\mathbb{Z}_1, \mathbb{Z}_2, \mathbb{Z}_2 \times \mathbb{Z}_2$, or G.   ∎

We shall now see how the ideas in this chapter can be used to construct many groups from smaller component groups. *The groups in the remainder of this section are assumed to be finite* (even though some of the statements are also true for infinite groups).

A group G is said to be an *extension* of a group A by a group B if G contains a subgroup N such that

$$A \approx N \lhd G \quad \text{and} \quad G/N \approx B.$$

When this is so, we can think of G as being built up from component groups A and B. Because $|A| = |N|$ and $|G/N| = |G|/|N| = |B|$, we must have $|G| = |A| \cdot |B|$.

Example 23.4.   We saw in Example 22.1 that S_3 has a normal subgroup of order 3, namely $N = \langle (1 \quad 2 \quad 3) \rangle$. Also, $|S_3/N| = 2$. Every group of prime order p is isomorphic to $\mathbb{Z}_p$ (Theorem 19.3). Therefore,

$$\mathbb{Z}_3 \approx N \lhd S_3 \quad \text{and} \quad \frac{S_3}{N} \approx \mathbb{Z}_2.$$

Thus S_3 is an extension of $\mathbb{Z}_3$ by $\mathbb{Z}_2$.   ∎

Example 23.5.   For any groups A and B, there is always at least one extension of A by B, namely $A \times B$. This is because

$$A \approx A \times \{e\} \lhd A \times B \quad \text{and} \quad \frac{A \times B}{A \times \{e\}} \approx B$$

(Problem 23.10).   ∎

If we apply Example 23.5 with $A = \mathbb{Z}_3$ and $B = \mathbb{Z}_2$, we see that $\mathbb{Z}_3 \times \mathbb{Z}_2$ is an extension of $\mathbb{Z}_3$ by $\mathbb{Z}_2$. But Example 23.4 showed that S_3 is also an extension of $\mathbb{Z}_3$ by $\mathbb{Z}_2$, and $\mathbb{Z}_3 \times \mathbb{Z}_2 \not\approx S_3$ ($\mathbb{Z}_3 \times \mathbb{Z}_2$ is Abelian, but S_3 is non-Abelian). Thus there are nonisomorphic extensions of $\mathbb{Z}_3$ by $\mathbb{Z}_2$. This leads to the following general problem: *Given groups A and B, determine all (isomorphism classes of) extensions of A by B*. This problem was solved in part by Otto Hölder in the 1890s, and more completely by Otto Schreier in the 1920s. Their results show how to construct a collection of groups (Cayley tables if you like), from A and B, with the property that any extension of A by B will be isomorphic to some group in that collection. The details are complicated and will not be given here. The important point is that, in theory, there is a procedure for determining all extensions of A by B.

Some groups can be constructed from smaller groups by extension, and some cannot. For example, each group $\mathbb{Z}_p$ with p a prime has only the two obvious (normal) subgroups of orders 1 and p; therefore, $\mathbb{Z}_p$ can be thought of as an extension in only a trivial way—either as $\mathbb{Z}_1$ by $\mathbb{Z}_p$, or as $\mathbb{Z}_p$ by $\mathbb{Z}_1$. A group G having no normal subgroup other than $\{e\}$ and G itself is called a *simple* group. An Abelian simple group must be isomorphic to $\mathbb{Z}_p$ for some prime p (Problem 23.11).

All finite groups can be constructed from simple groups by forming repeated extensions. This can be seen as follows. Assume that G is finite. Let G_1 denote a maximal proper normal subgroup of $G : G_1 \neq G$ (proper), $G_1 \lhd G$ (normal), and G has no normal subgroup strictly between G_1 and G (maximal normal). It can be shown that G/G_1 is a simple group (Problem 23.17). Now let G_2 denote a maximal proper normal subgroup of G_1. Then G_1/G_2 is simple also (Problem 23.17 again). Continuing in this way, with G denoted by G_0, we eventually arrive at a series of subgroups

$$G = G_0 \rhd G_1 \rhd G_2 \rhd \cdots \rhd G_{k-2} \rhd G_{k-1} \rhd G_k = \{e\}$$

such that each factor group G_{i-1}/G_i is simple ($1 \leq i \leq k$). Such a series is called a *composition series* of G. We see from this that G can be constructed by a succession of extensions by simple groups: extend G_{k-1} by G_{k-2}/G_{k-1} to get G_{k-2}; extend G_{k-2} by G_{k-3}/G_{k-2} to get G_{k-3}; and so on until we arrive at G. Moreover, it can be proved that although G may have more than one composition series, the factor groups arising from any two composition series can be paired (after rearrangement, perhaps) in such a way that corresponding factor groups are isomorphic (*Jordan-Hölder Theorem*).

These remarks about composition series show that, with the problem of how to determine extensions effectively solved, the place to direct attention is the class of simple groups. If all finite simple groups could be determined, then, in theory, all finite groups would be determined, just by constructing all successive extensions by simple groups. We close our discussion with some general remarks about this problem.

We have already observed that groups of prime order are simple; they are the only Abelian simple groups. These groups of prime order are simple in the technical sense, defined previously, and they are also simple in the sense of being uncomplicated. Other simple groups (technical sense) can be quite complicated. The smallest non-Abelian simple group is A_5, the alternating group of degree 5, which is of order 60. (Alternating groups are discussed in Section 7.) In general, each alternating group A_n is simple for $n \geq 5$. Thus there is a non-Abelian simple group of order $\frac{1}{2}(n!)$ for each $n \geq 5$. These groups account for 5 of the 56 non-Abelian simple groups of order less than 1,000,000.

The most comprehensive easy-to-state theorem about non-Abelian simple groups is a celebrated theorem proved in the 1960s by the American group theorists Walter Feit and John Thompson: *There are no non-Abelian simple groups of odd order*. The classification of all simple groups was completed in the early 1980s. The original proof of the classification involved some 500 journal articles covering approximately 10,000 printed pages.

PROBLEMS

Find all homomorphic images of each of the following groups.

23.1. $\mathbb{Z}_6$ **23.2.** $\mathbb{Z}_4$

23.3. $\mathbb{Z}_5$ **23.4.** S_3

23.5. Find two nonisomorphic extensions of Z_2 by Z_2.

23.6. Find two nonisomorphic extensions of Z_4 by Z_2.

23.7. With θ, ϕ, and η as in the paragraph following the proof of the Fundamental Homomorphism Theorem, prove that $\phi \circ \eta = \theta$.

23.8. Use the Fundamental Homomorphism Theorem to prove that if G is any group with identity e, then $G/\{e\} \approx G$.

23.9. Verify the isomorphism $G/\langle \mu_3 \rangle \approx \mathbb{Z}_2 \times \mathbb{Z}_2$, from Example 23.3. (It is sufficient to check that the factor group on the left has no element of order 4. Why?)

23.10. Verify the isomorphisms in Example 23.5. (*Suggestion*: Use the Fundamental Homomorphism Theorem and the homomorphism $\pi_2 : A \times B \to B$ from Problem 21.2.)

23.11. Prove that if G is a simple Abelian group, then $G \approx \mathbb{Z}_p$ for some prime p.

23.12. Prove that $\mathbb{Z}_{18}/\langle [3] \rangle \approx \mathbb{Z}_3$.

23.13. Prove that if k and n are positive integers and k is a divisor of n, then $\mathbb{Z}_n/\langle[k]\rangle \approx \mathbb{Z}_k$.

23.14. Find all homomorphic images of the quaternion group Q_8.

23.15. Find all homomorphic images of Z_n (n a positive integer).

23.16. Prove that if θ is a homomorphism of G onto H, $B \lhd H$, and $A = \{g \in G : \theta(g) \in B\}$, then $A \lhd G$.

23.17. Prove that if $N \lhd G$, $N \neq G$, and G has no normal subgroup strictly between N and G, then G/N is simple. (*Suggestion*: Use the natural homomorphism $\eta : G \to G/N$ and Problem 23.16.)

23.18. If G is a simple group, then any homomorphic image of G is either isomorphic to G or of order one. Why?

23.19. Let A denote the group of all mappings $\alpha_{a,b} : \mathbb{R} \to \mathbb{R}$, defined as in Example 5.8. Let $B = \{\alpha_{1,b} : b \in \mathbb{R}\}$, and let $\mathbb{R}^\#$ denote the multiplicative group of $\mathbb{R}$. Define $\theta : A \to \mathbb{R}^\#$ by $\theta(\alpha_{a,b}) = a$ for each $\alpha_{a,b} \in A$.
(a) Verify that θ is a homomorphism with $\operatorname{Ker}\theta = B$.
(b) Explain why $B \lhd A$ and $A/B \approx \mathbb{R}^\#$.
(c) Explain why A is an extension of $\mathbb{R}$ (operation addition) by $\mathbb{R}^\#$ (operation multiplication).
(d) Give an example of a group that is an extension of $\mathbb{R}$ (operation addition) by $\mathbb{R}^\#$ (operation multiplication) and is not isomorphic to A.

23.20. (First Isomorphism Theorem) Assume that H and K are subgroups of a group G and $K \lhd G$.
(a) Prove that $HK = \{hk : h \in H \text{ and } k \in K\}$ is a subgroup of G, and that $K \lhd HK$. (Compare Problem 23.22.)
(b) Define $\theta : H \to HK/K$ by $\theta(h) = Kh$ for each $h \in H$. Verify that θ is a homomorphism of H onto HK/K.
(c) Verify that $\operatorname{Ker}\theta = H \cap K$. Now explain why
$$\frac{H}{H \cap K} \approx \frac{HK}{K}.$$

23.21. (Second Isomorphism Theorem) Assume that H and K are subgroups of a group G, and that $K \lhd H$, $K \lhd G$, and $H \lhd G$. Prove that $H/K \lhd G/K$ and that $(G/K)/(H/K) \approx G/H$. [*Suggestion*: Consider $\theta : G/K \to G/H$ defined by $\theta(Kg) = Hg$. Verify that θ is well-defined and is a homomorphism, and apply the Fundamental Homomorphism Theorem.]

23.22. Give an example to show that if A and B are subgroups of a group G, then $AB = \{ab : a \in A \text{ and } b \in B\}$ need not be a subgroup of G. [*Suggestion*: Try $G = S_3$. If $A \lhd G$ or $B \lhd G$, then AB is a subgroup; see Problem 23.20(a).]

NOTES ON CHAPTER V

1.   Aschbacher, M., *Finite Group Theory*, Cambridge University Press, Cambridge, England, 1994.
2.   Curtis, R., and R. A. Wilson, eds., *The Atlas of Finite Groups: Ten Years On*, Cambridge University Press, Cambridge, England, 1998.
3.   Feit, W., and J. G. Thompson, Solvability of groups of odd order, *Pacific Journal of Mathematics* **13** (1963) 755–1029.
4.   Hurley, J. F., and A. Rudvalis, Finite simple groups, *American Mathematical Monthly* **84** (1977) 693–714.
5.   Gallian, J. A., The search for finite simple groups, *Mathematics Magazine* **49** (1976) 163–179.
6.   Gorenstein, D., *Finite Simple Groups: An Introduction to Their Classification*, Plenum Press, New York, 1982.

CHAPTER VI

INTRODUCTION TO RINGS

In considering the integers as a group, we have used addition but not multiplication. In doing that, we have ignored more than multiplication. We have also ignored properties that combine the two operations, such as the law $a(b + c) = ab + ac$. The same has happened with other groups—the rational numbers and the integers mod n, for example. The concept of a *ring* covers all of these systems, as well as many others with two operations. Rings, just like groups, arise in widely varying contexts. Some of the most basic ideas about rings are used in the next chapter to analyze the integers and other familiar number systems. Rings also play a role in all of the remaining chapters except XIII and XIV.

SECTION 24   DEFINITION AND EXAMPLES

A ring consists of a set with two operations, which are nearly always written as a sum $(a + b)$ and a product (ab). For rings whose elements are numbers, this notation always has its usual meaning unless the contrary is explicitly stated. In some cases it is necessary to specify what is meant by the two operations. In reading the axioms for a ring, which follow, it may help to keep the integers in mind—they do form a ring. Also, do not confuse R (which may be any nonempty set) with $\mathbb{R}$ (which we reserve for the real numbers).

Definition.   A *ring* is a set R together with two operations on R, called *addition* $(a + b)$ and *multiplication* (ab), such that each of the following axioms is satisfied:

R with addition is an Abelian group

$$a + (b + c) = (a + b) + c \quad \text{for all} \quad a, b, c \in R,$$

there is an element $0 \in R$ such that

$$a + 0 = 0 + a = a \quad \text{for each } a \in R,$$

for each $a \in R$ there is an element $-a \in R$ such that

$$a + (-a) = (-a) + a = 0,$$
$$a + b = b + a \quad \text{for all} \quad a, b \in R,$$

multiplication is associative

$$a(bc) = (ab)c \quad \text{for all} \quad a, b, c \in R, \text{ and}$$

distributive laws

$$a(b + c) = ab + ac \quad \text{and} \quad (a + b)c = ac + bc \quad \text{for all} \quad a, b, c \in R.$$

The group formed by R with addition is referred to as the *additive group* of R. Its identity element, 0, is called the *zero* of the ring; the context will usually make clear whether this or the integer zero is meant. In expressions such as $ab + ac$, multiplications are to be performed first, just as in elementary arithmetic; that is, $ab + ac$ means $(ab) + (ac)$.

Example 24.1.   The integers form a ring with respect to the usual addition and multiplication. The same is true for the rational numbers, the real numbers, and also the even integers. ∎

Example 24.2.   For each positive integer n, $\mathbb{Z}_n$, the integers mod n, forms a ring with respect to the operations $\oplus$ and $\odot$ introduced in Section 11. Theorem 11.1 shows that $\mathbb{Z}_n$ with $\oplus$ is a group, and Problem 11.7 shows that this group is Abelian. Associativity of $\odot$ was recorded in Lemma 11.3. Here is a verification of the first distributive law:

$$
\begin{aligned}
[a] \odot ([b] \oplus [c]) &= [a] \odot [b + c] & &\text{definition of } \oplus \\
&= [a(b + c)] & &\text{definition of } \odot \\
&= [ab + ac] & &\text{distributivity for } +, \cdot \\
&= [ab] \oplus [ac] & &\text{definition of } \oplus \\
&= ([a] \odot [b]) \oplus ([a] \odot [c]) & &\text{definition of } \odot .
\end{aligned}
$$

The proof of the other distributive law is similar (Problem 21.2). ∎

Example 24.3.   Recall that $M(2, \mathbb{Z})$ denotes the set of all 2×2 matrices with integers as entries. With matrix addition and multiplication, $M(2, \mathbb{Z})$ is a ring (Problem 24.4). Appendix D has more about groups and rings of matrices, which are very important in linear algebra. ∎

Example 24.4.   Let $\mathbb{Z}[\sqrt{2}]$ denote the set of all numbers $a + b\sqrt{2}$ with $a, b \in \mathbb{Z}$. The sum of two numbers in $\mathbb{Z}[\sqrt{2}]$ is also in $\mathbb{Z}[\sqrt{2}]$:

$$(a + b\sqrt{2}) + (c + d\sqrt{2}) = (a + c) + (b + d)\sqrt{2},$$

and $a + c \in \mathbb{Z}$ and $b + d \in \mathbb{Z}$ if $a, b, c, d \in \mathbb{Z}$. The set $\mathbb{Z}[\sqrt{2}]$ is also closed under multiplication:

$$
\begin{aligned}
(a + b\sqrt{2})(c + d\sqrt{2}) &= ac + ad\sqrt{2} + bc\sqrt{2} + bd\sqrt{2}\sqrt{2} \\
&= (ac + 2bd) + (ad + bc)\sqrt{2}.
\end{aligned}
$$

With these operations $\mathbb{Z}[\sqrt{2}]$ is a ring (Problem 24.5). ∎

Example 24.5.   Let F denote the set $M(\mathbb{R})$ of all functions (mappings) $f: \mathbb{R} \to \mathbb{R}$. We can define $f + g$ and fg for $f, g \in F$ in a way that will give a ring. If $f + g$ is to be in F

then it must be a function from $\mathbb{R}$ to $\mathbb{R}$. Thus we must specify $(f + g)(x)$ for each $x \in \mathbb{R}$. Similarly for fg. The definitions are

$$(f + g)(x) = f(x) + g(x) \quad \text{for each } x \in \mathbb{R}$$

and $\qquad\qquad\qquad\qquad\qquad\qquad\qquad\qquad\qquad\qquad\qquad\qquad$ (24.1)

$$(fg)(x) = f(x)g(x) \quad \text{for each } x \in \mathbb{R}$$

To verify that this operation $+$ is associative, $f + (g + h) = (f + g) + h$, we observe that since each side is a function with domain $\mathbb{R}$, what must be shown is that for all $f, g, h \in F$

$$[f + (g + h)](x) = [(f + g) + h](x) \quad \text{for each } x \in \mathbb{R}.$$

To do this, write

$$
\begin{aligned}
[f + (g + h)](x) &= f(x) + (g + h)(x) && \text{definition of } + \text{ on } F \\
&= f(x) + [g(x) + h(x)] && \text{definition of } + \text{ on } F \\
&= [f(x) + g(x)] + h(x) && \text{associativity of } + \text{ on } \mathbb{R} \\
&= (f + g)(x) + h(x) && \text{definition of } + \text{ on } F \\
&= [(f + g) + h](x) && \text{definition of } + \text{ on } F.
\end{aligned}
$$

The 0 (identity element for $+$) for this ring is the function defined by $0(x) = 0$ for each $x \in \mathbb{R}$, where the 0 on the right is the zero of $\mathbb{R}$: if $f \in F$, then

$$(f + 0)(x) = f(x) + 0(x) = f(x) + 0 = f(x)$$

for each $x \in \mathbb{R}$, so $f + 0 = f$. The negative of a function f is the function $-f$ defined by $(-f)(x) = -f(x)$ for each $x \in \mathbb{R}$. Verification of the remaining axioms is left to Problem 24.6.

Notice that the product fg in this example is *not* $f \circ g$. See Problem 24.10 for what happens when $f \circ g$ is used. ∎

Example 24.6. Let R and S be rings, and $R \times S$ the Cartesian product of R and S, that is, the set of all ordered pairs (r, s) with $r \in R$ and $s \in S$. Then $R \times S$ becomes a ring with the following operations:

$$(r_1, s_1) + (r_2, s_2) = (r_1 + r_2, s_1 + s_2)$$

and

$$(r_1, s_1)(r_2, s_2) = (r_1 r_2, s_1 s_2)$$

for all $r_1, r_2 \in R$, $s_1, s_2 \in S$. The discussion of direct products of groups in Section 15 carries over to show that $R \times S$ is a group with respect to addition: the additive identity is $(0, 0) = (0_R, 0_S)$, and the negative of (r, s) is $(-r, -s)$. Here is verification of one of the distributive laws:

$$
\begin{aligned}
(r_1, s_1)((r_2, s_2) + (r_3, s_3)) &= (r_1, s_1)(r_2 + r_3, s_2 + s_3) \\
&= (r_1(r_2 + r_3), s_1(s_2 + s_3)) \\
&= (r_1 r_2 + r_1 r_3, s_1 s_2 + s_1 s_3) \\
&= (r_1 r_2, s_1 s_2) + (r_1 r_3, s_1 s_3) \\
&= (r_1, s_1)(r_2, s_2) + (r_1, s_1)(r_3, s_3).
\end{aligned}
$$

The remaining details are left to Problem 24.11. This ring is called the *direct sum* of R and S. ∎

When first working with any abstract concept such as *group* or *ring*, there is bound to be uncertainty over what is and is not allowed. For example, in a group written multiplicatively the left cancellation law holds: If $ab = ac$, then $b = c$ [Theorem 14.1(a)]. Because a ring is (among other things) a group with respect to addition, this can be translated into a statement about rings: If $a + b = a + c$, then $b = c$. But what about left cancellation for *multiplication* in a ring? In the ring of integers $ab = ac$ implies $b = c$ (for $a \neq 0$), but what about other rings? We shall come to see that sometimes it is safe to cancel and sometimes it is not. Only experience can guide us in such matters. Once it has been determined that there are enough important examples of a concept to make that concept worth studying, we set about trying to discover theorems about it. Not only do examples tell us whether the concept is worth studying, they are also the source of ideas for theorems. Given a property of a specific ring, for example, we can ask whether that property holds for all rings. Very often, the answer will be No, and it will be another example—a counterexample—that will give us that answer. But sometimes the answer will be Yes, and we then have another piece of the theory surrounding the general concept. In this sense, the example of the integers is a good one to keep in mind when first studying rings. Even though not everything true about the integers is true about all rings, we can gradually improve our perspective by comparing each theorem and example with this familiar and important special case.

With these remarks and some examples behind us, we now turn to some elementary theorems. Because a ring is a group with respect to addition, the elementary properties of rings that involve only addition are obtained by simply translating the elementary properties of groups into additive notation. An example is the cancellation law mentioned earlier. And the law $(a^{-1})^{-1} = a$ becomes $-(-a) = a$. The following theorem gives a summary of such properties. The last part uses the conventions that if n is a positive integer, then $na = a + a + \cdots + a$ (n terms) and $(-n)a = -(na)$.

Theorem 24.1.   *Let R be a ring and $a, b, c \in R$.*

(a) *The zero element of R is unique.*

(b) *Each element of R has a unique negative.*

(c) *If $a + b = a + c$, then $b = c$ (left cancellation law).*

(d) *If $b + a = c + a$, then $b = c$ (right cancellation law).*

(e) *Each of the equations $a + x = b$ and $x + a = b$ has a unique solution.*

(f) $-(-a) = a$ *and* $-(a + b) = (-a) + (-b)$.

(g) *If m and n are integers, then $(m + n)a = ma + na$, $m(a + b) = ma + mb$, and $m(na) = (mn)a$.*

The proof of Theorem 24.1 is left to Problem 24.12. The next theorem concerns properties involving multiplication in a ring. Here, and elsewhere, $a - b$ means $a + (-b)$.

Theorem 24.2.   *Let R be a ring, 0 the zero of R, and $a, b, c \in R$.*

(a) $0a = a0 = 0$

(b) $a(-b) = (-a)b = -(ab)$.

(c) $(-a)(-b) = ab$.

(d) $a(b - c) = ab - ac$ and $(a - b)c = ac - bc$.

PROOF.   (a) $0a + 0a = (0 + 0)a = 0a = 0a + 0$, and therefore, by canceling $0a$ from the left of the first and last expressions, $0a = 0$. The proof for $a0 = 0$ is similar.

(b) The equation $x + ab = 0$ has $x = -(ab)$ as a solution. But $a(-b) + ab = a(-b + b) = a0 = 0$; hence $x = a(-b)$ is also a solution. By uniqueness of the solution [Theorem 24.1(e)], it follows that $-(ab) = a(-b)$. The proof for $-(ab) = (-a)b$ is similar.

(c) Using part (b) and Theorem 24.1(f), we have $(-a)(-b) = (-(-a))(b) = ab$.

(d) Using part (b), we can write $a(b - c) = a(b + (-c)) = ab + a(-c) = ab + (-(ac)) = ab - ac$. The proof for $(a - b)c = ac - bc$ is similar.   ■

In the definition of *ring*, nothing is assumed about multiplication except associativity and the connection of multiplication and addition through the distributive laws. Other assumptions are considered in the next section when we look at some special types of rings. We close this section by mentioning two of the assumptions that arise most frequently.

An element e in a ring R is called a *unity* (or *identity* or *unit element*) for the ring if $ea = ae = a$ for each $a \in R$. Thus a unity is simply an identity for multiplication. The number 1 is a unity for the ring of integers. The ring of even integers has no unity.

A ring R is said to be *commutative* if $ab = ba$ for all $a, b \in R$. The rings in Examples 24.1, 24.2, 24.4, and 24.5 are commutative. If $ab \neq ba$ for some $a, b \in R$, the ring is *noncommutative*. The ring $M(2, \mathbb{Z})$ (Example 24.3) is noncommutative. The ring $R \times S$ in Example 24.6 will be commutative iff both R and S are commutative.

PROBLEMS

24.1. Compute $[3] \odot ([4] \oplus [5])$ and $([3] \odot [4]) \oplus ([3] \odot [5])$ in $\mathbb{Z}_6$. Show each step.

24.2. Prove that $([a] \oplus [b]) \odot [c] = ([a] \odot [c]) \oplus ([b] \odot [c])$ for all $[a], [b], [c] \in \mathbb{Z}_n$ (Example 24.2).

24.3. Compute $\begin{bmatrix} 3 & 1 \\ -1 & 2 \end{bmatrix} \left(\begin{bmatrix} 0 & 4 \\ 5 & -2 \end{bmatrix} + \begin{bmatrix} -1 & -2 \\ 0 & 3 \end{bmatrix} \right)$ in $M(2, \mathbb{Z})$.

24.4. Prove that $M(2, \mathbb{Z})$ is a ring with the usual matrix operations (Example 24.3). Verify that this ring is noncommutative.

24.5. Prove that $\mathbb{Z}\left[\sqrt{2}\right]$ is a ring (Example 24.4).

24.6. Complete the verification that $F = M(\mathbb{R})$ is a ring (Example 24.5). Also verify that it is commutative and has a unity.

24.7. Which of the following properties hold in every ring R? What about every commutative ring R?
(a) $a^m a^n = a^{m+n}$ for all $a \in R$, $m, n \in \mathbb{N}$.
(b) $(a^m)^n = a^{mn}$ for all $a \in R$, $m, n \in \mathbb{N}$.
(c) $(ab)^m = a^m b^m$ for all $a \in R$, $m \in \mathbb{N}$. (The powers occurring here are defined as in Section 14.)

24.8. Prove that if R is a ring, $a, b \in R$, and $ab = ba$, then $a(-b) = (-b)a$ and $(-a)(-b) = (-b)(-a)$. (Use Theorem 24.2.)

24.9. Show that if R is a ring and S is a nonempty set, then the set of all mappings from S into R can be made into a ring by using operations like those defined in (Example 24.5).

24.10. Consider Example 24.5 with fg, as defined there, replaced by $f \circ g$. Verify that this does not define a ring.

24.11. Complete the verification that Example 24.6 is a ring.

24.12. Prove Theorem 24.1. (This can be done by referring to proofs that have already been carried out for groups.)

24.13. Prove that if R is a ring, then each of the following properties holds for all $a, b, c \in R$.
(a) $a0 = 0$. (This is part of Theorem 24.2.)
(b) $-(ab) = (-a)b$. (This is part of Theorem 24.2.)
(c) $(a - b)c = ac - bc$. (This is part of Theorem 24.2.)
(d) $-(a + b) = (-a) + (-b)$
(e) $(a - b) + (b - c) = a - c$.

24.14. Prove that a ring has at most one unity.

24.15. Let E denote the set of even integers. Prove that with the usual addition, and with multiplication defined by $m * n = (1/2)mn$, E is a ring. Is there a unity?

24.16. Prove that $a^2 - b^2 = (a + b)(a - b)$ for all a, b in a ring R iff R is commutative.

24.17. Prove that $(a + b)^2 = a^2 + 2ab + b^2$ for all a, b in a ring R iff R is commutative.

24.18. Verify that if A is an Abelian group, with addition as operation, and an operation $*$ is defined on A by $a * b = 0$ for all $a, b \in A$, then A is a ring with respect to $+$ and $*$.

24.19. In the ring of integers, if $ab = ac$ and $a \neq 0$, then $b = c$. Is this true in all rings? (*Suggestion*: Look carefully at the rings in Examples 24.2, 24.3, 24.5, and 24.6.)

24.20. Verify that if R is a ring and $a, b \in R$, then

$$(a + b)^3 = a^3 + aba + ba^2 + b^2a + a^2b + ab^2 + bab + b^3.$$

Which ring axioms do you need?

24.21. Prove that if R is a commutative ring, $a, b \in R$, and n is a positive integer, then $(a + b)^n$ can be computed by the Binomial Theorem. (See Appendix C.)

24.22. For each set S, let $\mathcal{P}(S)$ denote the set of all subsets of S. For $A, B \in \mathcal{P}(S)$, define $A + B$ and AB by

$$A + B = (A \cup B) \backslash (A \cap B) = \{x : x \in A \cup B \quad \text{and} \quad x \notin A \cap B\}$$

and

$$AB = A \cap B.$$

Verify that with these operations, $\mathcal{P}(S)$ is a ring. [$\mathcal{P}(S)$ is called the *power set* of A. See Appendix A for facts about sets. *Suggestion*: Make generous use of Venn diagrams in this problem.]

SECTION 25   INTEGRAL DOMAINS. SUBRINGS

To isolate what is unique about the integers—something we first promised in Section 18— we focus here on the pertinent abstract ring properties. Commutativity and the existence of a unity are examples of these properties. After introducing one more such property we define a special class of rings known as integral domains. This class contains the ring of integers and brings us very near to a characterization of that ring. The characterization will be completed in Section 29.

We begin by recalling a fact relating to zero in the ring of integers: If a and b are integers and $ab = 0$, then either $a = 0$ or $b = 0$. This is not true in some rings. In $\mathbb{Z}_6$, for example, $[2] \odot [3] = [0]$, but $[2] \neq [0]$ and $[3] \neq [0]$. The following definition singles out such examples.

Definition.   An element $a \neq 0$ in a commutative ring R is called a *zero divisor* in R if there exists an element $b \neq 0$ in R such that $ab = 0$.

Thus the ring of integers has no zero divisors. But both [2] and [3] are zero divisors in $\mathbb{Z}_6$. Notice that the definition is restricted to elements in a commutative ring; for what can happen in a noncommutative ring see Problem 25.9. By the definition that follows, zero divisors are forbidden in an integral domain.

Definition.   A commutative ring with unity $e \neq 0$ and no zero divisors is called an *integral domain*.

Example 25.1.   The ring of integers, the ring of rational numbers, and the ring of real numbers are all integral domains. The ring of even integers is not an integral domain because it has no unity.   ∎

Example 25.2.   The ring $\mathbb{Z}_6$ is not an integral domain because, as we have seen, it has zero divisors. This happens because 6 is not a prime. More generally, if n is not a prime, and $n = rs$ with r and s each greater than 1, then, in $\mathbb{Z}_n$, $[r] \odot [s] = [rs] = [n] = [0]$ with $[r] \neq [0]$ and $[s] \neq [0]$. Thus $\mathbb{Z}_n$ is not an integral domain if n is not a prime. On the other hand, it can be proved that if n is a prime, then $\mathbb{Z}_n$ is an integral domain (Problem 25.10). For the special case $n = 5$ this can be seen from Table 11.2; every $\mathbb{Z}_n$ is commutative and has a unity, and Table 11.2 shows that the product of any two nonzero elements in $\mathbb{Z}_5$ is nonzero.   ∎

Of the other examples of rings in Section 24, $M(2, \mathbb{Z})$ is not an integral domain because it is noncommutative (Problem 24.4). The ring in Example 24.4 is an integral domain (Problem 25.11). The ring in Example 24.5 is not an integral domain (Problem 25.12). The ring $R \times S$ in Example 24.6 is an integral domain only if one of R or S is an integral domain and the other contains only a zero element (Problem 25.13).

The following theorem gives an easy-to-prove but important property of integral domains.

Theorem 25.1.   *If D is an integral domain, $a, b, c \in D$, $a \neq 0$, and $ab = ac$, then $b = c$ (left cancellation property).*

PROOF.   From $ab = ac$ we have $ab - ac = 0$, and thus $a(b - c) = 0$. Since a is a nonzero element of the integral domain D, it cannot be a zero divisor, so we must have $b - c = 0$ and $b = c$.   ∎

Because multiplication is commutative in an integral domain, the (left) cancellation property in Theorem 25.1 is equivalent to the *right cancellation property*: If $a \neq 0$ and $ba = ca$, then $b = c$. Moreover, for a commutative ring each of these cancellation properties implies that the ring has no zero divisors (Problem 25.14). Thus an alternative definition of an integral domain is this: *An integral domain is a commutative ring D with unity $e \neq 0$ such that if $a, b, c \in D$, $ab = ac$, and $a \neq 0$, then $b = c$.*

The notion of subring is the obvious analogue of the notion of subgroup.

Definition.   A subset S of a ring R is a *subring* of R if S is itself a ring with respect to the operations on R.

Example 25.3.   The ring of even integers is a subring of the ring of all integers. The ring of integers is a subring of the ring of rational numbers.   ∎

Theorem 7.1 gave a criterion for determining when a subset of a group is a subgroup. Here is the corresponding criterion for subrings.

Theorem 25.2.   *A subset S of a ring R is a subring of R iff S is nonempty, S is closed under both the addition and the multiplication of R, and S contains the negative of each of its elements.*

PROOF.   Problem 25.15.   ∎

Example 25.4.   Let $F = M(\mathbb{R})$ denote the ring of all functions $f \colon \mathbb{R} \to \mathbb{R}$, introduced in Example 24.5. Let S denote the set of all $f \in F$ such that $f(1) = 0$. We shall use Theorem 25.2 to prove that S is a subring of F.

Certainly $0_F \in S$, since $0_F(x) = 0$ for all $x \in \mathbb{R}$ so that in particular $0_F(1) = 0$. Thus S is nonempty. If f and g are in S, then

$$(f + g)(1) = f(1) + g(1) = 0 + 0 = 0$$

and

$$(fg)(1) = f(1)g(1) = 0 \cdot 0 = 0$$

so that $f + g$ and fg are in S. Finally, if $f(1) = 0$, then $(-f)(1) = -f(1) = -0 = 0$; therefore, the negative of each element of S is also in S.   ∎

PROBLEMS

25.1. Which elements of $\mathbb{Z}_4$ are zero divisors?

25.2. Which elements of $\mathbb{Z}_{10}$ are zero divisors?

25.3. Verify that $([2], [0])$ is a zero divisor in $\mathbb{Z}_3 \times \mathbb{Z}_3$.

25.4. Which elements of $\mathbb{Z} \times \mathbb{Z}$ are zero divisors? (See Example 24.6.)

25.5. What is the smallest subring of $\mathbb{Z}$ containing 3?

25.6. What is the smallest subring of $\mathbb{R}$ containing 1/2?

Which of the following are subrings of $M(2, \mathbb{Z})$ (Example 24.3)?

25.7. $\left\{ \begin{bmatrix} a & b \\ 0 & 0 \end{bmatrix} : a, b \in \mathbb{Z} \right\}$

25.8. $\left\{ \begin{bmatrix} a & b \\ 0 & c \end{bmatrix} : a, b, c \in \mathbb{Z} \right\}$

25.9. Compute $\begin{bmatrix} 0 & 1 \\ 0 & 0 \end{bmatrix} \begin{bmatrix} 1 & 0 \\ 0 & 0 \end{bmatrix}$ and $\begin{bmatrix} 1 & 0 \\ 0 & 0 \end{bmatrix} \begin{bmatrix} 0 & 1 \\ 0 & 0 \end{bmatrix}$ in $M(2, \mathbb{Z})$.

25.10. Prove that if p is a prime, then $\mathbb{Z}_p$ is an integral domain.

25.11. Explain why $\mathbb{Z}[\sqrt{2}]$ (Example 24.4) is an integral domain. (Assume that $\mathbb{R}$ is an integral domain.)

25.12. Prove that the ring $F = M(\mathbb{R})$ in Example 24.5 is not an integral domain.

25.13. Prove that $R \times S$ (Example 24.6) is an integral domain iff one of R or S is an integral domain and the other contains only a zero element.

25.14. Prove that for a commutative ring the cancellation property of Theorem 25.1 implies that there are no zero divisors.

25.15. Prove Theorem 25.2.

25.16. Let $C(\mathbb{R})$ denote the set of all continuous functions from $\mathbb{R}$ into $\mathbb{R}$. Prove that $C(\mathbb{R})$ is a subring of the ring $F = M(\mathbb{R})$ in Example 24.5. What properties of continuous functions are required? What happens if $C(\mathbb{R})$ is replaced by the set of differentiable functions from $\mathbb{R}$ into $\mathbb{R}$?

25.17. Prove that if C denotes any collection of subrings of a ring R, then the intersection of all of the rings in C is also a subring of R. (Compare Theorem 15.1.)

25.18. State and prove a theorem for rings that is analogous to Theorem 15.2 for groups. (Use Problem 25.17.)

25.19. Prove or disprove that if C denotes a collection of subrings of a ring R, and each ring in C is an integral domain, then the intersection of all of the rings in C is an integral domain.

25.20. For which $n \in \mathbb{Z}$ is the smallest subring containing n an integral domain?

25.21. Verify that $\{a + b\sqrt[3]{2} + c\sqrt[3]{4} : a, b, c \in \mathbb{Z}\}$ is a subring of $\mathbb{R}$. Is this subring an integral domain?

25.22. The *center* of a ring R is defined to be $\{c \in R: cr = rc$ for every $r \in R\}$. Prove that the center of a ring is a subring. What is the center of a commutative ring?

25.23. What is the center of $M(2, \mathbb{Z})$? (See Example 24.3 and Problem 25.22.)

25.24. What is the center of $M(\mathbb{R})$? (See Example 24.5 and Problem 25.22.)

25.25. State and prove a theorem giving a necessary and sufficient condition for a subset of an integral domain to be an integral domain. (Compare Theorem 25.2.)

SECTION 26   FIELDS

Because an integral domain has no zero divisors, its set of nonzero elements is closed with respect to multiplication. Therefore, multiplication is an operation on this set of nonzero elements, and it is natural to ask if it yields a group. Because multiplication is required to be associative in a ring, the operation is associative. Because an integral domain has a unity, the operation has an identity element. Thus, with respect to multiplication, the set of nonzero elements of an integral domain can fail to be a group only because of the absence of inverse elements. In the integral domain of integers, for instance, the nonzero elements do not form a group with respect to multiplication because only 1 and -1 have inverses. In the integral domain of rational numbers, however, each nonzero element does have an inverse relative to multiplication. Such integral domains are singled out by the following definition.

Definition.   A commutative ring in which the set of nonzero elements forms a group with respect to multiplication is called a *field*.

Alternatively, a field can be defined as an integral domain in which each nonzero element has an inverse relative to multiplication (Problem 26.11). Another example of a field, besides the ring of rational numbers, is the ring of real numbers. Fields are indispensable to much of mathematics. For example, the field of real numbers is basic in calculus and its applications. In Chapter XV we shall see that there are also useful finite fields.

Here are some relationships between classes of rings:

$$\text{fields} \subset \text{integral domains} \subset \text{commutative rings} \subset \text{rings}.$$

Each class is contained in, but different from, the class that follows it. If we restrict attention to rings having only finitely many elements, however, then the first two classes are the same, because of the following theorem.

Theorem 26.1.   *Every finite integral domain is a field.*

PROOF.   Let D be a finite integral domain. We must show that each nonzero element $a \in D$ has an inverse relative to multiplication; that is, if $a \in D$ and $a \neq 0$, then there is an element $b \in D$ such that $ab = e$. This means we must show that e is among the set of elements ax for $x \in D$. To show this, assume that $a \neq 0$ and consider the mapping λ_a: $D \rightarrow D$ defined by $\lambda_a(x) = ax$ for each $x \in D$. If this mapping is onto, then in particular $\lambda_a(x) = e$ for some $x \in D$, say $x = b$, and then $\lambda_a(b) = ab = e$. Thus it suffices to show that λ_a is onto.

Because λ_a is a mapping of a finite set to itself, it suffices to establish that λ_a is one-to-one, for it will then necessarily be onto. To show that λ_a is one-to-one, assume that $\lambda_a(x_1) = \lambda_a(x_2)$. Then $ax_1 = ax_2$, and therefore, by Theorem 25.1, $x_1 = x_2$. Thus λ_a is one-to-one, as required.   ∎

Corollary. $\mathbb{Z}_n$ is a field iff n is a prime.

PROOF.   From the remarks in Example 25.2 we know that $\mathbb{Z}_n$ is an integral domain iff n is a prime. That makes this corollary an immediate consequence of Theorem 26.1.   ∎

Example 26.1. Tables 26.1 and 26.2 show operations on $\{0, e, a, b\}$ that produce a field.

Table 26.1

+	0	e	a	b
0	0	e	a	b
e	e	0	b	a
a	a	b	0	e
b	b	a	e	0

Table 26.2

	0	e	a	b
0	0	0	0	0
e	0	e	a	b
a	0	a	b	e
b	0	b	e	a

Once 0 has been chosen for the zero element, and e for the unity, there is no choice about how to complete the table for multiplication (Problem 26.15). The table for addition must produce a group of order 4; Problem 26.16 asks for verification that the additive group here is isomorphic to $\mathbb{Z}_2 \times \mathbb{Z}_2$ (with operation $\oplus$ on each $\mathbb{Z}_2$). This example shows that there is a field of order 4, even though $\mathbb{Z}_4$ (with $\oplus$ and $\odot$) is not a field by the corollary of Theorem 26.1. It can be proved that there is a finite field of order n iff n is a power of a prime (see Section 43). ∎

Definition. A subset K of a field F is a *subfield* of F if K is itself a field with respect to the operations on F.

Theorem 26.2. *A subset K of a field F is a subfield of F iff*

(a) *K contains the zero and unity of F,*

(b) *if $a, b \in K$, then $a + b \in K$ and $ab \in K$,*

(c) *if $a \in K$, then $-a \in K$, and*

(d) *if $a \in K$ and $a \neq 0$, then $a^{-1} \in K$.*

PROOF. Problem 26.19. ∎

Example 26.2. The ring $\mathbb{Z}[\sqrt{2}] = \{a + b\sqrt{2} : a, b \in \mathbb{Z}\}$, considered in Example 24.4, is a subring of $\mathbb{R}$. Although $\mathbb{Z}[\sqrt{2}]$ is an integral domain (see Problem 25.11), it is not a field. For instance, $-2 + \sqrt{2} \in \mathbb{Z}[\sqrt{2}]$, but $(-2 + \sqrt{2})^{-1} = -1 - \frac{1}{2}\sqrt{2} \notin \mathbb{Z}[\sqrt{2}]$. If, however, $\mathbb{Z}$ is replaced by $\mathbb{Q}$, then we do get a field: $\mathbb{Q}[\sqrt{2}] = \{a + b\sqrt{2} : a, b \in \mathbb{Q}\}$ is a subfield of $\mathbb{R}$ (Problem 26.14). ∎

If the requirement of commutativity is dropped from the definition of a field, what is left is the definition of a *division ring*: a ring in which the set of nonzero elements forms a group with respect to multiplication. Thus a commutative division ring is a field. Problem 32.19 gives an example of a division ring that is not commutative.

PROBLEMS

In each of Problems 26.1–26.10, give an example of a ring satisfying the given conditions or, if there is no example, so state.

26.1. An integral domain that is not a field.

26.2. A finite integral domain that is not a field.

26.3. A commutative ring with unity that is not an integral domain.

26.4. An infinite integral domain.

26.5. A finite field.

26.6. A commutative ring with a zero divisor.

26.7. A field that is not an integral domain.

26.8. A field that contains a zero divisor.

26.9. A commutative ring without zero divisors that is not an integral domain.

26.10. A noncommutative ring without a unity.

26.11. Prove that an integral domain is a field iff each nonzero element has an inverse relative to multiplication.

26.12. Prove that an integral domain D is a field iff each equation $ax = b$ ($a, b \in D$ and $a \neq 0$) has a unique solution in D.

26.13. Let $\mathbb{Z}_n^\#$ denote the nonzero elements of $\mathbb{Z}_n$. For which n is $\mathbb{Z}_n^\#$ a group with respect to $\odot$? (Compare Examples 11.4 and 11.5.)

26.14. Verify that $\mathbb{Q}[\sqrt{2}] = \{a + b\sqrt{2} : a, b \in \mathbb{Q}\}$ is a subfield of the field of real numbers (Example 26.2).

26.15. Prove that if $\{0, e, a, b\}$ is to be a field with 0 as zero element and e as unity, then the multiplication must be as defined in Example 26.1.

26.16. Prove that the additive group of the field in Example 26.1 is isomorphic to $\mathbb{Z}_2 \times \mathbb{Z}_2$.

26.17. Show that the ring $\mathbb{Z}_2 \times \mathbb{Z}_2$ is not a field. Why is this not in conflict with Problem 26.16?

26.18. Prove that a direct sum of two or more fields is never a field.

26.19. Prove Theorem 26.2.

26.20. Prove that if C denotes any collection of subfields of a field F, then the intersection of all fields in C is also a subfield of F. (Compare Problem 25.17.).

26.21. State and prove a theorem for fields that is analogous to Theorem 15.2 for groups. (Use Problem 26.20, and compare Problem 25.18.)

26.22. What is the smallest subfield of $\mathbb{R}$ containing $\mathbb{Z}$? (There is such a subfield by Problem 26.21.)

26.23. (a) An element a in a commutative ring R with unity e is said to be *invertible* if there is an element $b \in R$ such that $ab = e$. Prove that if R is a commutative ring with unity, then the invertible elements form a group with respect to the multiplication of the ring.
(b) What is the group of invertible elements in a field?
(c) What is the group of invertible elements in $\mathbb{Z}$?

(d) What is the group of invertible elements in $\mathbb{Z}_4$?

(e) What is the group of invertible elements in $\mathbb{Z}_n$?

26.24. Prove that a zero divisor in a commutative ring with unity cannot be invertible. (See Problem 26.23.)

26.25. For each r in a ring R define $\rho_r : R \rightarrow R$ by $\rho_r(a) = ar$ for each $a \in R$. For R commutative, explain why Ker $\rho_r \neq \{0\}$ iff $r = 0$ or r is a zero divisor in R. For R a field and $r \neq 0$, explain why ρ_r is an isomorphism of the additive group of R onto itself.

SECTION 27   ISOMORPHISM. CHARACTERISTIC

In Section 18 we met the idea of isomorphism for groups and learned that isomorphic groups are essentially the same—they differ at most in the nature of their elements and operations. A similar idea applies to rings.

Definition.   Let R and S be rings. An *isomorphism* of R onto S is a mapping $\theta : R \rightarrow S$ that is one-to-one and onto and satisfies

$$\theta(a + b) = \theta(a) + \theta(b)$$

and

$$\theta(ab) = \theta(a)\theta(b)$$

for all $a, b \in R$. If there is an isomorphism of R onto S, then R and S are said to be *isomorphic* and we write $R \approx S$.

In the conditions $\theta(a + b) = \theta(a) + \theta(b)$ and $\theta(ab) = \theta(a)\theta(b)$, the operations on the left in each equation are, of course, those of R, and the operations on the right are those of S. Notice that because of the first of these conditions a ring isomorphism is necessarily an isomorphism of the additive group of R and S. It follows that $\theta(0) = 0$ and $\theta(-a) = -\theta(a)$ for each $a \in R$, by translation into additive notation of parts (a) and (b) of Theorem 18.2.

The following example shows that an isomorphism between the additive groups of two rings is not necessarily a ring isomorphism.

Example 27.1.   Let θ be the mapping from the ring of integers to the ring of even integers defined by $\theta(n) = 2n$ for each n. We verified in Example 18.3 that this is an isomorphism between additive groups. But it is not a ring isomorphism because it does not preserve multiplication: $\theta(mn) = 2mn$ but $\theta(m)\theta(n) = (2m)(2n) = 4mn$.

Although this mapping θ is not a ring isomorphism, we might ask whether some other mapping from the ring of integers to the ring of even integers can be a ring isomorphism. The answer is No. For example, the ring of integers has a unity, but the ring of even integers does not, and if one of two isomorphic rings has a unity, then the other must as well (Problem 27.1). Notice, then, that although the integers and even integers cannot be distinguished as groups (Example 18.3), they can be distinguished as rings.   ∎

Example 27.2.   Consider the ring $\mathbb{Z}[\sqrt{2}]$ defined in Example 24.4, and define $\theta\colon \mathbb{Z}[\sqrt{2}] \to$ $\mathbb{Z}[\sqrt{2}]$ by $\theta(a + b\sqrt{2}) = a - b\sqrt{2}$. This mapping is clearly one-to-one and onto. It also preserves both ring operations: For addition,

$$\theta((a + b\sqrt{2}) + (c + d\sqrt{2})) = \theta((a + c) + (b + d)\sqrt{2})$$
$$= (a + c) - (b + d)\sqrt{2}$$

and also

$$\theta(a + b\sqrt{2}) + \theta(c + d\sqrt{2}) = (a - b\sqrt{2}) + (c - d\sqrt{2})$$
$$= (a + c) - (b + d)\sqrt{2}.$$

For multiplication,

$$\theta((a + b\sqrt{2})(c + d\sqrt{2})) = \theta((ac + 2bd) + (bc + ad)\sqrt{2})$$
$$= (ac + 2bd) - (bc + ad)\sqrt{2}.$$

and also

$$\theta(a + b\sqrt{2})\theta(c + d\sqrt{2}) = (a - b\sqrt{2})(c - d\sqrt{2})$$
$$= (ac + 2bd) - (bc + ad)\sqrt{2}.$$

Thus θ is an isomorphism of $\mathbb{Z}[\sqrt{2}]$ onto $\mathbb{Z}[\sqrt{2}]$. An isomorphism like this, of a ring onto itself, is called an *automorphism*.   ■

Example 27.3.   We can prove that $\mathbb{Z}_6 \approx \mathbb{Z}_2 \times \mathbb{Z}_3$ by using the mapping $\theta\colon \mathbb{Z}_6 \to \mathbb{Z}_2 \times \mathbb{Z}_3$ defined by $\theta([a]_6) = ([a]_2, [a]_3)$. The following string of equivalent statements shows that θ is both well-defined and one-to-one.

$$
\begin{array}{llll}
[a]_6 = [b]_6 & \text{iff} & 6 \mid (a - b) \\
& \text{iff} & 2(a - b) \text{ and } 3 \mid (a - b) \\
& \text{iff} & [a]_2 = [b]_2 \text{ and } [a]_3 = [b]_3 \\
& \text{iff} & ([a]_2, [a]_3) = ([b]_2, [b]_3).
\end{array}
$$

(The "only if" portion shows that θ is well-defined; the "if" portion shows that it is one-to-one.) Because $|\mathbb{Z}_6| = |\mathbb{Z}_2 \times \mathbb{Z}_3|$, θ is onto since it is one-to-one. Next, θ preserves addition:

$$\theta([a]_6 \oplus [b]_6) = \theta([a + b]_6)$$
$$= ([a + b]_2, [a + b]_3)$$
$$= ([a]_2 \oplus [b]_2, [a]_3 \oplus [b]_3)$$
$$= ([a]_2, [a]_3) + ([b]_2, [b]_3)$$
$$= \theta([a]_6) + \theta([b]_6)$$

Similarly, θ preserves multiplication (Problem 27.17). Therefore, $\mathbb{Z}_6 \approx \mathbb{Z}_2 \times \mathbb{Z}_3$, as claimed. Problem 27.16 suggests how to show that $\mathbb{Z}_4 \not\approx \mathbb{Z}_2 \times \mathbb{Z}_2$. Problem 27.20 asks you to show that $\mathbb{Z}_{mn} \approx \mathbb{Z}_m \times \mathbb{Z}_n$ iff m and n are relatively prime.   ■

If one of two isomorphic groups is Abelian, then the other must also be Abelian (Theorem 18.1). In the same way, if one of two isomorphic rings is commutative, then the other must also be commutative (Problem 27.2). Other properties shared by isomorphic rings include the existence of a unity, existence of a zero divisor, that of being an integral domain, and that of being a field (see the problems). The most common method of showing that two rings are not isomorphic is by finding some such property that one of the rings has but the other does not.

The next concept will help in determining what is unique about the ring of integers. It is also especially useful in the study of fields. Recall that if n is a positive integer and a is a ring element, then $na = a + a + \cdots + a$ (n terms).

Definition. Let R be a ring. If there is a positive integer n such that $na = 0$ for each $a \in R$, then the least such integer is called the *characteristic* of R. If there is no such positive integer, then R is said to have *characteristic* 0 (zero).

If a ring has a unity e and characteristic $n \neq 0$, then in particular $ne = 0$. On the other hand, if $ne = 0$ and $a \in R$, then $na = n(ea) = (ne)a = 0a = 0$. *Thus, for a ring with unity e, the characteristic can be defined alternatively as the least positive integer n such that $ne = 0$, if there is such an integer; otherwise the ring has characteristic 0.*

Example 27.4.
  (a) The ring of integers has characteristic 0, for there is no positive integer n such that $n \cdot 1 = 0$. For the same reason, the ring of rational numbers and the ring of real numbers also have characteristic 0.
  (b) The characteristic of $\mathbb{Z}_n$ is n, because $n[1] = [n] = [0]$, but $k[1] = [k] \neq [0]$ for $0 < k < n$. ∎

In Example 25.2 we observed that if a ring $\mathbb{Z}_n$ is an integral domain, then n is a prime. Thus, in view of the last example, if a ring $\mathbb{Z}_n$ is an integral domain, then its characteristic is a prime. Here is a more general statement.

Theorem 27.1. *If D is an integral domain, then the characteristic of D is either 0 or a prime.*

  PROOF. Assume that D is an integral domain with characteristic $n \neq 0$. Let e denote the unity of D. Since $1e = e \neq 0$, we must have $n > 1$. We shall prove that n must be a prime. Assume otherwise. Then $n = rs$ for some integers r and s with $1 < r < n$ and $1 < s < n$. From $ne = 0$ we have $(rs)e = 0$ and $(rs)(ee) = (re)(se) = 0$. But D, being an integral domain, has no zero divisors, so $(re)(se) = 0$ implies that either $re = 0$ or $se = 0$. Since $1 < r < n$ and $1 < s < n$, either possibility contradicts the fact that n is the characteristic of D. Thus n must be a prime. ∎

Theorem 27.2. *If D is an integral domain of characteristic 0, then D contains a subring isomorphic to $\mathbb{Z}$.*

  PROOF. Let e denote the unity of D, and define $\theta \colon \mathbb{Z} \to D$ by $\theta(n) = ne$ for each $n \in \mathbb{Z}$. We shall prove that θ is one-to-one and that it preserves both ring operations.
    If $\theta(m) = \theta(n)$, then $me = ne$, $me - ne = 0$, and $(m - n)e = 0$. Thus $m = n$ because D has characteristic 0. Therefore θ is one-to-one. If $m, n \in \mathbb{Z}$, then

$$\theta(m + n) = (m + n)e = me + ne = \theta(m) + \theta(n)$$
$$\theta(mn) = (mn)e = (me)(ne) = \theta(m)\theta(n).$$

The image of θ is a subring of D (Problem 27.10), and θ is an isomorphism of $\mathbb{Z}$ onto that subring. This completes the proof.   ∎

Theorem 27.3.   *If D is an integral domain of prime characteristic p, then D contains a subring isomorphic to* $\mathbb{Z}_p$.

PROOF.   The proof of this theorem is similar to that of Theorem 27.2. The relevant mapping is $\theta\colon \mathbb{Z}_p \to D$ defined by $\theta([k]) = ke$ for each $[k] \in \mathbb{Z}_p$. The details are left as an exercise (Problem 27.11).   ∎

If a ring R contains a subring isomorphic to a ring S, then it is said that S can be *embedded* in R. Using this terminology, Theorem 27.2 becomes: *The ring of integers can be embedded in every integral domain of characteristic 0.* And Theorem 27.3 becomes: *The ring* $\mathbb{Z}_p$ *can be embedded in every integral domain of prime characteristic p.*

PROBLEMS

27.1. Prove that if R and S are rings, $\theta\colon R \to S$ is an isomorphism, and e is a unity of R, then $\theta(e)$ is a unity of S.

27.2. Prove that if R and S are isomorphic rings and R is commutative, then S is commutative.

27.3. Prove that if R and S are isomorphic rings and R is an integral domain, then S is an integral domain.

27.4. Prove that if R and S are isomorphic rings and R is a field, then S is a field.

Assume that R is a ring with unity e, that $a, b \in R$, *and that* $m, n \in \mathbb{Z}$. *Prove each statement in Problems 27.5–27.8.*

27.5. $m(ab) = (ma)b = a(mb)$

27.6. $(mn)e = (me)(ne)$

27.7. $(mn)a = m(na)$

27.8. $m(ea) = (me)a$

27.9. Prove that if E and F are fields, $\theta\colon E \to F$ is an isomorphism, and $a \in E$, $a \neq 0$, then $\theta(a^{-1}) = \theta(a)^{-1}$.

27.10. Prove that if R and S are rings, $\theta\colon R \to S$, and θ preserves both ring operations, then $\theta(R)$ is a subring of S.

27.11. Complete the proof of Theorem 27.3. Prove, in particular, that the mapping θ is well-defined.

27.12. Prove that isomorphism is an equivalence relation on the class of all rings.

27.13. List five ring properties that hold for each ring isomorphic to $\mathbb{Z}$ but not for every ring. (By a *ring property* is meant a property of a ring R that is shared by every ring isomorphic to R.)

27.14. What can be said about the characteristic of a ring R in which $x = -x$ for each $x \in R$?

27.15. Prove that if R and S are isomorphic rings, then their characteristics are equal.

27.16. Use Problem 27.15 to explain why $\mathbb{Z}_4 \not\approx \mathbb{Z}_2 \times \mathbb{Z}_2$.

27.17. Verify that the mapping θ in Example 27.3 preserves multiplication.

27.18. Give a counterexample to the following statement: If R is a ring, n is a positive integer, and $na = 0$ for some $a \in R$, then $nr = 0$ for all $r \in R$.

27.19. Give an example of a ring of characteristic 3 that is not a field.

27.20. (a) Prove that if $(m, n) = 1$, then the rings $\mathbb{Z}_{mn}$ and $\mathbb{Z}_m \times \mathbb{Z}_n$ are isomorphic.

(b) Prove that if $(m, n) \neq 1$, then $\mathbb{Z}_{mn} \not\approx \mathbb{Z}_m \times \mathbb{Z}_n$.

27.21. A ring R is called a *Boolean ring* if $x^2 = x$ for each $x \in R$. Prove that every Boolean ring R is commutative and satisfies $2x = 0$ for each $x \in R$. Give an example of such a ring.

27.22. Prove that if R is a finite ring, then the characteristic of R is a divisor of $|R|$. (Section 17 is relevant.)

27.23. Let R denote the subfield $\{a + b\sqrt{2} : a, b \in \mathbb{Q}\}$ of $\mathbb{R}$, and let S denote the subfield $\{a + b\sqrt{3}: a, b \in \mathbb{Q}\}$ of $\mathbb{R}$. Verify that $\theta : R \to S$ defined by $\theta(a + b\sqrt{2}) = a + b\sqrt{3}$ is not a ring isomorphism.

27.24. (a) Prove that if R and S are rings with unities e and f respectively, $R \approx S$, and $x^2 = e + e$ has a solution in R, then $x^2 = f + f$ has a solution in S. (See Problem 27.1.)

(b) Verify that $x^2 = 2$ has a solution in the ring R of Problem 27.23.

(c) The rings R and S in Problem 27.23 are not isomorphic. Give a reason. (Notice that the solution of Problem 27.23 shows that a *particular* mapping from R to S is not an isomorphism. The solution of this problem shows that *no* mapping from R to S is an isomorphism.)

27.25. Every ring R can be embedded in a ring with a unity. Prove this by verifying the following steps.

(a) The set $\mathbb{Z} \times R$ is a ring with respect to the operations

$$(m, a) + (n, b) = (m + n, a + b)$$

and

$$(m, a)(n, b) = (mn, na + mb + ab).$$

(Notice that this is not the direct sum of the rings $\mathbb{Z}$ and R.)

(b) The ring in part (a) has unity $(1, 0)$.

(c) $R' = \{(0, a): a \in R\}$ is a subring of the ring in part (a), and $R \approx R'$

27.26. Prove that every ring with prime characteristic p can be embedded in a ring with unity and characteristic p.

27.27. Let A be an Abelian group, written additively. If α and β are homomorphisms from A to A, define $\alpha + \beta$ and $\alpha\beta$ by

$$(\alpha + \beta)(a) = \alpha(a) + \beta(a)$$

and

$$(\alpha\beta)(a) = a(\beta(a))$$

for all $a \in A$. Prove that $\alpha + \beta$ and $\alpha\beta$ are also homomorphisms from A to A. Also prove that with these operations the set of all homomorphisms from A to A is a ring. (A homomorphism from a group to itself is called an *endomorphism*. The ring in this problem is called the *ring of endomorphisms* of the Abelian group A. Compare Problem 24.10.)

27.28. Let R denote the ring of endomorphisms of $\mathbb{Z}$ (see Problem 27.27). Prove that $R \approx \mathbb{Z}$ (ring isomorphism). (*Suggestion:* For each $k \in \mathbb{Z}$, define $\alpha_k : \mathbb{Z} \to \mathbb{Z}$ by $\alpha_k(n) = kn$ for all $n \in \mathbb{Z}$. Verify that $\alpha_k \in R$ for each $k \in \mathbb{Z}$ and that $\theta : \mathbb{Z} \to R$ defined by $\theta(k) = \alpha_k$ is a ring isomorphism.)

NOTES ON CHAPTER VI

We shall return to rings in a number of later chapters. References [1], [2], and [4] are standard sources that go beyond what is in this book. Chapter 49 of [3] contains remarks on the history of ring theory.

1. Herstein, L. N., *Noncommutative Rings*, Carus Monograph Series, No. **15**, Mathematical Association of America, 1968.
2. Kaplansky, I., *Commutative Rings*, Polygonal Publishing House, 1994.
3. Kline, M., *Mathematical Thought from Ancient to Modern Times*, Oxford University Press, London, 1990.
4. McCoy, N. H., *Rings and Ideals*, Carus Monograph Series, No. **8**, Mathematical Association of America, 1948.

CHAPTER VII

THE FAMILIAR NUMBER SYSTEMS

This chapter shows what distinguishes each of the familiar number systems—integers, rational numbers, real numbers, complex numbers—in terms of its special properties as a ring or field. This requires that we introduce ideas relating to order ($<$) and to the existence of solutions of polynomial equations. Complete proofs are given or sketched in the case of the integers and rational numbers, but not in the case of the real or complex numbers. This chapter gives part of what is necessary to replace an intuitive understanding of the familiar number systems by an understanding based on a more solid logical foundation.

SECTION 28   ORDERED INTEGRAL DOMAINS

In this section and the one that follows we take the first steps in characterizing the ring of integers. The first definition given, that of an ordered integral domain, applies to the integers as well as to many other integral domains. It will lead to the ideas of *positive*, *negative*, *greater than*, and *less than*. In order to read the definition with the integers in mind as an example, think of D^p as being the set of positive integers.

Definition.   An integral domain D is said to be *ordered* if there is a subset D^p of D such that:

closure under addition

$$\text{if } a, b \in D^p, \text{ then } a + b \in D^p,$$

closure under multiplication

$$if \ a, b \in D^p, \text{ then } ab \in D^p,$$

law of trichotomy

if $a \in D$, then exactly one of the following is true:

$$a = 0, \quad a \in D^p, \quad \text{or } -a \in D^p.$$

The elements of D^p are called the *positive* elements of D. Elements that are neither zero nor positive are said to be *negative*.

Besides the integers, other ordered integral domains include the rational numbers and the real numbers, with the set of positive elements being the set of positive numbers in each case. We shall see that the integral domains $\mathbb{Z}_p$ are not ordered (regardless of what one tries to use for the set of positive elements). Assume in the remainder of this section that D is an ordered integral domain with unity e.

Lemma 28.1.   *If* $a \in D$ *and* $a \neq 0$, *then* $a^2 \in D^p$

PROOF.   By definition of D^p, since $a \neq 0$ we have either $a \in D^p$ or $-a \in D^p$. If $a \in D^p$, then $a \cdot a = a^2 \in D^p$ by closure of D^p under multiplication. If $-a \in D^p$, then $(-a)(-a) = (-a)^2 \in D^p$ by closure of D^p under multiplication; but $(-a)^2 = a^2$ in any ring, so that again $a^2 \in D^p$.   ■

Corollary. $e \in D^p$.

PROOF.   Since $e \in D$ and $e \neq 0$, $e^2 \in D^p$ by Lemma 28.1. But $e^2 = e$.   ■

Lemma 28.2.   *If* $a \in D^p$ *and* n *is a positive integer, then* $na \in D^p$.

PROOF.   The proof is by induction on n (which is reviewed in Appendix C). We are given $1a = a \in D^p$. Assume $ka \in D^p$. Then $(k + 1)a = ka + 1a \in D^p$ by closure of D^p under addition. Thus $na \in D^p$ for every positive integer n.   ■

Theorem 28.1.   *If* D *is an ordered integral domain, then* D *has characteristic 0.*

PROOF.   By the corollary of Lemma 28.1, $e \in D^p$, so $ne \in D^p$ for each positive integer n by Lemma 28.2. Thus $ne \neq 0$ for each positive integer n, since a positive element cannot be 0. Therefore the characteristic cannot be $n \neq 0$.   ■

Corollary.   *If* D *is an ordered integral domain, then* D *contains a subring isomorphic to* $\mathbb{Z}$.

PROOF.   Apply Theorems 28.1 and 27.2.   ■

Corollary.   *A finite integral domain cannot be ordered. In particular,* $\mathbb{Z}_p$ *(p a prime) cannot be ordered.*

PROOF.   This is a direct consequence of the preceding corollary.   ■

Definition.   Assume that D is an ordered integral domain and $a, b \in D$. Then $a > b$ will mean that $a - b \in D^p$. If $a > b$, we say that a is *greater than* b, and that b is *less than* a.

For the ring of integers (or rational numbers or real numbers), this is the customary meaning of $>$. As usual,

$$
\begin{array}{lll}
b < a & \text{means} & a > b \\
a \geq b & \text{means} & a > b \text{ or } a = b, \text{ and} \\
a \leq b & \text{means} & a < b \text{ or } a = b.
\end{array}
$$

The following theorem brings together many of the properties of the relation $>$.

Theorem 28.2.   *Let D be an ordered integral domain and let* $a, b, c \in D$.

(a) *If* $a > 0$ *and* $b > 0$, *then* $a + b > 0$.

(b) *If* $a > 0$ *and* $b > 0$, *then* $ab > 0$.

(c) *Exactly one of the following is true:* $a = b, a > b$, *or* $b > a$.

(d) *If* $a > b$, *then* $a + c > b + c$.

(e) *If* $a > b$ *and* $c > 0$, *then* $ac > bc$.

(f) *If* $a \neq 0$, *then* $a^2 > 0$

(g) *If* $a > b$ *and* $b > c$, *then* $a > c$.

PROOF.   Each property follows from the definitions or other properties already given in this section. To prove (e), for instance, suppose that $a > b$ and $c > 0$. Then $a - b \in D^p$ and $c \in D^p$ by the definition of $>$. Therefore $(a - b)c \in D^p$ by the closure of D^p under multiplication. But $(a - b)c = ac - bc$ so that $ac - bc \in D^p$. Therefore, applying the definition of $>$ once more, $ac > bc$.

The proofs of the other properties are left as exercises (Problem 28.7).   ■

PROBLEMS

Throughout this set of problems D denotes an ordered integral domain.

28.1. If e is the unity of D, then $x^2 + e = 0$ has no solution in D. Why?

28.2. Prove that if $a, b, c \in D, a > b$, and $c < 0$, then $ac < bc$.

28.3. Prove that if $a, b, c \in D, ac > bc$, and $c > 0$, then $a > b$.

28.4. Prove that if $a, b \in D$, and $a < 0$ and $b < 0$, then $ab > 0$.

28.5. Prove that if $a, b \in D$ and $a > b$, then $-a < -b$.

28.6. Prove that if $a, b \in D^p$ and $a > b$, then $a^2 > b^2$.

28.7. Complete the proof of Theorem 28.2.

28.8. Prove or disprove that if E is a subring of an ordered integral domain D, and E is also an integral domain, then E is ordered.

28.9. Let D denote the ring of integers. Prove that the only subset of D that satisfies the conditions on D^p in the definition of integral domain is the set of natural numbers.

28.10. A commutative ring R is said to be *ordered* if there is a subset R^p of R such that R^p satisfies the conditions on D^p in the definition of ordered integral domain.

(a) Verify that if *positive* has its usual meaning, then the ring of even integers is ordered. Make clear which properties of the even integers are used.

(b) Let $R = M(\mathbb{R})$, the ring in Example 24.5. Let $f \in R^p$ mean that $f(0) > 0$. Does this make R an ordered commutative ring? Justify your answer.

28.11. Begin with Lemma 28.1, continue through the two corollaries of Theorem 28.1, and in each case replace D by an ordered commutative ring R with unity. (See Problem 28.10 for the definition of ordered commutative ring.) Which of the lemmas, corollaries, and theorems are still true?

28.12. Prove or disprove that if $a, b \in D$ and $a > b$, then $a^2 > b^2$.

28.13. Prove or disprove that if $a, b \in D$ and $a > b$, then $a^3 > b^3$.

28.14. For $a \in D$, define $|a|$ by

$$|a| = \begin{cases} a & \text{if} & a \in D^p \\ 0 & \text{if} & a = 0 \\ -a & \text{if} & -a \in D^p \end{cases}$$

Prove that if $a, b \in D$, then:
(a) $|ab| = |a| \cdot |b|$
(b) $|a| \geq a \geq -|a|$ and $|b| \geq b \geq -|b|$
(c) $|a| + |b| \geq |a + b|$ [*Suggestion*: Add inequalities from part (b).]
(d) $|a - b| \geq ||a| - |b||$

SECTION 29   THE INTEGERS

Definition.   An element a in a subset S of an ordered integral domain D is a *least element* of S if $x > a$ for each $x \in S$ such that $x \neq a$.

Definition.   An ordered integral domain D is *well-ordered* if every nonempty subset of D^p has a least element.

The Least Integer Principle (Section 10) states that the integral domain of integers is well-ordered. (What we have called the Least Integer Principle is sometimes even called the *Well-Ordering Principle*.) The integral domain of rational numbers is not well-ordered, because the set of positive rational numbers has no least element (Problem 29.1). In fact, the integers form the "only" well-ordered integral domain. The following theorem makes this precise.

Theorem 29.1.   *If D is a well-ordered integral domain, then D is isomorphic to the ring of integers.*

The proof of the theorem will be easier to grasp if the following fact is proved separately.

Lemma 29.1.   *If D is a well-ordered integral domain with unity e, then e is the least element of D^p.*

PROOF.   Because D is well-ordered, D^p must have a least element; assume it to be $a \neq e$ (this will lead to a contradiction). Since $e \in D^p$ by the corollary of Lemma 28.1, and a is the least element of D^p by our assumption, we must have $e > a$. Now $e > a$ and $a > 0$ imply $a > a^2$, by Theorem 28.2(e). However, $a^2 \in D^p$ by Lemma 28.1. Thus we have $a^2 \in D^p$, and $a > a^2$, which contradicts the assumption that a is the least element of D^p. Thus the least element of D^p must be e.   ∎

PROOF OF THEOREM 29.1.   Assume that D is a well-ordered integral domain with unity e, and define $\theta: \mathbb{Z} \to D$ by $\theta(n) = ne$ for each $n \in \mathbb{Z}$. We showed in proving Theorem 27.2 that this θ is an isomorphism of $\mathbb{Z}$ onto $\theta(\mathbb{Z})$; therefore it suffices to prove that $\theta(\mathbb{Z}) = D$. We shall use an indirect proof for this.

Assume that θ is not onto, and let d denote an element such that $d \in D$ but $d \notin \theta (\mathbb{Z})$. Then also $-d \in D$ and $-d \notin \theta(\mathbb{Z})$: for if $-d \in \theta(\mathbb{Z})$, say $\theta(m) = -d$, then $\theta(m) = me = -d$ so that $\theta(-m) = (-m)e = -(me) = d$, implying $d \in \theta(\mathbb{Z})$, which is false. Since $d \notin \theta (\mathbb{Z})$ and $-d \notin \theta(\mathbb{Z})$, and either $d \in D^p$ or $-d \in D^p$, we conclude that there is a positive element in D that is not in $\theta(\mathbb{Z})$. Therefore the set of elements that are in D^p but not in $\theta (\mathbb{Z})$ is nonempty, and so since D is well-ordered there is a least such element—call it s.

Thus s is the least element of D^p that is not in $\theta (\mathbb{Z})$. Since $\theta(1) = 1e = e$, we have $e \in \theta (\mathbb{Z})$ so that $e \neq s$. Therefore, since e is the least element of D^p (Lemma 29.1), we must have $s > e, s - e > 0$, and $s - e \in D^p$. But $s > s - e$ [because $s - (s - e) = e \in D^p$], so that since s is the least element of D^p not in $\theta (\mathbb{Z})$, we must have $s - e \in \theta (\mathbb{Z})$. But if $\theta(k) = ke = s - e$ for $k \in \mathbb{Z}$, then $\theta(k + 1) = (k + 1)e = ke + e = (s - e) + e = s$, and hence $s \in \theta (\mathbb{Z})$, which is a contradiction. Therefore $\theta (\mathbb{Z}) = D$.   ∎

If we do not distinguish between isomorphic rings, we now have a characterization of the ring of integers: $\mathbb{Z}$ *is the unique well-ordered integral domain.*

Throughout this book we have assumed and used properties of the integers without proving them. In the book *Foundations of Analysis*, E. Landau begins with the Peano Postulates—a set of five postulates (axioms) for the system of natural numbers—and then works through a careful development of properties of the natural numbers, the integers, and each of the other number systems in this chapter. Such an axiomatic presentation, together with the uniqueness proved in Theorem 29.1, provides a more complete description of the integers than that provided by Theorem 29.1 alone.

PROBLEMS

29.1. Prove (without Theorem 29.1) that $\mathbb{Q}$ is not well-ordered.

29.2. Prove that the integral domain $\mathbb{Z}[\sqrt{2}]$ in Example 24.4 is not well-ordered. (*Suggestion*: Use Theorem 29.1 and Problem 27.24.)

29.3. Prove that if D is an ordered integral domain with unity e, and $a \in D$, then, $a > a - e$.

29.4. Prove that if D is a well-ordered integral domain, $a \in D, a \neq 0$, and $a \neq e$, then $a^2 > a$. Is the preceding statement true if D is merely ordered rather than well-ordered?

29.5. For D an ordered integral domain, let D^n denote the set of all nonzero elements not in D^p. Prove that if D is well-ordered, then every nonempty subset S of D^n has a greatest element, that is, an element $b \in S$ such that $x < b$ for each $x \in S$ such that $x \neq b$.

29.6. Prove that if n is an integer, then $n + 1$ is the least integer greater than n. (Everyone "knows" this, but prove it.)

29.7. Prove the Principle of Mathematical Induction (Appendix C) from the Least Integer Principle, that is, from the fact that $\mathbb{Z}$ is well-ordered. [*Suggestion*: Let $S = \{k: P(k)$ is false$\}$ and show that S must be empty.]

29.8. Show that if $\theta: \mathbb{Z} \to \mathbb{Z}$ is a ring isomorphism, then θ must be the identity mapping. Is there an additive group isomorphism $\theta: \mathbb{Z} \to \mathbb{Z}$ other than the identity mapping?

SECTION 30   FIELD OF QUOTIENTS. THE FIELD OF RATIONAL NUMBERS

If a and b are integers with $a \neq 0$, then the equation $ax = b$ may not have a solution in the integral domain of integers. The equation does have a solution in the field of rational numbers, however. Moreover, the field of rational numbers is just large enough to contain all such solutions, because every rational number has the form $a^{-1}b$ for a and b integers with $a \neq 0$, so that every rational number is a solution of an equation $ax = b$. We prove in this section that if D is *any* integral domain, then there is a "unique smallest" field "containing" D such that each equation $ax = b$, with $a, b \in D$ and $a \neq 0$, has a solution in that field. This field is called the *field of quotients* of D. The field of quotients of the integral domain of integers is the field of rational numbers. Since we have already characterized the integers, the uniqueness of its field of quotients gives us a characterization of the rational numbers.

Before starting through the formalities needed to construct a field of quotients, here is the basic idea as it applies to the integers. A fraction a/b gives us an ordered pair of integers (a, b) with the second component nonzero. Instead of thinking of the fraction, think of the ordered pair. To account for the fact that different fractions (such as $\frac{2}{3}$, $\frac{16}{24}$, and $\frac{-20}{-30}$) can represent the same rational number, we agree not to distinguish between ordered pairs if they correspond to such fractions. This is done with an equivalence relation: pairs (a, b) and (c, d) will be equivalent if the corresponding fractions are equal [thus $(2, 3)$ and $(-20, -30)$ will be equivalent]. The equivalence classes for this equivalence relation form a set, and we define two operations on this set that make it a field—the field of rational numbers. Next, we prove that this field contains an integral domain isomorphic to the integral domain of integers. Finally, we prove that any field containing an integral domain isomorphic to the integers must contain a field isomorphic to the rational numbers.

Why do all this if we already know about the rational numbers? First, it will tell us what is unique about the rational numbers in terms of the appropriate abstract ring properties. Second, the procedure used will apply to any integral domain, not just the integers.

Throughout the following discussion D denotes an integral domain with unity e, and D' denotes the set of all nonzero elements of D. The Cartesian product of D and D' is

$$D \times D' = \{(a, b): a, b \in D, b \neq 0\}.$$

For elements (a, b) and (c, d) in $D \times D'$ we write

$$(a, b) \sim (c, d) \quad \text{iff} \quad ad = bc.$$

Lemma 30.1.   *The relation $\sim$ is an equivalence relation on $D \times D'$.*

PROOF.   The verification of the reflexive and symmetric properties is left to Problem 30.5. To prove that $\sim$ is transitive, assume that $(a, b) \sim (c, d)$ and $(c, d) \sim (f, g)$, with each pair in $D \times D'$. Then $ad = bc$ and $cg = df$. By the first of these equations $adg = bcg$, and by the second $bcg = bdf$. From these last equations we conclude that $adg = bdf$, or $(ag)d = (bf)d$ (remember that D is commutative). But D is an integral domain and $d \neq 0$, so that by cancellation (Theorem 25.1) $ag = bf$. This proves that $(a, b) \sim (f, g)$, which establishes transitivity.   ∎

If $(a, b) \in D \times D'$, we shall denote the equivalence class to which (a, b) belongs relative to $\sim$ by $[a, b]$. Thus

$$[a, b] = \{(x, y) \in D \times D' : (a, b) \sim (x, y)\}.$$

The set of all such equivalence classes will be denoted by F_D. If we recall how fractions are added and multiplied $[(a/b) + (c/d) = (ad + bc)/bd$ and $(a/b)(c/d) = ac/bd]$, then we are led to define two operations on F_D as follows:

$$[a, b] + [c, d] = [ad + bc, bd] \qquad \text{and} \qquad [a, b] \cdot [c, d] = [ac, bd] \tag{30.1}$$

for all $[a, b], [c, d] \in F_D$. (Notice that each second component, bd, is in D' because $b \in D'$ and $d \in D'$ and D has no zero divisors.) The next lemma shows that these operations are well-defined.

Lemma 30.2. *If $[a_1, b_1] = [a_2, b_2]$ and $[c_1, d_1] = [c_2, d_2]$, then*

$$[a_1 d_1 + b_1 c_1, b_1 d_1] = [a_2 d_2 + b_2 c_2, b_2 d_2]$$

and

$$[a_1 c_1, b_1 d_1] = [a_2 c_2, b_2 d_2]$$

PROOF. The proof for multiplication is left to Problem 30.6. Here is the proof for addition. From $[a_1, b_1] = [a_2, b_2]$ we know that $(a_1, b_1) \sim (a_2, b_2)$, and so

$$a_1 b_2 = b_1 a_2. \tag{30.2}$$

Similarly, from $[c_1, d_1] = [c_2, d_2]$ we know that

$$c_1 d_2 = d_1 c_2. \tag{30.3}$$

We must show that

$$[a_1 d_1 + b_1 c_1, b_1 d_1] = [a_2 d_2 + b_2 c_2, b_2 d_2],$$

that is,

$$(a_1 d_1 + b_1 c_1)(b_2 d_2) = (b_1 d_1)(a_2 d_2 + b_2 c_2). \tag{30.4}$$

Using (30.2) and (30.3), and the commutativity of D, we can deduce (30.4) as follows:

$$\begin{aligned}
(a_1 d_1 + b_1 c_1)(b_2 d_2) &= a_1 d_1 b_2 d_2 + b_1 c_1 b_2 d_2 \\
&= a_1 b_2 d_1 d_2 + b_1 b_2 c_1 d_2 \\
&= b_1 a_2 d_1 d_2 + b_1 b_2 d_1 c_2 \\
&= b_1 d_1 a_2 d_2 + b_1 d_1 b_2 c_2 \\
&= b_1 d_1 (a_2 d_2 + b_2 c_2).
\end{aligned}$$ ∎

It is easy to verify that

$$\text{if } c \in D' \text{ and } [a, b] \in F_D, \text{ then } [a, b] = [ac, bc] = [ca, cb].$$

This will be used without explicit reference.

Lemma 30.3. F_D *with the operations defined in (30.1) is a field. The zero is* $[0, e]$; *the negative of* $[a, b]$ *is* $[-a, b]$; *the unity is* $[e, e]$; *and the inverse of* $[a, b] \neq [0, e]$ *is* $[b, a]$.

PROOF.   We shall verify that $[0, e]$ is a zero and that one of the two distributive laws is satisfied. The remainder of the proof is left to Problem 30.7.

As to zero: if $[a, b] \in F_D$, then $[a, b] + [0, e] = [ae + b0, be] = [ae, be] = [a, b]$, and $[0, e] + [a, b] = [0b + ea, eb] = [ea, eb] = [a, b]$.

If $[a, b], [c, d], [f, g] \in F_D$, then

$$[a, b], ([c, d] + [f, g]) = [a, b][cg + df, dg]$$
$$= [acg + adf, bdg]$$
$$= [b(acg + adf), b(bdg)]$$
$$= [(ac)(bg) + (bd)(af), (bd)(bg)]$$
$$= [ac, bd] + [af, bg]$$
$$= [a, b][c, d] + [a, b][f, g].$$

This establishes one of the two distributive laws.   ∎

Lemma 30.4.   *Let* D_1 *denote the subset of* F_D *consisting of all* $[a, e]$ *for* $a \in D$. *Then* D_1 *is a subring of* F_D *and* $D \approx D_1$.

PROOF.   Define $\theta \colon D \to F_D$ by $\theta(a) = [a, e]$ for each $a \in D$. The image of θ is clearly D_1. We shall prove that θ is one-to-one and preserves sums and products. It will follow from this that D_1 is a subring and that $D \approx D_1$. [The proof that D_1 is a subring is analogous to the proof of Theorem 18.2(d).]

If $\theta(a_1) = \theta(a_2)$, then $[a_1, e] = [a_2, e]$ so that $a_1 e = ea_2$ and $a_1 = a_2$. Thus θ is one-to-one. If $a, b \in D$, then $\theta(a + b) = [a + b, e] = [ae + eb, ee] = [a, e] + [b, e] = \theta(a) + \theta(b)$, and $\theta(ab) = [ab, e] = [ab, ee] = [a, e][b, e] = \theta(a)\theta(b)$.   ∎

The field F_D, constructed from the integral domain D in this way, is called the *field of quotients* of D. Recall that to say that a ring S can be embedded in a ring R means that R contains a subring isomorphic to S (Section 27). Using this terminology, what we have shown is that *any integral domain can be embedded in a field*—its field of quotients. But in general an integral domain can be embedded in more than one field. For example, the integral domain of integers can be embedded in the field of rational numbers and also the field of real numbers. However, the field of quotients is the smallest such field, in that it can be embedded in any field in which the given integral domain can be embedded. We put all of this together in the following theorem.

Theorem 30.1.   *If* D *is an integral domain, then there exists a field* F_D, *the field of quotients of* D, *such that*

(a) F_D *contains an integral domain isomorphic to* D, *and*

(b)   *if* K *is any field containing an integral domain isomorphic to* D, *then* K *contains a field isomorphic to* F_D.

PROOF.   Lemma 30.3 proves that F_D is a field. Lemma 30.4 proves property (a).

In proving property (b), we shall assume that D is actually a subring of K; this amounts to identifying D with an integral domain to which it is isomorphic, and allows us to get at the main idea without having it obscured by distractive notation. Because of the way in which the field of quotients F_D is constructed from the elements of D, we have a natural correspondence between elements $ab^{-1} \in K$, with $a, b \in D$ and $b \neq 0$, and elements $[a, b] \in F_D$. If we identify each element of F_D with the element to which it corresponds ($[a, b] \leftrightarrow ab^{-1}$), then we can think of F_D as a subring of K, which is what property (b) asserts. [If you prefer, the map $\phi: F_D \to K$ defined by $\phi([a, b]) = ab^{-1}$ is one-to-one and preserves both addition and multiplication, and thus is an isomorphism of F_D onto a subfield of K.]   ∎

We can now characterize the field of rational numbers among all sets with two operations: *The field of rational numbers is the (unique) field of quotients of the (unique) well-ordered integral domain.* Again, this assumes that we do not distinguish between isomorphic rings or fields.

If K is any field of characteristic 0, then K must contain a subring isomorphic to $\mathbb{Z}$, by Theorem 27.2. But then K must also contain a subfield isomorphic to $\mathbb{Q}$, by Theorem 30.1. This gives the following corollary.

Corollary.   If K is any field of characteristic 0, then K contains a subfield isomorphic to $\mathbb{Q}$.

PROBLEMS

30.1. Verify that Lemma 30.1 is not true if D is replaced by $\mathbb{Z}_4$. This will show that Lemma 30.1 is not true if D is assumed to be just a commutative ring with unity.

30.2. Prove or disprove: If D is a field in the definition preceding Lemma 30.1, then $(a, b) \sim (c, d)$ iff $(a, b) = (c, d)$.

30.3. Verify that the field of quotients of the integral domain $\mathbb{Z}[\sqrt{2}]$ (Example 24.4) is isomorphic to $\mathbb{Q}[\sqrt{2}] = \{a + b\sqrt{2} : a, b \in \mathbb{Q}\}$.

30.4. Let R denote the set of all fractions of the form $a/2^k$, where a is an integer and k is a nonnegative integer.
(a) Prove that R is an integral domain.
(b) Verify that the field of quotients of R is isomorphic to the field of rational numbers.

30.5. Complete the proof of Lemma 30.1.

30.6. Complete the proof of Lemma 30.2.

30.7. Complete the proof of Lemma 30.3.

30.8. The ring $\mathbb{Z}_6$ cannot be embedded in a field. Why?

30.9. The field of quotients of any field D is isomorphic to D. Why?

30.10. Assume that D, D_1, and F_D are as in Lemma 30.4. Show that each element of F_D is a solution of some equation $ax = b$ with $a, b \in D_1$

30.11. State and prove the analogue of the corollary of Theorem 30.1 with characteristic p in place of characteristic 0.

30.12. In place of the operation $+$ defined on F_D in (30.1), consider $\boxplus$ "defined" by $[a, b] \boxplus [c, d] = [a + c, b + d]$. (That is, try to define the sum of two fractions so that the numerator is the sum of the numerators and the denominator is the sum of the denominators, an idea often used by students new to fractions.) Verify that $\boxplus$ is not well-defined.

SECTION 31   ORDERED FIELDS. THE FIELD OF REAL NUMBERS

By moving from the integral domain of integers to the field of rational numbers, we have obtained solutions to all equations $ax = b$ (a and b integers, $a \neq 0$). But other deficiencies remain. As we shall prove in Theorem 31.1, for instance, there is no rational number x such that $x^2 = 2$. In other words, $\sqrt{2}$ is *irrational*, that is, not rational. This was first discovered by the Pythagoreans, in the fifth century B.C., in its geometric form: there is no rational number that will measure the hypotenuse of a right triangle with each leg of unit length (see Figure 31.1). In terms of a number line, this means that if two points are chosen on a straight line and labeled 0 and 1, and if other points are then made to correspond to the rational numbers in the obvious way, there will be no number corresponding to the point "$\sqrt{2}$ units" from 0 in the positive direction. In fact, there will be many points not corresponding to rational numbers. The basic assumption of coordinate geometry is that this problem can be overcome by using real numbers in place of (just) rational numbers. This important use of real numbers emphasizes the fact that the deficiency of the rational numbers that the real numbers corrects has more to do with order than with the solution of equations (such as $x^2 = 2$). This should become clearer with the discussion of real numbers in this section and complex numbers in the next.

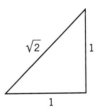

Figure 31.1

We now prove the irrationality of $\sqrt{2}$. The usual notation a/b ($a, b \in \mathbb{Z}$) will be used in place of the ordered pair notation (a, b) of Section 30, and also to denote the equivalence class $[a, b]$.

Theorem 31.1. *There is no rational number x such that $x^2 = 2$.*

PROOF.   The proof is by contradiction, and so we begin by assuming the theorem to be false. Thus assume that there are integers a and b, with $b \neq 0$, such that $(a/b)^2 = 2$. We also assume a/b reduced to lowest terms, so that a and b have no common factor except ± 1. From $(a/b)^2 = 2$ we have $a^2 = 2b^2$. The right side of this equation, $2b^2$, is even; therefore, the left side, a^2, must also be even. But if a^2 is even, then a must be even, and hence $a = 2k$ for some integer k. Substituting $a = 2k$ in $a^2 = 2b^2$, we obtain $(2k)^2 = 2b^2$, or $2k^2 = b^2$ Since $2k^2$ is even, b^2 is even, so that b must be even. Thus we have deduced that a and b are both even, and they therefore have 2 as a common factor. This contradicts the assumption that a/b was reduced to lowest terms, and completes the proof.   ∎

The next theorem shows that the rational numbers form an *ordered field*, that is, an ordered integral domain that is also a field. Again, the notation a/b is used in place of the ordered pair notation (a, b) of Section 30. (The proof of this theorem should be considered optional.)

Theorem 31.2.   *Let $\mathbb{Q}$ denote the field of rational numbers, and let $\mathbb{Q}^p$ denote the set of all elements of $\mathbb{Q}$ with representations a/b such that $ab > 0$. Then $\mathbb{Q}$ is an ordered field with $\mathbb{Q}^p$ as its set of positive elements.*

PROOF.   We must first verify that the condition $(ab > 0)$ for an element of $\mathbb{Q}$ to be in $\mathbb{Q}^p$ is independent of the particular fraction chosen to represent it. To this end, assume that $a/b = c/d$. Then $ad = bc$. Making use of this, and $b^2 > 0$ and $d^2 > 0$, together with properties of the relation $>$, we can deduce that if $ab > 0$, then $abd^2 > 0$, $adbd > 0$, $bcbd > 0$, $cdb^2 > 0$, and finally $cd > 0$. Similarly, if $cd > 0$, then $ab > 0$. Thus $ab > 0$ iff $cd > 0$, as required.

It remains to verify the three properties in the definition of ordered integral domain (Section 28).

If a/b and c/d represent elements of $\mathbb{Q}^p$, then $ab > 0$ and $cd > 0$, and thus $abd^2 > 0$ and $cdb^2 > 0$. This implies that $(ad + bc)bd = abd^2 + cdb^2 > 0$, so that $(a/b) + (c/d) = (ad + bc)/bd \in \mathbb{Q}^p$.

As to closure of $\mathbb{Q}^p$ under multiplication, if $a/b \in \mathbb{Q}^p$ and $c/d \in \mathbb{Q}^p$, then $ab > 0$ and $cd > 0$, so that $(ac)(bd) = (ab)(cd) > 0$ and $(a/b)(c/d) = (ac)/(bd) \in \mathbb{Q}^p$.

Finally, if a/b represents a nonzero element of $\mathbb{Q}$, then $a \neq 0$ and $b \neq 0$; therefore $ab \neq 0$. Therefore $ab > 0$ or $0 > ab$, and, correspondingly, $a/b \in \mathbb{Q}^p$ or $-(a/b) = (-a)/b \in \mathbb{Q}^p$. This establishes the law of trichotomy.   ∎

An element u of an ordered field F is said to be an *upper bound* for a subset S of F if $u \geq x$ for each $x \in S$. For example, any positive rational number is an upper bound for the set of all negative rational numbers. The set of integers has no upper bound in the field of rational numbers (or in any other field). An element u of an ordered field F is said to be a *least upper bound* for a subset S of F provided

1. u is an upper bound for S, and
2. if $v \in F$ is an upper bound for S, then $v \geq u$.

Thus 0 is a least upper bound for the set of negative rational numbers.

If S denotes the set of all rational numbers r such that $r^2 < 2$, then S has an upper bound in the field of rational numbers (1.5, for instance); but it does not have a least upper bound in the field of rational numbers. However, S does have a least upper bound in the field of real numbers—namely $\sqrt{2}$. This leads to the following definition, which isolates the property distinguishing the field of real numbers from the field of rational numbers.

Definition.   An ordered field F is said to be *complete* if every nonempty subset of F having an upper bound in F has a least upper bound in F.

Theorem 31.3. *There exists a complete ordered field. Any two such fields are isomorphic, and any such field contains a subfield isomorphic to the field of rational numbers.*

The field of real numbers is a complete ordered field. Theorem 31.3 shows that such a field exists and is essentially unique. The fact that it must contain a subfield isomorphic to $\mathbb{Q}$ follows quickly from results in Sections 28 and 30 (Problem 31.15). We shall not prove the other parts of the theorem. (For details of the construction of the real numbers, see, for example, the book by Landau referred to in Section 29.)

In applications the real numbers are usually thought of as the numbers having decimal representations. Examples are

$$\tfrac{1}{2} = 0.5 \qquad\qquad \tfrac{1}{3} = 0.\overline{3}$$

$$-12.138 \qquad\qquad \tfrac{11}{7} = 1.\overline{571428}$$

$$\sqrt{2} = 1.414213\ldots \qquad \pi = 3.141592\ldots,$$

where the lines over 3 and 571428 mean that they repeat without end. It can be shown that *the decimal numbers representing rational numbers are precisely those that either terminate or become periodic* (Problems 31.10 and 31.12). The number $0.1010010001\ldots$, where the number of 0's between 1's increases each time, is irrational. Each number whose decimal representation terminates (such as 0.5 or -12.138) also has a representation with 9 repeating on the end. For example, $1.0 = 0.\overline{9}$ (Problem 31.7). The numbers π and e (the base for natural logarithms) are both irrational, but the proofs are more difficult than that for $\sqrt{2}$.

PROBLEMS

In Problems 31.1 and 31.2, assume that F is an ordered field, $a, b \in F$, and 1 is the unity of F.

31.1. Prove $b > a > 0$ implies $a^{-1} > b^{-1} > 0$.

31.2. Prove $0 > b > a$ implies $0 > a^{-1} > b^{-1}$.

31.3. Prove that if a is rational and b is irrational, then $a + b$ is irrational.

31.4. Prove that if a is rational, $a \neq 0$, and b is irrational, then ab is irrational.

31.5. Prove that if u is a least upper bound for a subset S of $\mathbb{R}$, then $2u$ is a least upper bound for $\{2x : x \in S\}$.

31.6. Prove that if u is a least upper bound for a subset S of $\mathbb{R}$, then $3 + u$ is a least upper bound for $\{3 + x : x \in S\}$.

31.7. Explain why $0.\overline{9} = 1$.

31.8. Express $1.9\overline{35}$ as a fraction.

31.9. Prove that a decimal number that terminates represents a rational number.

31.10. Prove that a decimal number that becomes periodic represents a rational number. (See Problems 31.7 and 31.8.)

31.11. Determine the decimal representations of each of the following numbers.

(a) $\frac{3}{11}$

(b) $\frac{1984}{7}$

(c) $\frac{9}{8}$

(d) $\frac{8}{9}$

31.12. Explain why the decimal representation of a rational number must terminate or become periodic. (*Suggestion*: In computing the decimal representation of a/b by long division, there are only b possible remainders. Look at what happens in a special case.)

31.13. Prove that if p is a prime, then $\sqrt{p}$ is irrational. (Problem 13.16 gives a more general statement.)

31.14. (a) Write definitions of *lower bound* and *greatest lower bound* for a subset S of an ordered field F.

(b) Prove that if F is a complete ordered field, then every nonempty subset of F having a lower bound in F has a greatest lower bound in F.

(c) Is the converse of the statement in part (b) true?

31.15. Prove the last part of Theorem 31.3, that is, prove that each complete ordered field contains a subfield isomorphic to $\mathbb{Q}$. (Use Sections 28 and 30.)

31.16. Prove that if a and b are two distinct positive real numbers, then $(a + b)/2 > \sqrt{ab}$ ("The arithmetic mean is greater than the geometric mean." *Suggestion*: First explain why $(\sqrt{a} - \sqrt{b})^2 > 0$.)

31.17. Prove that a subset of an ordered field has at most one least upper bound in the field.

31.18. Prove that if a and b are positive real numbers, then there exists an integer n such that $na > b$. (This is called the *Archimedean Property* of $\mathbb{R}$. Suppose that the statement is false, so that b is an upper bound for $\{na: n \in \mathbb{Z}\}$, and then deduce a contradiction by using the completeness of $\mathbb{R}$.)

31.19. Prove that if $a, b \in \mathbb{Q}$, and $a > b$, then there are infinitely many $x \in \mathbb{Q}$ such that $a > x > b$. [*Suggestion*: If c and d are rational, then so is $(c + d)/2$.]

31.20. Prove that the statement in Problem 31.19 is true if $\mathbb{Q}$ is replaced by any other ordered field.

31.21. Prove that if $a, b \in \mathbb{R}$, and $a > b$, then there exists a rational number m/n such that $a > m/n > b$. Compare Problem 31.22. [*Suggestion*: By Problem 31.18 there is a positive integer n such that $n(a - b) > 1$, or $(a - b) > 1/n$. Let m be the least integer such that $m > nb$. Then $(m - 1)/n \le b$, and so $m/n = (m - 1)/n + 1/n < b + (a - b) = a$.]

31.22. Assume $a, b \in \mathbb{R}$ and $a > b$.

(a) Prove that

$$a > b + \frac{a - b}{\sqrt{2}} > b$$

(b) Use part (a) to prove that between any two rational numbers there is an irrational number. Compare Problem 31.21.

31.23. True or false: If a is irrational, then a^{-1} is irrational.

31.24. Prove that every real number is a least upper bound of some set of rational numbers. (For $u \in \mathbb{R}$, let $S = \{r \in \mathbb{Q}: r \le u\}$. Use Problem 31.21 to explain why u must be a least upper bound for S.)

31.25. Give examples to show that if a and b are irrational, then ab may be either rational or irrational, depending on a and b.

31.26. Prove that the order on $\mathbb{Q}$ given by Theorem 31.2 is the only one that will make $\mathbb{Q}$ an ordered field.

SECTION 32   THE FIELD OF COMPLEX NUMBERS

There is no real number x such that $x^2 = -1$, because the square of any nonzero element in an ordered integral domain must be positive (Lemma 28.1). The field of complex numbers, which contains the field of real numbers as a subfield, overcomes this deficiency. It does much more than that, in fact, as can be seen from the following theorem.

Theorem 32.1 (Fundamental Theorem of Algebra).   *Every polynomial equation*

$$a_n x^n + a_{n-1} x^{n-1} + \cdots + a_1 x + a_0 = 0, \tag{32.1}$$

which is of degree at least one and whose coefficients $a_n, a_{n-1}, \ldots, a_1, a_0$ are complex numbers, has at least one solution in the field of complex numbers.

Notice the implications of this theorem. To have solutions for all equations $ax = b$ (coefficients integers), we extended the integers to the rational numbers. To have a solution for $x^2 = 2$, we went outside the rational numbers to the real numbers. To have a solution for $x^2 = -1$, we are extending the real numbers to the complex numbers. The Fundamental Theorem of Algebra asserts that in looking for solutions to polynomial equations there will be no need to extend further, because any such equation with complex numbers as coefficients will have a solution in the field of complex numbers. We shall not prove the Fundamental Theorem of Algebra, but we say more about polynomial equations in Chapter XI.

We now give a description of the complex numbers. The rational numbers were constructed using equivalence classes of ordered pairs of integers. The complex numbers will be constructed using ordered pairs of real numbers. (*Suggestion*: After studying the statement of Theorem 32.2, pass over the proof and read through Example 32.2; then return to the proof. This should make the operations in the theorem seem more natural.)

Theorem 32.2.   *Let $\mathbb{C}$ denote the set of all ordered pairs (a, b) with $a, b \in \mathbb{R}$. Define addition and multiplication of these pairs by*

$$(a, b) + (c, d) = (a + c, b + d) \quad \text{and} \quad (a, b)(c, d) = (ac - bd, ad + bc)$$

for all $a, b, c, d \in \mathbb{R}$. With these operations, $\mathbb{C}$ is a field. The subset of $\mathbb{C}$ consisting of all $(a, 0)$ with $a \in \mathbb{R}$ forms a subfield of $\mathbb{C}$, isomorphic to $\mathbb{R}$.

PROOF.   Most of the details will be left as an exercise, including the verification that $(0, 0)$ serves as an identity element for addition and $(-a, -b)$ serves as a negative of (a, b).

To prove that multiplication is associative, assume that $(a, b), (c, d), (e, f) \in \mathbb{C}$. Using the definition of multiplication, and properties of the real numbers, we can write

$$
\begin{aligned}
(a, b)[(c, d)(e, f)] &= (a, b)(ce - df, cf + de) \\
&= (a(ce - df) - b(cf + de), a(cf + de) + b(ce - df)) \\
&= (ace - adf - bcf - bde, acf + ade + bce - bdf) \\
&= ((ac - bd)e - (ad + bc)f, (ac - bd)f + (ad + bc)e) \\
&= (ac - bd, ad + bc)(e, f) \\
&= [(a, b)(c, d)](e, f),
\end{aligned}
$$

as required.

To prove that multiplication is commutative, write

$$(a, b)(c, d) = (ac - bd, ad + bc)$$
$$= (ca - db, cb + da)$$
$$= (c, d)(a, b).$$

The unity is $(1, 0)$:

$$(a, b)(1, 0) = (a \cdot 1 - b \cdot 0, a \cdot 0 + b \cdot 1)$$
$$= (a, b).$$

Because of the commutativity of multiplication we need not check separately that $(1, 0)(a, b) = (a, b)$. A similar remark applies to verification of inverse elements and the distributive laws.

Assume that (a, b) is different from $(0, 0)$, the zero of $\mathbb{C}$. Then $a \neq 0$ or $b \neq 0$, so that $a^2 > 0$ or $b^2 > 0$, and $a^2 + b^2 > 0$. Thus $(a/(a^2 + b^2), -b/(a^2 + b^2))$ is an element of $\mathbb{C}$, and it is the inverse of (a, b) relative to multiplication:

$$(a, b)\left(\frac{a}{(a^2 + b^2)}, \frac{-b}{(a^2 + b^2)}\right) = \left(\frac{(a^2 + b^2)}{(a^2 + b^2)}, \frac{(-ab + ba)}{(a^2 + b^2)}\right) = (1, 0)$$

The remainder of the proof that $\mathbb{C}$ is a field is left to Problem 32.9.

To prove that $\{(a, 0) : a \in \mathbb{R}\}$ is a subfield isomorphic to $\mathbb{R}$, consider the mapping $\theta : \mathbb{R} \to \mathbb{C}$ defined by $\theta(a) = (a, 0)$ for each $a \in \mathbb{R}$. The mapping θ preserves both operations:

$$\theta(a + b) = (a + b, 0) = (a, 0) + (b, 0) = \theta(a) + \theta(b)$$

and

$$\theta(ab) = (ab, 0) = (a, 0)(b, 0) = \theta(a)\theta(b)$$

for all $a, b \in \mathbb{R}$. Also, θ is one-to-one because if $\theta(a) = \theta(b)$, then $(a, 0) = (b, 0)$ so that $a = b$. Thus θ is an isomorphism of $\mathbb{R}$ onto $\{(a, 0): a \in \mathbb{R}\}$; the proof that the latter is a field is left to Problem 32.10. ∎

In light of the last part of this theorem it is natural to identify $a \in \mathbb{R}$ with $(a, 0) \in \mathbb{C}$. In this way $\mathbb{R}$ actually becomes a subset of $\mathbb{C}$, so that every real number is a complex number. The element $(0, 1)$ of $\mathbb{C}$ is customarily denoted by i, and then each element $(0, b)$ by bi. This leads to the notation $a + bi$ for the element (a, b) of $\mathbb{C}$. For $a, b, c, d \in \mathbb{R}$,

$$a + bi = c + di \quad \text{iff} \quad a = c \quad \text{and} \quad b = d.$$

The rules for addition and multiplication become

$$(a + bi) + (c + di) = (a + c) + (b + d)i$$

and

$$(a + bi)(c + di) = (ac - bd) + (ad + bc)i.$$

In particular, $i^2 = (0 + 1i)(0 + 1i) = -1 + 0i = -1$, and hence i is a solution in $\mathbb{C}$ to the equation $x^2 = -1$.

To compute with elements of $\mathbb{C}$, simply apply the various associative, commutative, and distributive laws, and replace i^2 by -1 wherever it occurs. In this way any expression involving complex numbers can be reduced to the form $a + bi$ with $a, b \in \mathbb{R}$. When such a number has $b \neq 0$ it is said to be *imaginary*.

Example 32.1.
   (a) $(1 + i)^2 = 1 + 2i + i^2 = 1 + 2i - 1 = 2i$
   (b) $i^4 = (i^2)^2 = (-1)^2 = 1$
   (c) $(-i)^2 = (-1)^2(i)^2 = i^2 = -1$
   (d) $i(1 - i) + 2(3 + i) = i - i^2 + 6 + 2i = i - (-1) + 6 + 2i = i + 1 +$
       $6 + 2i = 7 + 3i$ ■

The number $a - bi$ is called the *conjugate* of $a + bi$. To simplify a fraction with an imaginary number $a + bi$ in the denominator, multiply both numerator and denominator by this conjugate, making use of $(a + bi)(a - bi) = a^2 - (bi)^2 = a^2 + b^2$.

Example 32.2.
   (a) $\dfrac{1}{1 + i} = \dfrac{1}{i + i} \cdot \dfrac{1 - i}{1 - i} = \dfrac{1 - i}{2} = \dfrac{1}{2} - \dfrac{1}{2}i$

   (b) $\dfrac{2 + i}{2 - i} = \dfrac{2 + i}{2 - i} \cdot \dfrac{2 + i}{2 + i} = \dfrac{4 + 4i - 1}{4 + 1} = \dfrac{3}{5} + \dfrac{4}{5}i$ ■

More will be said about calculations with complex numbers in the next section. Following now is a concise discussion of the ideas needed to characterize the field $\mathbb{C}$.

If E and F are fields, then E is said to be an *extension* of F if E contains a subfield isomorphic to F; for convenience, F is often thought of as actually being a subfield of E. Thus $\mathbb{R}$ is an extension of $\mathbb{Q}$, and $\mathbb{C}$ is an extension of both $\mathbb{R}$ and $\mathbb{Q}$. Any field is an extension of itself.

Assume that E is an extension of F. An element $a \in E$ is said to be *algebraic* over F if a is a solution of some polynomial equation (32.1) with coefficients in F. For example, $\sqrt{2}$ is algebraic over $\mathbb{Q}$ because it is a solution of $x^2 - 2 = 0$. Neither π nor e is algebraic over $\mathbb{Q}$, but these facts are not easy to prove. A field E is an *algebraic extension* of F if every $a \in E$ is algebraic over F. The remarks about π and e show that $\mathbb{R}$ is not an algebraic extension of $\mathbb{Q}$. However, $\mathbb{C}$ is an algebraic extension of $\mathbb{R}$ (Problem 32.12).

A field F is *algebraically closed* if every polynomial equation (32.1) with coefficients in F has a solution in F. By the Fundamental Theorem of Algebra, $\mathbb{C}$ is algebraically closed. But neither $\mathbb{Q}$ nor $\mathbb{R}$ is algebraically closed. (Why?) A field E is an *algebraic closure* of a field F if

1. E is an algebraic extension of F, and
2. E is algebraically closed.

It can be proved that every field has an algebraic closure. Moreover, this algebraic closure is essentially unique: If E_1 and E_2 are algebraic closures of F, then E_1 must be isomorphic to E_2.

Because $\mathbb{C}$ is an algebraic extension of $\mathbb{R}$, and $\mathbb{C}$ is also algebraically closed, we see that $\mathbb{C}$ is an algebraic closure of $\mathbb{R}$. If we put all of this together with what we know about $\mathbb{R}$, and agree not to distinguish between isomorphic fields, we arrive at the following characterization: *The field of complex numbers is the (unique) algebraic closure of the (unique) complete ordered field.*

In the introductory remarks of Section 31 it was stated that the necessity for extending $\mathbb{Q}$ to $\mathbb{R}$ had more to do with order than with algebra. We can now put this in better focus. The question of order was covered in Section 31. Regarding algebra, the algebraic closure of $\mathbb{Q}$ cannot be $\mathbb{R}$ because $\mathbb{R}$ is not algebraically closed. Also, the algebraic closure of $\mathbb{Q}$ cannot be $\mathbb{C}$ because $\mathbb{C}$ is not an algebraic extension of $\mathbb{Q}$. The algebraic closure of $\mathbb{Q}$ is, in fact, a subfield of $\mathbb{C}$ known as the *field of algebraic numbers*. This field consists precisely of those elements of $\mathbb{C}$ that are algebraic over $\mathbb{Q}$. If we were to begin with $\mathbb{Q}$, and concern ourselves only with finding solutions for polynomial equations, we could work wholly within the field of algebraic numbers. We would not need all of $\mathbb{C}$, and, although we would need some elements outside $\mathbb{R}$, we would not need all of $\mathbb{R}$. Questions about algebraic numbers have been important throughout the history of modern algebra; more is said about this in Chapter IX.

Further questions about polynomial equations and field extensions are dealt with in Chapters X, XI, and XII.

PROBLEMS

Express each of the following in the form $a + bi$, with $a, b \in \mathbb{R}$.

32.1. (a) $(2 - i)(1 + i)$ (b) i^3 (c) $\dfrac{1}{1 + 2i}$

32.2. (a) $(-i)^3$ (b) $(1 + i)^3$ (c) $\dfrac{1 + i}{2 - 3i}$

32.3. Explain why $\mathbb{Q}$ is not algebraically closed.

32.4. Explain why $\mathbb{R}$ is not algebraically closed.

32.5. Prove that $\mathbb{Z}_2$ is not an algebraically closed field.

32.6. Prove that if p is a prime, then the field $\mathbb{Z}_p$ is not algebraically closed.

32.7. Determine a pair of complex numbers $z = a + bi$ and $w = c + di$ giving a solution of the system

$$3z - 2w = -i$$
$$iz + 2iw = -5.$$

(The usual methods of solving systems of equations over $\mathbb{R}$ also work over $\mathbb{C}$. Why?)

32.8. Repeat Problem 32.7 for the system

$$z + iw = 1$$
$$-2iz + w = -1.$$

32.9. Complete the proof that $\mathbb{C}$ is a field, in the proof of Theorem 32.1.

32.10. Prove that $\{\theta(a, 0): a \in \mathbb{R}\}$ is a subfield of $\mathbb{C}$ (Theorem 32.1).

32.11. (a) Verify that in $\mathbb{C}$, thought of as $\{(a, b) : a, b \in \mathbb{R}\}$

$$(0, 1)(0, 1) = (-1, 0).$$

(b) What is $(0, 1)^4$ in $\mathbb{C}$?

32.12. Prove that $\mathbb{C}$ is an algebraic extension of $\mathbb{R}$. [*Suggestion:* Consider the equation $[x - (a + bi)][x - (a - bi)] = 0$.]

32.13. Explain why $\mathbb{C}$ cannot be an ordered field.

32.14. Prove or disprove that the mapping $\theta \colon \mathbb{C} \to \mathbb{C}$ defined by $\theta(a + bi) = a - bi$ is a ring isomorphism. (Compare Problem 29.8.)

32.15. Find two complex numbers that are solutions of $x^2 = -4$.

32.16. Let z^* denote the conjugate of the complex number z, that is, $(a + bi)^* = a - bi$. Prove that each of the following is true for each $z \in \mathbb{C}$.
   (a) $(z^*)^* = z$ (b) $z + z^* \in \mathbb{R}$
   (c) $z = z^*$ iff $z \in \mathbb{R}$ (d) $(z^{-1})^* = (z^*)^{-1}$

32.17. Prove that if θ is an isomorphism of $\mathbb{C}$ onto $\mathbb{C}$ and $\theta(a) = a$ for each $a \in \mathbb{R}$, then either θ is the identity mapping or θ maps each element of $\mathbb{C}$ to its conjugate, that is, $\theta(a + bi) = a - bi$ for each $a + bi \in \mathbb{C}$. [*Suggestion*: First prove that either $\theta(i) = i$ or $\theta(i) = -i$.]

32.18. Verify that

$$\theta(a + bi) = \begin{bmatrix} a & b \\ -b & a \end{bmatrix}$$

defines an isomorphism of $\mathbb{C}$ onto a subring of $M(2, \mathbb{R})$, the ring of 2×2 matrices over $\mathbb{R}$.

32.19. Let z^* denote the conjugate of z, as in Problem 32.16. Let Q denote the set of all matrices in $M(2, \mathbb{C})$ that have the form

$$\begin{bmatrix} z & w \\ -w^* & z^* \end{bmatrix}. \tag{32.2}$$

For example,

$$\begin{bmatrix} 1 + 2i & 2 + i \\ -2 + i & 1 - 2i \end{bmatrix}$$

is in Q. Prove that Q is a division ring, that is, a ring with a unity in which each nonzero element has an inverse relative to multiplication. [*Suggestion*: Assume that $z = a + bi$ and $w = c + di$ in (32.2), and let $k = a^2 + b^2 + c^2 + d^2$. Then

$$\frac{1}{k} \begin{bmatrix} z^* & -w \\ w^* & z \end{bmatrix}$$

is an inverse in Q for the matrix in (32.2).] Also prove that Q is noncommutative, so that it is not a field. The division ring Q is called the ring of *Hamilton's quaternions*, after the Irish mathematician W. R. Hamilton (1805–1865).

SECTION 33   COMPLEX ROOTS OF UNITY

The other sections of this chapter have been concerned primarily with general properties and abstractions. This section has to do with computation. We look at some useful ways to represent complex numbers and show how they can be used to determine the complex roots of unity—the solutions of equations of the form $x^n = 1$. These roots of unity are useful for examples; they also arise often enough in other areas of mathematics to make their inclusion here worthwhile.

Just as the points on a line can be used to represent real numbers geometrically, the points in a plane can be used to represent complex numbers geometrically. A rectangular coordinate system is chosen for the plane, and then each complex number $a + bi$ is represented by the point with coordinates (a, b). Because

$$a + bi = c + di \quad \text{iff} \quad a = c \quad \text{and} \quad b = d$$

(for $a, b, c, d \in \mathbb{R}$), the correspondence

$$a + b \leftrightarrow (a, b)$$

is one-to-one between complex numbers and points of the plane. Figure 33.1 shows some examples.

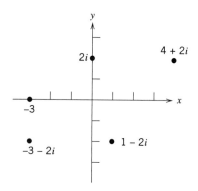

Figure 33.1

Addition of complex numbers corresponds to vector addition of points in the plane:

$$(a + bi) + (c + di) = (a + c) + (b + d)i \leftrightarrow (a + c, b + d) = (a, b) + (c, d).$$

To describe multiplication of complex numbers geometrically, we turn to polar coordinates. Recall that the polar representation of a point with rectangular coordinates (a, b) is (r, θ), where r denotes the distance between the origin and the given point, and θ denotes the angle from the positive x-axis to the ray from the origin through the given point, with the positive direction taken counterclockwise (Figure 33.2).

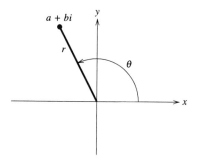

Figure 33.2

Thus $r = \sqrt{a^2 + b^2}$, $a = r\cos\theta$, $b = r\sin\theta$, and

$$a + bi = r(\cos\theta + i\sin\theta).$$

The latter is called the *polar* (or *trigonometric*) form of $a + bi$. The nonnegative number r appearing here is called the *absolute value* (or *modulus*) of $a + bi$ and is denoted by $|a + bi|$. The angle θ is called the *argument* (or *amplitude*) of $a + bi$. The absolute value of $a + bi$ is unique, but the argument is not: if θ is an argument of $a + bi$, then so is $\theta + 2n\pi$ for any integer n. If θ is restricted so that $0 \leq \theta < 2\pi$, then each nonzero complex number does have a unique argument.

Example 33.1.   The absolute value of $-2 + 2i$ is $|-2 + 2i| = \sqrt{(-2)^2 + 2^2} = 2\sqrt{2}$. The smallest positive argument is $\theta = 135° = 3\pi/4$. Thus the polar form of $-2 + 2i$ is $2\sqrt{2}[\cos(3\pi/4) + i\sin(3\pi/4)]$. ∎

The following theorem shows that polar form is especially well-suited for computing products.

Theorem 33.1.   *If* $z = r(\cos\theta + i\sin\theta)$ *and* $w = s(\cos\phi + i\sin\phi)$, *then*

$$zw = rs[\cos(\theta + \phi) + i\sin(\theta + \phi)].$$

Thus the absolute value of a product is the product of the absolute values, and the argument of a product is the sum of the arguments.

PROOF.   Recall the following two addition formulas from trigonometry:

$$\cos(\theta + \phi) = \cos\theta\cos\phi - \sin\theta\sin\phi$$

and

$$\sin(\theta + \phi) = \sin\theta\cos\phi + \cos\theta\sin\phi.$$

Using these, we can write

$$\begin{aligned}
zw &= r(\cos\theta + i\sin\theta) \cdot s(\cos\phi + i\sin\phi) \\
&= rs[(\cos\theta\cos\phi - \sin\theta\sin\phi) \\
&\quad + i(\sin\theta\cos\phi + \cos\theta\sin\phi)] \\
&= rs[\cos(\theta + \phi) + i\sin(\theta + \phi)].
\end{aligned}$$
∎

DeMoivre's Theorem.   *If* n *is a positive integer and* $z = r(\cos\theta + i\sin\theta)$, *then* $z^n = r^n(\cos n\theta + i\sin n\theta)$.

PROOF.   Use induction on n. For $n = 1$ the result is obvious. Assuming the theorem true for $n = k$, and using Theorem 33.1 with $w = z^k$, we have

$$\begin{aligned}
z^{k+1} = zz^k &= r(\cos\theta + i\sin\theta) \cdot r^k(\cos k\theta + i\sin k\theta) \\
&= r^{k+1}[\cos(k + 1)\theta + i\sin(k + 1)\theta],
\end{aligned}$$

as required. ∎

Example 33.2.   To compute $(-2 + 2i)^5$, begin with

$$-2 + 2i = 2\sqrt{2}[\cos(3\pi/4) + i\sin(3\pi/4)]$$

from Example 33.1. Now apply DeMoivre's Theorem. We have

$$(2\sqrt{2})^5 = 128\sqrt{2} \quad \text{and} \quad 5(3\pi/4) = 15\pi/4 = 2\pi + 7\pi/4.$$

Therefore

$$\begin{aligned}
(-2 + 2i)^5 &= 128\sqrt{2}[\cos(7\pi/4) + i\sin(7\pi/4)] \\
&= 128\sqrt{2}(\sqrt{2}/2 - i\sqrt{2}/2) \\
&= 128 - 128i.
\end{aligned}$$
∎

For each integer $n \geq 1$, there can be at most n distinct complex numbers that are solutions of $x^n = 1$. (This will be a corollary of Theorem 44.1.) There are, in fact, exactly n solutions, called the complex *n*th *roots of unity.* They can be determined by using DeMoivre's Theorem, as follows.

Theorem 33.2.   *For each integer $n \geq 1$, the n complex nth roots of unity are*

$$\cos\frac{2k\pi}{n} + i\sin\frac{2k\pi}{n}, \quad k = 0, 1, \dots n - 1. \tag{33.1}$$

PROOF.   By DeMoivre's Theorem, the argument of the nth power of each number in (33.1) is $n(2k\pi/n) = 2k\pi$, and the absolute value is $1^n = 1$. Thus each of the numbers is an nth root of unity. The n numbers in (33.1) are distinct because the numbers $2k\pi/n$ are distinct and $0 \leq 2k\pi/n < 2\pi$ for $k = 0, 1, \dots, n - 1$. Thus the numbers in (33.1) represent all the nth roots of unity.   ∎

The nth roots of unity are represented geometrically by n equally spaced points on the circle with center at the origin and radius 1, with one of the points being 1. Figure 33.3 shows the case $n = 6$.

Theorem 33.2 can be extended to give a formula for the nth roots of any complex number (see Problems 33.27 and 33.28).

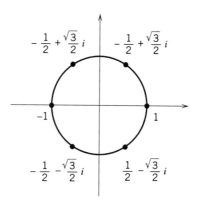

Figure 33.3

PROBLEMS

Write each of the following complex numbers in the form $a + bi$.

33.1. The number with absolute value 2 and argument $\pi/6$.

33.2. The number with absolute value $\frac{1}{2}$ and argument $5\pi/3$.

33.3. The number with absolute value 5 and argument $9\pi/4$.

33.4. The number with absolute value 3 and argument π.

Use DeMoivre's Theorem to write each of the following numbers in the form $a + bi$.

33.5. $(1 + i)^{10}$ **33.6.** $(\sqrt{3} + i)^5$

33.7. $(1 - i)^6$ **33.8.** $(-i)^{10}$

33.9. $\left(\frac{1}{2} - \frac{1}{2}i\right)^4$ **33.10.** $\left(-\sqrt{2} - \sqrt{2}i\right)^{12}$

33.11. Determine all complex 8th roots of unity, and represent them geometrically.

33.12. Determine all complex 5th roots of unity, and represent them geometrically.

Express each of the following complex numbers in polar form.

33.13. $1 + i$ **33.14.** $\sqrt{3} - i$

33.15. -5 **33.16.** $-2i$

33.17. $2 - 2i$ **33.18.** $2i + 2\sqrt{3}$

33.19. Prove that if $z = r(\cos\theta + i\sin\theta)$ and $z \neq 0$, then
$$z^{-1} = r^{-1}[\cos(-\theta) + i\sin(-\theta)]$$

33.20. State and prove DeMoivre's Theorem for negative integers. (For $n = -1$ see Problem 33.19.)

33.21. Prove that if n is a positive integer, then the set of all nth roots of unity forms a cyclic group of order n with respect to multiplication. [Any generator of this group is called a *primitive root of unity*. One such root is $\cos(2\pi/n) + i\sin(2\pi/n)$.]

33.22. Prove that for each integer $n > 1$ the sum of the nth roots of unity is 0. (*Suggestion*: They form a geometric progression, which can be summed by a formula in Appendix C.)

33.23. Let n be a positive integer. What is the product of all the nth roots of unity?

33.24. Prove that the set of all roots of unity forms a group with respect to multiplication.

33.25. Prove that with respect to multiplication, the set of all complex numbers of absolute value 1 forms a group that is isomorphic to the group of all rotations of the plane about a fixed point p (Example 5.7).

33.26. Let z^* denote the conjugate of the complex number z, that is, $(a + bi)^* = a - bi$. Prove that the following are true for each $z \in \mathbb{C}$.
 (a) $|z^*| = |z|$ (b) $zz^* = |z|^2$
 (c) $z^{-1} = z^*/|z|^2$ if $z \neq 0$.

33.27. Prove that for each integer $n \geq 1$, the n complex nth roots of
$$z = r(\cos\theta + i\sin\theta)$$

are
$$r^{1/n} = \left(\cos\frac{\theta + 2k\pi}{n} + i\sin\frac{\theta + 2k\pi}{n}\right), \quad k = 0, 1, \ldots, n - 1,$$

where $r^{1/n}$ is the positive real nth root of r.

33.28. Prove that for each integer $n \geq 1$, the n complex nth roots of

$$z = r(\cos \theta + i \sin \theta)$$

are

$$v, vw, vw^2, \ldots, vw^{n-1},$$

where

$$v = r^{1/n} \left[\cos \frac{\theta}{n} + i \sin \frac{\theta}{n} \right]$$

and

$$w = \cos \frac{2\pi}{n} + i \sin \frac{2\pi}{n}.$$

33.29. Use Problem 33.27 to find the complex cube roots of $-2i$.

33.30. Use Problem 33.28 to find the complex cube roots of $-2i$.

33.31. Use Problem 33.27 to find the complex fourth roots of -5.

33.32. Use Problem 33.28 to find the complex fourth roots of -5.

33.33. Let $\mathbb{C}^{\#}$ denote the multiplicative group of nonzero complex numbers and $\mathbb{R}^p$ the multiplicative group of positive real numbers. Why is $\theta : \mathbb{C}^{\#} \to \mathbb{R}^p$, defined by $\theta(z) = |z|$, a homomorphism? What is Ker θ?

33.34. Verify that $\alpha : \mathbb{R} \to \mathbb{C}$ defined by $\alpha(x) = \cos(2\pi x) + i \sin(2\pi x)$ is a homomorphism of the additive group of $\mathbb{R}$ onto the group of all complex numbers of absolute value 1. What is Ker α? Interpret α geometrically. (See Problem 33.25.)

CHAPTER VIII

POLYNOMIALS

This chapter presents the facts about polynomials that are necessary for studying fields (Chapter X) and polynomial equations (Chapter XI). Polynomials with real numbers as coefficients will be familiar from elementary algebra and calculus; now we must allow for the possibility that the coefficients are from some ring other than $\mathbb{R}$. A similarity between the theorems in this chapter and those in Sections 12 and 13 will be obvious; more is said about this similarity in the last section of this chapter.

SECTION 34 DEFINITION AND ELEMENTARY PROPERTIES

If R is a commutative ring and $a_0, a_1, \ldots, a_n \in R$, then an expression of the form

$$a_0 + a_1 x + a_2 x^2 + \cdots + a_n x^n \tag{34.1}$$

is called a polynomial in x: it is a finite sum of terms, each of which is some element of R times a nonnegative integral power of x. We become acquainted with such expressions, and how to add and multiply them, in elementary algebra. Here we want to consider polynomials in the context of commutative rings.

Our first problem is that if x is not an element of R, then terms such as $a_1 x$ and $a_n x^n$, as well as "sums" of such terms, may not have a predetermined meaning. One way around this is to consider not (34.1), but rather the sequence $(a_0, a_1, \ldots, a_n, 0, \ldots)$ of elements of R arising from (34.1), and to define appropriate ring operations on the set of all these sequences. This procedure is outlined in the appendix to this section. It has the advantage that it avoids questions about the precise meaning of expressions such as that in (34.1), but it is not the way polynomials are handled in practice. For most purposes the discussion that follows will be more satisfactory.

Definition. Let R be a commutative ring. A *polynomial in indeterminate x over R* is an expression of the form (34.1) where the *coefficients* $a_0, a_1, \ldots, a_n$ are elements of R. If $a_n \neq 0$, then the integer n is the *degree* of the polynomial, and a_n is its *leading coefficient*. A polynomial over a field is said to be *monic* if its leading coefficient is the unity of the field. Two polynomials in x are *equal* iff the coefficients of like powers of x are equal. The set of all polynomials in x over R will be denoted $R[x]$.

The indeterminate x used in constructing $R[x]$ can be any element such that an expression of the form (34.1) equals the zero element of R iff $a_0 = a_1 = \cdots = a_n = 0$. This requirement on x is equivalent to the requirements that two polynomials in x are equal iff the coefficients of like powers of x are equal.

Polynomials are added by adding coefficients of like powers of x. They are multiplied by assuming that the laws of a commutative ring apply to all symbols present (the elements of R, the powers of x, the $+$ sign, and the juxtaposition of the coefficients with powers of x). Before stating the formal definition that follows from this assumption, let us look at an example.

Example 34.1.   In $\mathbb{Z}[x]$,

$$(2x + 5x^2) + (1 - 3x^2 - x^3)$$
$$= (0 + 2x + 5x^2 + 0x^3) + (1 + 0x + (-3)x^2 + (-1)x^3)$$
$$= (0 + 1) + (2 + 0)x + (5 - 3)x^2 + (0 - 1)x^3$$
$$= 1 + 2x + 2x^2 - x^3$$

and

$$(2x + 5x^2)(1 - 3x^2 - x^3) = 2x(1 - 3x^2 - x^3) + 5x^2(1 - 3x^2 - x^3)$$
$$= (2x - 6x^3 - 2x^4) + (5x^2 - 15x^4 - 5x^5)$$
$$= 2x + 5x^2 - 6x^3 - 17x^4 - 5x^5. \qquad \blacksquare$$

Definition.   Let

$$p(x) = a_0 + a_1 x + \cdots + a_m x^m$$

and

$$q(x) = b_0 + b_1 x + \cdots + b_n x^n$$

be polynomials over a commutative ring R. Then

$$p(x) + q(x) = (a_0 + b_0) + (a_1 + b_1)x + \cdots + (a_n + b_n)x^n$$
$$+ a_{n+1}x^{n+1} + \cdots + a_m x^m, \quad \text{for } m \geq n, \qquad (34.2)$$

with a similar formula if $m < n$. And

$$p(x)q(x) = a_0 b_0 + (a_0 b_1 + a_1 b_0)x + \cdots + a_m b_n x^{m+n}, \qquad (34.3)$$

the coefficient of x^k being

$$a_0 b_k + a_1 b_{k-1} + a_2 b_{k-2} + \cdots + a_k b_0.$$

Example 34.2.   In $\mathbb{Z}_4[x]$,

$$([2] + [2]x) + ([2] + [3]x - [1]x^2)$$
$$= ([2] \oplus [2]) + ([2] \oplus [3])x + (-[1])x^2$$
$$= [0] + [1]x + [-1]x^2$$
$$= [1]x + [3]x^2$$

and

$$([2] + [2]x)([2] + [3]x - [1]x^2)$$
$$= ([2] \odot [2]) + ([2] \odot [3] \oplus [2] \odot [2])x$$
$$+([2] \odot [-1] \oplus [2] \odot [3])x^2$$
$$+([2] \odot [-1])x^3$$
$$= [0] + [2]x + [0]x^2 + [-2]x^3$$
$$= [2]x + [2]x^3.$$

We shall use the convention that $-ax^n$ means $(-a)x^n$. Thus, above, $-[1]x^2 = [-1]x^2 = [3]x^2$. ∎

In working with polynomials over any ring $\mathbb{Z}_n$, we shall frequently omit brackets from the coefficients, write $+$ instead of $\oplus$, use juxtaposition instead of $\odot$, and rely on the context to remind us that all calculations are to be done modulo n. With this abbreviated notation the results of the preceding calculations would be written

$$(2 + 2x) + (2 + 3x - x^2) = x + 3x^2$$

and

$$(2 + 2x)(2 + 3x - x^2) = 2x + 2x^3.$$

Theorem 34.1.   *If R is a commutative ring, then $R[x]$ is a commutative ring with respect to the operations defined by (34.2) and (34.3). If R is an integral domain, then $R[x]$ is an integral domain.*

PROOF.   The zero of $R[x]$ is the polynomial having all coefficients equal to the zero of R. (This polynomial is not assigned a degree.) The details of proving that $R[x]$ is a commutative ring are left as an exercise (Problem 34.7).

Assume that R is an integral domain with unity e. Then it is easy to verify that the polynomial of degree zero with coefficient e is a unity for $R[x]$. Also, if $p(x)$ and $q(x)$ are nonzero elements of $R[x]$, with leading coefficients $a_m \neq 0$ and $b_n \neq 0$, then (34.3) shows that $p(x)q(x)$ has leading coefficient $a_m b_n \neq 0$, and thus $p(x)q(x)$ is also nonzero. Therefore $R[x]$ has no zero divisors. This proves that $R[x]$ is an integral domain. ∎

Notice, in particular, that $F[x]$ is an integral domain if F is a field. However, $F[x]$ is not a field, no matter what F is (Problem 34.8). The ring $R[x]$ is called the *ring of polynomials in x over R*.

APPENDIX TO SECTION 34

Here is an alternative way to define the ring of polynomials in an indeterminate over a commutative ring R.

If R is a commutative ring, then a *sequence* of elements of R is a mapping from the set of nonnegative integers to R. Such sequences can be denoted by $(a_0, a_1, \ldots, a_n, \ldots)$, where a_n, the nth term of the sequence, is the ring element corresponding to the nonnegative integer n.

Definition.   Let R be a commutative ring. A *polynomial over R* is a sequence

$$(a_0, a_1, \ldots, a_n, \ldots) \tag{34.1A}$$

of elements of R such that only finitely many of its terms are different from the zero element of R. If $a_n \neq 0$, but all terms with subscripts greater than n are zero, then the polynomial is said to be of *degree n*; the terms $a_0, a_1, \ldots, a_n$ are called the *coefficients* of the polynomial, with a_n the *leading coefficient*. Denote the set of all polynomials over R by $R[X]$.

By the definition of equality of mappings (Section 1), two polynomials over R are equal iff they have the same degree and their corresponding terms are equal. (This uses the fact that polynomials are sequences, that is, mappings from the set of nonnegative integers to R.)

Definition.   Let

$$p = (a_0, a_1, \ldots, a_n, \ldots)$$

and

$$q = (b_0, b_1, \ldots, b_n, \ldots)$$

be polynomials over the commutative ring R. Then

$$p + q = (a_0 + b_0, a_1 + b_1, \ldots, a_n + b_n, \ldots) \tag{34.2A}$$

and

$$pq = (a_0 b_0, a_0 b_1 + a_1 b_0, \ldots), \tag{34.3A}$$

where the kth term of pq is

$$a_0 b_k + a_1 b_{k-1} + \cdots + a_k b_0.$$

Because the sequences p and q each have only finitely many nonzero terms, the same is true for both $p + q$ and pq. Therefore, both $p + q$ and pq are members of $R[X]$.

Theorem 34.1A.   *If R is a commutative ring then $R[X]$ is a commutative ring with respect to the operations defined by (34.2A) and (34.3A). If R is an integral domain, then $R[X]$ is an integral domain.*

PROOF.   The zero of $R[X]$ is the sequence $(0, 0, \ldots, 0, \ldots)$. If R has a unity e, then $R[X]$ has a unity $(e, 0, 0, \ldots, 0, \ldots)$. If R is an integral domain, and p and q are nonzero elements of $R[X]$ having degrees m and n, then pq has degree mn, and thus is also nonzero. The other details of the proof are left as an exercise (Problem 34.15).   ∎

The mapping that assigns the "polynomial" in (34.1) to the "polynomial" in (34.1A), with all terms after a_n assumed to be zero in (34.1A), is an isomorphism of the ring $R[x]$ onto the ring $R[X]$ (Problem 34.16).

Assume that R has unity e, and let X denote the sequence $(0, e, ..0, \ldots)$. Then it follows from (34.3A) that $X^2 = XX = (0, 0, e, \ldots, 0, \ldots)$, $X^3 = (0, 0, 0, e, \ldots, 0, \ldots)$, and, in general, $X^k = (0, 0, \ldots, 0, e, 0, \ldots)$, the sequence (element of $R[X]$) having the

$(k + 1)$st term e and all other terms zero. If the product of an element $a \in R$ and a sequence $(b_0, b_1, \ldots, b_n, \ldots)$ is defined to be $(ab_0, ab_1, \ldots, ab_n, \ldots)$, then it follows that

$$a_0 + a_1 X + \cdots + a_n X^n = (a_0, a_1, \ldots, a_n, 0, \ldots).$$

This can be interpreted as giving a more precise meaning to the left side than that given by calling it an "expression," which was the way we referred to $a_0 + a_1 x + \cdots + a_n x^n$.

PROBLEMS

34.1. There are four different polynomials of degree 2 in $\mathbb{Z}_2[x]$. List them all.

34.2. There are nine different monic polynomials of degree 2 in $\mathbb{Z}_3[x]$. List them all.

34.3. The following are polynomials over $\mathbb{Z}$. Express each in the form (34.1).
(a) $(1 + 2x) + (2 - x + 2x^2)$
(b) $(2x + x^3) + (x + 2x^4)$
(c) $(1 + 2x)(2 - x + 2x^2)$
(d) $(2x + x^3)(x + 2x^4)$
(e) $(2x + x^2)^3$

34.4. (a) to (e). Interpret the polynomials in Problem 34.3 as being over $\mathbb{Z}_3$ rather than $\mathbb{Z}$, and write each in the form (34.1). (See the remark on notation following Example 34.2.)

34.5. (a) to (e). Repeat Problem 34.4 with $\mathbb{Z}_4$ in place of $\mathbb{Z}_3$.

34.6. (a) In $\mathbb{Z}_n[x]$, how many different polynomials are there of degree $\leq d$?
(b) In $\mathbb{Z}_n[x]$, how many different monic polynomials are there of degree d?
(c) In $\mathbb{Z}_n[x]$, how many different polynomials are there of degree d?

34.7. Complete the proof of Theorem 34.1 (that is, prove that $R[x]$ is a commutative ring).

34.8. Explain why a polynomial ring $R[x]$ cannot be a field.

34.9. (a) True or false: The degree of the sum of two polynomials is at least as large as the degree of each of the two polynomials.
(b) True or false: The degree of the product of two polynomials is the sum of the degrees of the two polynomials.

34.10. Prove that if R is a commutative ring and $R[x]$ is an integral domain, then R must be an integral domain. (Compare the last sentence of Theorem 34.1.)

34.11. Prove that if R is any commutative ring, then the characteristic of $R[x]$ is equal to the characteristic of R.

34.12. Prove that if R and S are commutative rings, and $R \approx S$, then $R[x] \approx S[x]$.

34.13. The *formal derivative* of a polynomial

$$p(x) = a_0 + a_1 x + \cdots + a_n x^n$$

over R is defined to be

$$p'(x) = a_1 + 2a_2 x + \cdots + na_n x^{n-1}.$$

Prove that

$$[p(x) + q(x)]' = p'(x) + q'(x)$$

and

$$[p(x)q(x)]' = p(x)q'(x) + p'(x)q(x).$$

(An application of formal derivatives will be given in Problem 44.9.)

34.14. Let D be an ordered integral domain (Section 28), and let $D[x]^P$ consist of all nonzero polynomials in $D[x]$ that have leading coefficient in D^P, the set of positive elements of D.
   (a) Prove that this makes $D[x]$ an ordered integral domain with $D[x]^P$ as the set of positive elements.
   (b) Prove that the polynomial 1 is a least positive element of $\mathbb{Z}[x]$.
   (c) Prove that $\mathbb{Z}[x]$ is not well-ordered.

34.15. Complete the proof of Theorem 34.1A.

34.16. Prove that for any commutative ring R, $R[x] \approx R[X]$. (See the paragraph following the proof of Theorem 34.1A.)

SECTION 35   THE DIVISION ALGORITHM

In the next two sections we concentrate on rings of polynomials over fields, proving divisibility and factorization theorems for these rings that are analogous to the divisibility and factorization theorems that were proved in Section 12 and 13 for the ring of integers. We use deg $f(x)$ to denote the degree of a polynomial $f(x)$.

Division Algorithm. *If $f(x)$ and $g(x)$ are polynomials over a field F, with $g(x) \neq 0$, then there exist unique polynomials $q(x)$ and $r(x)$ over F such that*

$$f(x) = g(x)q(x) + r(x), \quad \text{with} \quad r(x) = 0 \quad \text{or} \quad \text{deg } r(x) < \text{deg } g(x).$$

The polynomials $q(x)$ and $r(x)$ are called, respectively, the *quotient* and *remainder* in the division of $f(x)$ by $g(x)$. The following example illustrates how they can be computed. The same idea is used in the proof that follows the example.

Example 35.1.   Let $f(x) = 2x^4 + x^2 - x + 1$ and $g(x) = 2x - 1$.

$$
\begin{array}{r}
x^3 + \frac{1}{2}x^2 + \frac{3}{4}x - \frac{1}{8} \\
\hline
2x - 1 \overline{)\,2x^4 \qquad\quad + x^2 - x + 1} \\
2x^4 - x^3 \\
\hline
x^3 + x^2 - x + 1 \\
x^3 - \frac{1}{2}x^2 \\
\hline
\frac{3}{2}x^2 - x + 1 \\
\frac{3}{2}x^2 - \frac{3}{4}x \\
\hline
-\frac{1}{4}x + 1 \\
-\frac{1}{4}x + \frac{1}{8} \\
\hline
\frac{7}{8}
\end{array}
$$

Therefore, $q(x) = x^3 + \frac{1}{2}x^2 + \frac{3}{4}x - \frac{1}{8}$ and $r(x) = \frac{7}{8}$. Notice that at each step we eliminated the highest remaining power of $f(x)$ by subtracting an appropriate multiple of $g(x)$. For the proof we simply show that this can be done no matter what $f(x)$ and $g(x)$ are.   ∎

PROOF OF THE DIVISION ALGORITHM.   Let $f(x) = a_0 + a_1 x + \cdots + a_m x^m$ and $g(x) = b_0 + b_1 x + \cdots + b_n x^n$. Since $g(x) \neq 0$, we can assume that $b_n \neq 0$ so that $\deg g(x) = n$. The theorem is trivial for $f(x) = 0$ [use $q(x) = 0$ and $r(x) = 0$]. Therefore we also assume that $a_m \neq 0$ so that $\deg f(x) = m$.

We first prove the existence of $q(x)$ and $r(x)$ using induction on m. If $m < n$, then $f(x) = g(x) \cdot 0 + f(x)$ gives the required representation; that is, we can take $q(x) = 0$ and $r(x) = f(x)$. Thus assume that $m \geq n$. If $m = 0$, then $f(x) = a_0$ and $g(x) = b_0$. In this case $a_0 = b_0 \cdot b_0^{-1} a_0 + 0$; hence, we can take $q(x) = b_0^{-1} a_0$ and $r(x) = 0$.

It remains to prove the statement for $\deg f(x) = m$, on the basis of the induction hypothesis that it is true whenever $f(x)$ is replaced by a polynomial of degree less than m. Let $f_1(x) = f(x) - a_m b_n^{-1} x^{m-n} g(x)$. Then $\deg f_1(x) < \deg f(x)$ (Problem 35.11). Therefore, by the induction hypothesis, there exist polynomials $q_1(x)$ and $r_1(x)$ such that

$$f_1(x) = g(x) q_1(x) + r_1(x), \qquad \text{with} \qquad r_1(x) = 0 \quad \text{or} \quad \deg r_1(x) < \deg g(x).$$

This implies that

$$f(x) - a_m b_n^{-1} x^{m-n} g(x) = g(x) q_1(x) + r_1(x)$$
$$f(x) = g(x)[a_m b_n^{-1} x^{m-n} + q_1(x)] + r_1(x).$$

Thus we can take $q(x) = a_m b_n^{-1} x^{m-n} + q_1(x)$ and $r(x) = r_1(x)$. This proves the existence of $q(x)$ and $r(x)$.

To prove that the polynomials $q(x)$ and $r(x)$ are unique, assume that $q^*(x)$ and $r^*(x)$ are also polynomials over F, and that

$$f(x) = g(x) q^*(x) + r^*(x), \quad \text{with} \quad r^*(x) = 0 \quad \text{or} \quad \deg r^*(x) < \deg g(x).$$

Then

$$g(x) q(x) + r(x) = g(x) q^*(x) + r^*(x)$$

and

$$g(x)[q(x) - q^*(x)] = r^*(x) - r(x).$$

The right side of this equation is zero or of degree less than $\deg g(x)$. Since the left side is zero or of degree at least $g(x)$, this forces $q(x) = q^*(x)$. Then we must also have $r^*(x) = r(x)$.   ∎

Example 35.2.   Let

$$f(x) = [2]x^4 + [1]x^2 + [-1]x + [1] \in \mathbb{Z}_5[x]$$

and

$$g(x) = [2]x + [-1] \in \mathbb{Z}_5[x].$$

(Compare Example 35.1.) If we use the abbreviated notation a for $[a]$, $+$ for $\oplus$, and juxtaposition for $\odot$, then, for instance, since we are in $\mathbb{Z}_5[x]$, $2^{-1} = 3$, $3^{-1} = 2$, and $4^{-1} = 4$. The division of $f(x)$ by $g(x)$ follows.

$$
\begin{array}{r}
x^3 + 3x^2 + 2x + 3 \\
\hline
2x - 1\,\big|\,2x^4 \quad\quad\; + x^2 \; - \; x + 1 \\
2x^4 - x^3 \\
\hline
x^3 + x^2 - x + 1 \\
x^3 - 3x^2 \\
\hline
4x^2 - x + 1 \\
4x^2 - 2x \\
\hline
x + 1 \\
x - 3 \\
\hline
4
\end{array}
$$

Therefore $q(x) = [1]x^3 + [3]x^2 + [2]x + [3]$, and $r(x) = [4]$. ∎

If the indeterminate x in a polynomial

$$f(x) = a_0 + a_1 x + \cdots + a_n x^n \in F[x]$$

is replaced by an element $c \in F$, the result is an element of F:

$$f(c) = a_0 + a_1 c + \cdots + a_n c^n \in F$$

We say that $f(c)$ results from $f(x)$ by *substitution* of c for x. If $f(x) = g(x)$ in $F[x]$, then $f(c) = g(c)$ in F.

Example 35.3.
 (a) If $f(x) = x^3 - 2x^2 + 2 \in \mathbb{R}[x]$, then

$$f(3) = 3^3 - 2 \cdot 3^2 + 2 = 11 \in \mathbb{R}.$$

 (b) If $f(x) = [3] + [1]x + [-3]x^4 \in \mathbb{Z}_5[x]$, then

$$
\begin{aligned}
f([2]) &= [3] \oplus ([1] \odot [2]) \oplus ([-3] \odot [2]^4) \\
&= [3] \oplus [2] \oplus ([2] \odot [1]) = [7] = [2] \in \mathbb{Z}_5.
\end{aligned}
$$
∎

Remainder Theorem. *If $f(x) \in F[x]$ and $c \in F$, then the remainder in the division of $f(x)$ by $x - c$ is $f(c)$.*

PROOF. Because $\deg(x - c) = 1$, the remainder in the division of $f(x)$ by $x - c$ must be either 0 or of degree 0. Thus, for some $q(x) \in F[x]$,

$$f(x) = (x - c)q(x) + r, \quad \text{with} \quad r \in F.$$

Substitution of c for x yields

$$f(c) = (c - c)q(c) + r = r.$$
∎

Example 35.4.

(a) Divide $f(x) = x^3 - 2x^2 + 2 \in \mathbb{R}[x]$ by $x - 3$. The quotient is $x^2 + x + 3$ and the remainder is 11:

$$f(x) = (x - 3)(x^2 + x + 3) + 11.$$

Also, $f(3) = 11$, as we saw in Example 35.3(a).

(b) Divide $f(x) = [3] + [1]x + [-3]x^4 \in \mathbb{Z}_5[x]$ by $[1]x + [-2]$. The quotient is $[2]x^3 + [4]x^2 + [3]x + [2]$ and the remainder is $[2]$:

$$f(x) = ([1]x + [-2])([2]x^3 + [4]x^2 + [3]x + [2]) + [2].$$

Also, $f([2]) = [2]$, as we saw in Example 35.3(b).   ∎

If $f(x), g(x) \in F[x]$, with $g(x) \neq 0$, then $f(x)$ is *divisible* by $g(x)$ over F if $f(x) = g(x)q(x)$ for some $q(x) \in F[x]$. Thus $f(x)$ is divisible by $g(x)$ if the remainder in the division of $f(x)$ by $g(x)$ is zero. If $f(x)$ is divisible by $g(x)$ (over F), then we also say that $g(x)$ is a *factor* of $f(x)$ (over F).

Factor Theorem.   *If $f(x) \in F[x]$ and $c \in F$, then $x - c$ is a factor of $f(x)$ iff $f(c) = 0$.*

PROOF.   This is an immediate corollary of the Remainder Theorem.   ∎

An element $c \in F$ is called a *root* (or *zero*) of a polynomial $f(x) \in F[x]$ if $f(c) = 0$. By the Factor Theorem, c is a root of $f(x)$ iff $x - c$ is a factor of $f(x)$.

PROBLEMS

For each pair of polynomials $f(x)$ and $g(x)$ in Problems 35.1–35.6, determine $q(x)$ and $r(x)$, the quotient and remainder in the division of $f(x)$ by $g(x)$.

35.1. $f(x) = x^3 + x + 1$, $g(x) = x - 1$, both in $\mathbb{Q}[x]$.

35.2. $f(x) = x^4 - 1$, $g(x) = -x^2 + 2$, both in $\mathbb{Q}[x]$.

35.3. $f(x) = x^3 - 3$, $g(x) = x^4 + x$, both in $\mathbb{Z}_3[x]$.

35.4. $f(x) = x^2 + 2$, $g(x) = x - 1$, both in $\mathbb{Z}_3[x]$.

35.5. $f(x) = 3x^4 + 2x^2 - 1$, $g(x) = 2x^2 + 4x$, both in $\mathbb{Z}_5[x]$.

35.6. $f(x) = x^4 + ix^2 + 1$, $g(x) = ix^2 + 1$, both in $\mathbb{C}[x]$.

35.7. Use the Remainder Theorem to determine the remainder when $2x^5 - 3x^3 + 2x + 1 \in \mathbb{R}[x]$ is divided by $x - 2 \in \mathbb{R}[x]$.

35.8. Use the Remainder Theorem to determine the remainder when $2x^5 - 3x^3 + 2x + 1 \in \mathbb{R}[x]$ is divided by $x + 3 \in \mathbb{R}[x]$.

35.9. What is the remainder when $2x^5 - 3x^3 + 2x + 1 \in \mathbb{Z}_7[x]$ is divided by $x - 2 \in \mathbb{Z}_7[x]$?

35.10. What is the remainder when $ix^9 + 3x^7 + x^6 - 2ix + 1 \in \mathbb{C}[x]$ is divided by $x + i \in \mathbb{C}[x]$?

35.11. Verify that $\deg f_1(x) < \deg f(x)$ in the proof of the Division Algorithm.

35.12. Use the Factor Theorem to answer each of the following questions.

(a) Is $x - 3 \in \mathbb{Q}[x]$ a factor of $3x^3 - 9x^2 - 7x + 21 \in \mathbb{Q}[x]$?

(b) Is $x + 2 \in \mathbb{R}[x]$ a factor of $x^3 + 8x^2 + 6x - 8 \in \mathbb{R}[x]$?

(c) For which $k \in \mathbb{Q}$ is $x - 1$ a factor of $x^3 + 2x^2 + x + k \in \mathbb{Q}[x]$?

(d) Is $x - 2 \in \mathbb{Z}_5[x]$ a factor of $2x^5 - 3x^4 - 4x^3 + 3x \in \mathbb{Z}_5[x]$?

(e) For which $k \in \mathbb{C}$ is $x + i$ a factor of $ix^9 + 3x^7 + x^6 - 2ix + k \in \mathbb{C}[x]$?

35.13. Find all odd primes p for which $x - 2$ is a factor of $x^4 + x^3 + x^2 + x$ in $\mathbb{Z}_p$.

35.14. Prove that for each prime p there is an infinite field of characteristic p. (*Suggestion*: Consider the field of quotients of $\mathbb{Z}_p[x]$. Fields of quotients are discussed in Section 30.)

35.15. Construct an example to show that the Division Algorithm is not true if F is replaced by the integral domain $\mathbb{Z}$. (Compare Problem 35.16.)

35.16. Prove that if the Division Algorithm is true for polynomials over an integral domain D, then D is a field. (Compare Problem 35.15.)

35.17. Use the Factor Theorem to construct a single polynomial $f(x) \in \mathbb{Z}_5[x]$ such that every element of $\mathbb{Z}_5$ is a root of $f(x)$.

35.18. Prove that if p is a prime, then each element of $\mathbb{Z}_p$ is a root of $x^p - x$. (*Suggestion*: If $[a] \neq [0]$, then $[a]^{p-1} = [1]$ because $\mathbb{Z}_p^{\#}$ is a group with respect to $\odot$ by the corollary of Theorem 26.1.)

35.19. (Lagrange's interpolation formula). Assume that $a_0, a_1, \ldots, a_n$ are distinct elements of a field F, and that $b_0, b_1, \ldots, b_n \in F$. Let

$$f(x) = \sum_{k=0}^{n} \frac{b_k (x - x_0) \cdots (x - a_{k-1})(x - a_{k+1}) \cdots (x - a_n)}{(a_k - a_0) \cdots (a_k - a_{k-1})(a_k - a_{k+1}) \cdots (a_k - a_n)} \tag{35.1}$$

Verify that $f(a_j) = b_j$ for $0 \leq j \leq n$. (This shows that there exists a polynomial of degree at most n that takes on given values at $n + 1$ distinct given points of F. Problem 44.8 will show that such a polynomial is unique.)

35.20. (a) Use Problem 35.19 to write a polynomial $f(x) \in \mathbb{R}[x]$ such that $f(1) = 2$, $f(2) = 3$, and $f(3) = -1$.

(b) Use Problem 35.19 to write a polynomial $f(x) \in \mathbb{Z}_5[x]$ such that $f([1]) = [2]$, $f([2]) = [3]$, and $f([3]) = [-1]$.

SECTION 36   FACTORIZATION OF POLYNOMIALS

The next theorem is a direct parallel of Theorem 12.1.

Theorem 36.1.   *If $a(x)$ and $b(x)$ are polynomials over a field F, not both the zero polynomial, then there is a unique monic polynomial $d(x)$ over F such that*

(a) $d(x) \mid a(x)$ and $d(x) \mid b(x)$, and

(b) if $c(x)$ is a polynomial such that $c(x) \mid a(x)$ and $c(x) \mid b(x)$, then $c(x) \mid d(x)$.

The polynomial $d(x)$ in the theorem is called the *greatest common divisor* of $a(x)$ and $b(x)$. Just as in the case of the integers, the existence of the greatest common divisor is shown using the *Euclidean Algorithm*—this time for polynomials.

PROOF.   First assume $b(x) \neq 0$. By the Division Algorithm there are unique polynomials $q_1(x)$ and $r_1(x)$ such that

$$a(x) = b(x)q_1(x) + r_1(x), \qquad \text{with} \qquad r_1(x) = 0 \qquad \text{or} \qquad \deg r_1(x) < \deg b(x).$$

If $r_1(x) = 0$, then $b(x) \mid a(x)$; thus $d(x) = b(x)$ satisfies parts (a) and (b). If $r_1(x) \neq 0$, then we apply the Division Algorithm repeatedly, just as in the proof of Theorem 12.1:

$$
\begin{aligned}
a(x) &= b(x)q_1(x) + r_1(x), & \deg r_1(x) &< \deg b(x) \\
b(x) &= r_1(x)q_2(x) + r_2(x), & \deg r_2(x) &< \deg r_1(x) \\
r_1(x) &= r_2(x)q_3(x) + r_3(x), & \deg r_3(x) &< \deg r_2(x) \\
&\qquad\qquad \vdots \\
r_{k-2}(x) &= r_{k-1}(x)q_k(x) + r_k(x), & \deg r_k(x) &< \deg r_{k-1}(x) \\
r_{k-1}(x) &= r_k(x)q_{k+1}(x).
\end{aligned}
$$

Here we must eventually get the zero polynomial as a remainder because $\deg r_1(x) > \deg r_2(x) > \deg r_3(x) > \cdots$. Let $r_k(x)$ denote the last nonzero remainder. The proof that $r_k(x)$ satisfies both of the requirements (a) and (b) for $d(x)$ is similar to the proof of Theorem 12.1 (Problem 36.13). If the leading coefficient of $r_k(x)$ is r, then $r^{-1} \cdot r_k(x)$ is a *monic* polynomial satisfying (a) and (b).

If $b(x) = 0$, and a_m is the leading coefficient of $a(x)$, then $a_m^{-1} \cdot a(x)$ is a monic polynomial satisfying both requirements (a) and (b) for $d(x)$. Thus we have proved the existence of a greatest common divisor.

The proof of uniqueness relies on the requirement that the greatest common divisor be monic. It is similar to the uniqueness proof in Theorem 12.1 and is left as an exercise (Problem 36.14). ∎

Example 36.1.   Here the Euclidean Algorithm is applied to compute the greatest common divisor of $a(x) = x^4 - x^3 - x^2 + 1$ and $b(x) = x^3 - 1$, considered as polynomials over the field of rationals.

$$
\begin{aligned}
x^4 - x^3 - x^2 + 1 &= (x^3 - 1)(x - 1) + (-x^2 + x) \\
x^3 - 1 &= (-x^2 + x)(-x - 1) + (x - 1) \\
-x^2 + x &= (x - 1)(-x)
\end{aligned}
$$

Therefore, the greatest common divisor is $x - 1$. ∎

The next theorem follows from the proof of Theorem 36.1 in the same way that Theorem 12.2 follows from the proof of Theorem 12.1 (Problem 36.15).

Theorem 36.2.   *If $a(x)$ and $b(x)$ are polynomials over a field F, not both the zero polynomial, and $d(x)$ is their greatest common divisor, then there exist polynomials $u(x)$ and $v(x)$ over F such that*

$$d(x) = a(x)u(x) + b(x)v(x).$$

Example 36.2.   Consider Example 36.1. From the first equation there,

$$-x^2 + x = (x^4 - x^3 - x^2 + 1) - (x^3 - 1)(x - 1).$$

Using this with the second equation, we get

$$x - 1 = (x^3 - 1) - (-x^2 + x)(-x - 1)$$
$$= (x^3 - 1) - [(x^4 - x^3 - x^2 + 1) - (x^3 - 1)(x - 1)](-x - 1)$$
$$= (x^4 - x^3 - x^2 + 1)(x + 1) + (x^3 - 1)[1 + (x - 1)(-x - 1)]$$
$$= (x^4 - x^3 - x^2 + 1)(x + 1) + (x^3 - 1)(-x^2 + 2).$$

Thus $d(x) = x - 1 = a(x)u(x) + b(x)v(x)$ for $u(x) = x + 1$ and $v(x) = -x^2 + 2$.   ∎

Two polynomials $f(x)$ and $g(x)$ over a field F are said to be *associates* if $f(x) = c \cdot g(x)$ for some nonzero element c of F. For example, $2x^2 - 1$ and $6x^2 - 3 = 3(2x^2 - 1)$ are associates over $\mathbb{Q}$. Notice that each nonzero polynomial has precisely one monic polynomial among its associates. Notice also that each polynomial of degree at least one has two obvious sets of divisors: its associates and the polynomials of degree zero. If a polynomial of degree at least one has no other divisors, then it is said to be *irreducible* (or *prime*). Thus, if $f(x) = g(x)h(x)$, and $f(x)$ is irreducible, then one of $g(x)$ and $h(x)$ is of degree zero and the other is an associate of $f(x)$. If a polynomial is not irreducible, then it is said to be *reducible*.

The property of being irreducible depends on the field F. For example, $x^2 - 2$ is irreducible over the field of rational numbers, but not over the field of real numbers: $x^2 - 2 = (x + \sqrt{2})(x - \sqrt{2})$. The irreducible polynomials over a field F play the same role for $F[x]$ that the prime numbers do for the ring of integers. Here is the first indication of that. (Compare Lemma 13.2.)

Corollary. *If F is a field, $a(x), b(x), p(x) \in F[x]$, $p(x)$ is irreducible, and $p(x) \mid a(x)b(x)$, then $p(x) \mid a(x)$ or $p(x) \mid b(x)$.*

PROOF. If $p(x) \nmid a(x)$, then the greatest common divisor of $p(x)$ and $a(x)$ is e, the polynomial of degree zero with the coefficient the unity of F. Thus, if $p(x) \nmid a(x)$, then by Theorem 36.2 there are polynomials $u(x)$ and $v(x)$ such that $e = u(x)p(x) + v(x)a(x)$. Multiplication of both sides of this equation by $b(x)$ leads to $b(x) = u(x)p(x)b(x) + v(x)a(x)b(x)$. Because $p(x) \mid p(x)$ and $p(x) \mid a(x)b(x)$, we conclude that $p(x) \mid [u(x)p(x)b(x) + v(x)a(x)b(x)]$, and therefore $p(x) \mid b(x)$. Thus, if $p(x) \nmid a(x)$, then $p(x) \mid b(x)$, which proves the corollary.   ∎

Corollary. *If $p(x), a_1(x), a_2(x), \ldots, a_n(x)$ are polynomials over F, with $p(x)$ irreducible and $p(x) \mid a_1(x)a_2(x) \cdots a_n(x)$, then $p(x) \mid a_i(x)$ for some i $(1 \le i \le n)$.*

PROOF. Use the preceding corollary and induction on n (Problem 36.17).   ∎

Unique Factorization Theorem. *Each polynomial of degree at least one over a field F can be written as an element of F times a product of monic irreducible polynomials over F, and, except for the order in which these irreducible polynomials are written, this can be done in only one way.*

PROOF. Let S denote the set of those polynomials over F that are of degree at least one and that *cannot* be written as stated. We shall prove that S is empty. If not, then by the Least Integer Principle (applied to the set of degrees of polynomials in S) there is a polynomial of least positive degree in S; let $a(x)$ denote such a polynomial and assume $\deg a(x) = n$. Then $a(x)$ is not irreducible, so it can be factored as $a(x) = a_1(x)a_2(x)$, where $1 < \deg a_1(x) < n$ and $1 < \deg a_2(x) < n$. By the choice of $a(x)$, $a_1(x)$, and $a_2(x)$, we know $a_1(x) \notin S$ and $a_2(x) \notin S$. Therefore $a_1(x)$ and $a_2(x)$ can each be written as an element of F times a product of monic irreducible polynomials, so the same is true of $a(x) = a_1(x)a_2(x)$. This contradicts the fact that $a(x) \in S$, and we therefore conclude that S must be empty, as stated.

The proof of the last part of the theorem is similar to the last part of the proof of the Fundamental Theorem of Arithmetic (Section 13) and is left as an exercise (Problem 36.18).   ∎

Example 36.3.   The polynomial $3x^4 - 3x^2 - 6$ can be factored as

$$3(x^2 - 2)(x^2 + 1) \quad \text{in} \quad \mathbb{Q}[x],$$
$$3(x + \sqrt{2})(x - \sqrt{2})(x^2 + 1) \quad \text{in} \quad \mathbb{R}[x],$$

and

$$3(x + \sqrt{2})(x - \sqrt{2})(x + i)(x - i) \quad \text{in} \quad \mathbb{C}[x].$$

Each factor is irreducible in its context.   ∎

PROBLEMS

Use the Euclidean Algorithm to compute the greatest common divisors of the following pairs of polynomials over $\mathbb{Q}$. Also express each greatest common divisor as a linear combination of the two given polynomials (as in Theorem 36.2).

36.1. $x^3 - 3x^2 + 3x - 2$ and $x^2 - 5x + 6$

36.2. $x^4 + 3x^2 + 2$ and $x^5 - x$

36.3. $x^3 + x^2 - 2x - 2$ and $x^4 - 2x^3 + 3x^2 - 6x$

36.4. $x^5 + 4x$ and $x^3 - x$.

36.5. Determine the monic associate of $2x^3 - x + 1 \in \mathbb{Q}[x]$.

36.6. Determine the monic associate of $1 + x - ix^2 \in \mathbb{C}[x]$.

36.7. Determine the monic associate of $2x^5 - 3x^2 + 1 \in \mathbb{Z}_7[x]$.

36.8. Determine the monic associate of $2x^5 - 3x^2 + 1 \in Z_5[x]$.

36.9. Verify that $x^3 - 3 \in \mathbb{Z}_7[x]$ is irreducible.

36.10. Verify that $x^4 + x^2 + 1 \in \mathbb{Z}_5[x]$ is irreducible.

36.11. Write $x^3 + 3x^2 + 3x + 4 \in \mathbb{Z}_5[x]$ as a product of irreducible polynomials.

36.12. Write $x^5 + x^4 + x^2 + 2x \in Z_3[x]$ as a product of irreducible polynomials.

36.13. Prove that the polynomial $r_k(x)$ in the proof of Theorem 36.1 satisfies both of the requirements (a) and (b) of the theorem.

36.14. Prove the uniqueness of $d(x)$ in Theorem 36.1.

36.15. Prove Theorem 36.2. (The remark preceding it suggests the method.)

36.16. Explain why each nonzero polynomial has precisely one monic polynomial among its associates.

36.17. Write the proof of the second corollary of Theorem 36.2.

36.18. Complete the proof of the Unique Factorization Theorem.

36.19. (a) Prove that $(x - 1) \mid f(x)$ in $\mathbb{Z}_2[x]$ iff $f(x)$ has an even number of nonzero coefficients.
 (b) Prove that if $\deg f(x) > 1$ and $f(x)$ is irreducible over $\mathbb{Z}_2$, then $f(x)$ has constant term 1 and an odd number of nonzero coefficients.
 (c) Determine all irreducible polynomials of degree 4 or less over $\mathbb{Z}_2$.
 (d) Write each polynomial of degree 3 over $\mathbb{Z}_2$ as a product of irreducible factors.

36.20. (a) By counting the number of distinct possibilities for $(x - a)(x - b)$, verify that there are $p(p + 1)/2$ monic reducible polynomials of degree 2 over $\mathbb{Z}_p$ (p a prime).
 (b) How many monic irreducible polynomials of degree 2 over $\mathbb{Z}_p$ are there?

36.21. State and prove a theorem establishing the existence of a unique *least common multiple* for every pair of polynomials, not both the zero polynomial, over a field F. (Compare Theorem 12.3 and Theorem 36.1.)

36.22. (Eisenstein's irreducibility criterion) *Assume that p is a prime, $f(x) = a_0 + a_1x + \cdots + a_nx^n \in \mathbb{Z}[x]$, $p \mid a_i$ $(0 \leq i \leq n - 1)$, $p^2 \nmid a_0$ and $p \nmid a_n$. Then $f(x)$ is irreducible in $\mathbb{Z}[x]$.* Give an indirect proof of this by justifying each of the following statements.
 (a) Assume that $f(x) = (b_0 + b_1x + \cdots + b_ux^u)(c_0 + c_1x + \cdots + c_vx^v)$ Then p does not divide both b_0 and c_0.
 (b) But p divides one of b_0 and c_0. Assume that $p \nmid b_0$ and $p \mid c_0$.
 (c) Since $p \nmid c_v$ (why?), not all c_j are divisible by p. Let k be the smallest integer such that $p \nmid c_k$ and $p \mid c_j$ for $0 \leq j < k$.
 (d) Because $a_k = b_kc_0 + b_{k-1}c_1 + \cdots + b_0c_k$, we can conclude that $p \mid b_0c_k$, which is a contradiction. (For applications, see Problems 36.23 and 36.24.)

36.23. Use Eisenstein's irreducibility criterion (Problem 36.22) to show that each of the following polynomials is irreducible in $\mathbb{Z}[x]$.
 (a) $x^3 + 6x^2 + 3x + 3$ (b) $x^5 - 5x^3 + 15$

36.24. Use Eisenstein's irreducibility criterion (Problem 36.22) to prove that if p is a prime, then the polynomial

$$f(x) = \frac{x^p - 1}{x - 1} = x^{p-1} + x^{p-2} + \cdots + x + 1$$

is irreducible in $\mathbb{Z}[x]$. (*Suggestion*: Replace x by $y + 1$, and let $g(y)$ denote the result. Show that $g(y)$ is irreducible in $\mathbb{Z}[y]$, and explain why this implies that $f(x)$ is irreducible in $\mathbb{Z}[x]$. The polynomial $f(x)$ is called a *cyclotomic polynomial*; if p is odd, its roots are the imaginary pth roots of unity.)

SECTION 37   UNIQUE FACTORIZATION DOMAINS

The similarity between the factorization theorems for integers and polynomials has been emphasized throughout this chapter. We are concerned now and again in Section 41 with the question of whether these theorems can be generalized to other rings. This question was first studied in the early to mid-1800s, and it is important to know that it was studied out of necessity: progress in solving several notable problems in number theory depended on understanding factorization in rings of numbers other than the integers. Examples 37.1 and 37.2 illustrate the kinds of rings that were involved, then Section 41 gives a more detailed account of the relevant history. Two points can be made now, however: first, the ideas we meet here and in Section 41 came not merely from abstraction for abstraction's sake, but from attempts to solve specific problems; second, these ideas, and the problems that brought them into focus, gave rise to algebraic number theory, one of the most interesting and challenging branches of modern mathematics.

The proofs in this section are similar to proofs already given for $\mathbb{Z}$ and $F[x]$, and will be left as exercises. Except for Section 41, the remainder of the book will be independent of this section.

Suppose that D is an integral domain. If $a, b \in D$, with $b \neq 0$, then a is *divisible* by b if $a = bc$ for some $c \in D$. If a is divisible by b, we also say that b *divides* a, or that b is a *factor* of a, and we write $b \mid a$.

An element of D is called a *unit* if it divides the unity, e, of D. The units of $\mathbb{Z}$ are 1 and -1. If F is a field, then the units of $F[x]$ are the polynomials of degree zero (that is, the nonzero constant polynomials).

Elements a and b of D are called *associates* if $a \mid b$ and $b \mid a$. It can be proved that a and b are associates iff $a = bu$ for some unit u of D (Problem 37.14). Elements $a, b \in \mathbb{Z}$ are associates iff $a = \pm b$. If F is a field, then polynomials $f(x), g(x) \in F[x]$ are associates iff $g(x) = c \cdot f(x)$ for some nonzero $c \in F$.

An element in an integral domain D is *irreducible* if it is not a unit of D and its only divisors in D are its associates and the units of D. The irreducible elements of $\mathbb{Z}$ are the primes and their negatives. The irreducible elements of $F[x]$ are the irreducible polynomials of $F[x]$.

Definition.   An integral domain D is a *unique factorization domain* provided that

(a)   if $a \in D$, $a \neq 0$, and a is not a unit, then a can be written as a product of irreducible elements of D, and

(b)   if $a \in D$ and

$$a = p_1 p_2 \cdots p_s = q_1 q_2 \cdots q_t,$$

where each p_i and each q_j is irreducible, then $s = t$ and there is a permutation π of $\{1, 2, \ldots, s\}$ such that p_i and $q_{\pi(i)}$ are associates for $1 \leq i \leq s$.

The ring of integers is a unique factorization domain (Section 13); so is the ring $F[x]$ of polynomials over any field F (Section 36). Following is an example of an integral domain that is not a unique factorization domain. We shall see in Section 41 that there is a historical connection between examples of this type and attempts to prove Fermat's Last Theorem (which will be stated in Section 41).

Example 37.1.   Let $\mathbb{Z}\left[\sqrt{-5}\right] = \{a + b\sqrt{-5} : a, b \in \mathbb{Z}\}$. It can be verified that $\mathbb{Z}\left[\sqrt{-5}\right]$ is a subring of $\mathbb{C}$ (Problem 37.6). Therefore, since $\mathbb{Z}\left[\sqrt{-5}\right]$ contains 1, it is an integral domain. In proving that $\mathbb{Z}\left[\sqrt{-5}\right]$ is not a unique factorization domain, we shall use the mapping N from $\mathbb{Z}\left[\sqrt{-5}\right]$ to the set of nonnegative integers defined by

$$N(a + b\sqrt{-5}) = |a + b\sqrt{-5}|^2 = a^2 + 5b^2.$$

This mapping N is called a *norm*. If $z, w \in \mathbb{Z}[\sqrt{-5}]$, then

(a) $N(z) \geq 0$,

(b) $N(z) = 0$ iff $z = 0$, and

(c) $N(zw) = N(z)N(w)$ (Problem 37.7).

To determine the units of $\mathbb{Z}[\sqrt{-5}]$, we first observe that if $zw = 1$, then $N(z)N(w) = N(zw) = N(1) = 1$. Therefore, if $z = a + b\sqrt{-5}$ is a unit, then $N(z) = a^2 + 5b^2 = 1$, so that $a = \pm 1$ and $b = 0$. Thus the units of $\mathbb{Z}[\sqrt{-5}]$ are ± 1. It follows that an element of $\mathbb{Z}[\sqrt{-5}]$ has two associates, itself and its negative.

Now observe that

$$9 = 3 \cdot 3 = (2 + \sqrt{-5})(2 - \sqrt{-5}),$$

and that $9, 3, 2 \pm \sqrt{-5} \in \mathbb{Z}[\sqrt{-5}]$. If we show that 3 and $2 \pm \sqrt{-5}$ are irreducible in $\mathbb{Z}[\sqrt{-5}]$, then we will have shown that $\mathbb{Z}[\sqrt{-5}]$ is not a unique factorization domain, because by the preceding paragraph 3 is not an associate of either $2 + \sqrt{-5}$ or $2 - \sqrt{-5}$. We shall prove that 3 is irreducible, and leave the proof for $2 \pm \sqrt{-5}$ as an exercise (Problem 37.8).

Assume that $3 = zw$, with $z, w \in \mathbb{Z}[\sqrt{-5}]$. Then $N(z)N(w) = N(zw) = N(3) = 9$; hence $N(z)$ is either 1, 3, or 9. If $N(z) = 1$, then z is a unit, and if $N(z) = 9$, then $N(w) = 1$ and w is a unit. Therefore, if 3 is to have a factor that is neither an associate of 3 nor a unit, then that factor must have norm 3. But $a^2 + 5b^2 \neq 3$ for all integers a and b, so $\mathbb{Z}[\sqrt{-5}]$ has no element of norm 3. Thus 3 is irreducible in $\mathbb{Z}[\sqrt{-5}]$.   ∎

If you examine the proofs of unique factorization in $\mathbb{Z}$ and $F[x]$, you will see that they both rely on the Division Algorithm. We shall now see that essentially the same method can work in other cases. Specifically, Theorem 37.1 shows that an integral domain is a unique factorization domain if it satisfies the following definition.

Definition.   An integral domain D is a *Euclidean domain* if for each nonzero element $a \in D$ there exists a nonnegative integer $d(a)$ such that

(a) if a and b are nonzero elements of D, then $d(a) \leq d(ab)$, and

(b) if $a, b \in D$, with $b \neq 0$, then there exist elements $q, r \in D$ such that $a = bq + r$, with $r = 0$ or $d(r) < d(b)$.

The ring of integers is a Euclidean domain with $d(a) = |a|$. If F is a field, then $F[x]$ is a Euclidean domain with $d(f(x)) = \deg f(x)$. The following example, first introduced by Gauss, has historical interest that will be discussed in Section 41.

Example 37.2.   Let $\mathbb{Z}[i] = \{a + bi : a, b \in \mathbb{Z}\}$. The elements of $\mathbb{Z}[i]$ are called *Gaussian integers*. It is easy to verify that $\mathbb{Z}[i]$ is an integral domain, and we shall prove that it is a Euclidean domain with respect to $d(a + bi) = |a + bi|^2 = a^2 + b^2$.

If z and w are nonzero elements of $\mathbb{Z}[i]$, then

$$d(z) = |z|^2 \leq |z|^2 |w|^2 = |zw|^2 = d(zw).$$

Thus d satisfies condition (a) of the definition.

To verify condition (b), assume that $z, w \in \mathbb{Z}[i]$ with $w \neq 0$. Then $zw^{-1} \in \mathbb{C}$, and in fact $zw^{-1} = a + bi$ with $a, b \in \mathbb{Q}$. Let m and n be integers such that $|a - m| \leq 1/2$ and $|b - n| \leq 1/2$. Then

$$zw^{-1} = a + bi = m + ni + [(a - m) + (b - n)i]$$

and

$$z = (m + ni)w + [(a - m) + (b - n)i]w$$
$$= qw + r,$$

where $q = m + ni$ and $r = [(a - m) + (b - n)i]w$. Here $r \in \mathbb{Z}[i]$ because $qw \in \mathbb{Z}[i]$ and $z \in \mathbb{Z}[i]$. It is now sufficient to show that $d(r) < d(w)$:

$$d(r) = d[(a - m) + (b - n)i]d(w)$$
$$= [(a - m)^2 + (b - n)^2]d(w)$$
$$\leq [(\tfrac{1}{4}) + (\tfrac{1}{4})]d(w)$$
$$< d(w).$$

It can be shown that the units in $\mathbb{Z}[i]$ are ± 1 and $\pm i$ (Problem 37.9). Notice that 2 can be factored in $\mathbb{Z}[i]$ as $2 = i(1 - i)^2$, where i is a unit and $1 - i$ is irreducible.

Because we have proved that $\mathbb{Z}[i]$ is a Euclidean domain, we can conclude that it is a unique factorization domain because of the following theorem.   ■

Theorem 37.1.   *Every Euclidean domain is a unique factorization domain.*

The proof of Theorem 37.1 is similar to the proofs of unique factorization in the special cases $\mathbb{Z}$ and $F[x]$. It can be carried out by first working through proofs of each of the following lemmas. The details will be left as exercises, but just by reading the statements of these results you can get an understanding of how ideas of factorization extend beyond the most familiar examples.

Lemma 37.1.   *If D is an integral domain and $a, b \in D$, then a and b are associates iff $a = bu$ for some unit u of D.*

Definition.   If a and b are elements of an integral domain D, not both zero, then an element $d \in D$ is a *greatest common divisor* of a and b provided that

(a) $d \mid a$ and $d \mid b$, and

(b) if $c \in D$, $c \mid a$, and $c \mid b$, then $c \mid d$.

A pair of elements need not have a *unique* greatest common divisor by this definition. For example, both ± 2 are greatest common divisors of 4 and 6 in $\mathbb{Z}$. And $x + 1, 2x + 2$, and $(1/2)x + (1/2)$ are all greatest common divisors of $x^2 - 1$ and $x^2 + 2x + 1$ in $\mathbb{Q}[x]$. The uniqueness of greatest common divisors in $\mathbb{Z}$ (Theorem 12.1) depended in part on the requirement of being positive. Similarly, the uniqueness of greatest common divisors in $F[x]$ (Theorem 36.1) depended in part on the requirement of being monic. For more general discussions of factorization we must sacrifice these strong forms of uniqueness and use the preceding definition. However, Lemma 37.3 will show that even in the general case we do not lose all uniqueness. Before stating that, however, we state the following important fact.

Lemma 37.2.   *Any two nonzero elements a and b of a Euclidean domain D have a greatest common divisor d in D, and d = ar + bs for some r, s ∈ D.*

Lemma 37.3.   *If d_1 and d_2 are both greatest common divisors of elements a and b in an integral domain D, then d_1 and d_2 are associates in D.*

Lemma 37.4.   *Assume that a and b are nonzero elements of a Euclidean domain D. Then $d(a) = d(ab)$ iff b is a unit in D.*

Lemma 37.5.   *Assume that $a, b,$ and c are nonzero elements of a Euclidean domain D with unity e. If a and b have greatest common divisor e, and $a \mid bc$, then $a \mid c$.*

Lemma 37.6.   *If $a, b,$ and p are nonzero elements of a Euclidean domain D, and p is irreducible and $p \mid ab$, then $p \mid a$ or $p \mid b$.*

Section 41 will show how the ideas in this section are connected with other ideas about rings that are, on the surface, unrelated to factorization. For references for the history related to this chapter, see the notes at the end of Chapter IX.

PROBLEMS

Assume that D is an integral domain throughout these problems.

37.1. Find all of the associates of $2x^5 - x + 1 \in Z_5[x]$.

37.2. Find all of the associates of $2 - i \in Z[i]$. (*Suggestion*: Use Problem 37.9.)

37.3. Find $q, r \in Z[i]$ such that $1 + 5i = (2i)q + r$ with $|r| < 2$. (See Example 37.2.)

37.4. Find $q, r \in Z[i]$ such that $1 - 5i = (1 + 2i)q + r$ with $|r| < 5$. (See Example 37.2.)

———————

37.5. Sketch a proof, along the lines of Example 37.1, that $Z[\sqrt{-3}]$ is not a unique factorization domain. [Suggestion: $4 = 2 \cdot 2 = (1 + \sqrt{-3})(1 - \sqrt{-3})$.]

37.6. Prove that $Z[\sqrt{-5}]$ is subring of $\mathbb{C}$ (Example 37.1).

37.7. Verify the properties (a), (b), and (c) of the norm in Example 37.1.

37.8. Prove that $2 \pm \sqrt{-5}$ are irreducible in $Z[\sqrt{-5}]$ (Example 37.1).

37.9. Show that the units in $\mathbb{Z}[i]$ are ± 1 and $\pm i$.

37.10. Prove that every field is a Euclidean domain.

37.11. (a) What are the units of $\mathbb{Z}[x]$?
(b) Prove or disprove that $\mathbb{Z}[x]$ is a Euclidean domain with $d(f(x)) = \deg f(x)$.

37.12. Prove that an element a of a Euclidean domain is a unit iff $d(a) = d(e)$.

37.13. Prove that if $a \in D$ and $a \neq 0$, then a is a greatest common divisor of a and 0.

37.14. Prove Lemma 37.1.   **37.15.** Prove Lemma 37.2.

37.16. Prove Lemma 37.3.   **37.17.** Prove Lemma 37.4.

37.18. Prove Lemma 37.5.   **37.19.** Prove Lemma 37.6.

37.20. Prove Theorem 37.1.

37.21. Prove that if $a, u \in D$, and u is a unit, then $u \mid a$.

37.22. Prove that the units of D form a group with respect to the multiplication in D.

CHAPTER IX

QUOTIENT RINGS

The first two sections of this chapter show that the definitions and basic theorems for ring homomorphisms and quotient rings directly parallel those for group homomorphisms and quotient groups. The third section develops some facts about polynomial rings that will be used later in studying questions about fields and polynomial equations. In the last section of the chapter we see that ideals, which are the kernels of ring homomorphisms, arise naturally in the study of factorization in rings.

SECTION 38   HOMOMORPHISMS OF RINGS. IDEALS

Definition.   If R and S are rings, then a mapping $\theta : R \to S$ is a (ring) *homomorphism* if

$$\theta(a + b) = \theta(a) + \theta(b)$$

and

$$\theta(ab) = \theta(a)\theta(b)$$

for all $a, b \in R$.

Thus a ring isomorphism (Section 27) is a ring homomorphism that is one-to-one and onto. In the conditions $\theta(a + b) = \theta(a) + \theta(b)$ and $\theta(ab) = \theta(a)\theta(b)$, the operations on the left in each case are those of R, and the operations on the right are those of S. Because of the first of these conditions, a ring homomorphism $\theta : R \to S$ is necessarily a group homomorphism from the additive group of R to the additive group of S. It follows that any statement about group homomorphisms translates into some statement about ring homomorphisms. For example, if $\theta : R \to S$ is a ring homomorphism, then $\theta(0_R) = 0_S$ and $\theta(-a) = -\theta(a)$ for all $a \in R$.

Example 38.1.   The mapping $\theta : \mathbb{Z} \to \mathbb{Z}_n$ defined by $\theta(a) = [a]$ was seen to be a homomorphism of additive groups in Example 21.1. But also $\theta(ab) = [ab] = [a] \odot [b]$ for all $a, b, \in \mathbb{Z}$, so θ is a ring homomorphism. ∎

Example 38.2.   Let R and S be rings, and let $R \times S$ be the direct sum of R and S (Example 24.6). The mappings $\pi_1 : R \times S \to R$ and $\pi_2 : R \times S \to S$ defined by $\pi_1((r, s)) = r$ and $\pi_2((r, s)) = s$ are ring homomorphisms (Problem 38.8). ∎

Definition.   If $\theta : R \to S$ is a ring homomorphism, then Ker θ, the *kernel* of θ, is the set of all elements $r \in R$ such that $\theta(r) = 0_S$.

Thus the kernel of a ring homomorphism θ is just the kernel of θ thought of as a homomorphism of the additive groups of the rings. Having observed this, we know at once that Ker θ is a subgroup of the additive group of R. Kernels of ring homomorphisms are, in fact, subrings. But just as kernels of group homomorphisms are special among all subgroups, being normal, kernels of ring homomorphisms are special among all subrings—they are ideals, in the following sense.

Definition.   A subring I of a ring R is an *ideal* of R if $ar \in I$ and $ra \in I$ for all $a \in I$ and all $r \in R$.

The important point here is that in the products ar and ra, r can be any element in R; r is not restricted to I. If R is commutative, then the conditions $ar \in I$ and $ra \in I$ are equivalent. Notice that $\mathbb{Z}$ is a subring of $\mathbb{Q}$, but $\mathbb{Z}$ is not an ideal of $\mathbb{Q}$. Other examples of subrings that are not ideals are given in the problems.

Theorem 38.1.   *Let $\theta : R \to S$ be a ring homomorphism.*

   (a)   *The image of θ is a subring of S.*
   (b)   *The kernel of θ is an ideal of R.*
   (c)   *θ is one-to-one iff Ker $\theta = \{0_R\}$.*

    PROOF.   (a) Because θ is an additive group homomorphism, it follows from Section 21 (after Example 21.4) that $\theta(R)$, the image of θ, is a subgroup of the additive group of S. Thus it suffices to prove that $\theta(R)$ is closed with respect to multiplication. To this end, assume that $s_1, s_2 \in \theta(R)$. Then $\theta(r_1) = s_1$ and $0(r_2) = s_2$ for some $r_1, r_2 \in R$, and $\theta(r_1 r_2) = \theta(r_1)\theta(r_2) = s_1 s_2$. Therefore $s_1 s_2 \in \theta(R)$, as required.
    (b) We have already observed that Ker θ is subgroup of the additive group of R. On the other hand, if $a \in$ Ker θ and $r \in R$, then $\theta(ar) = \theta(a)\theta(r) = 0 \cdot \theta(r) = 0$ and $\theta(ra) = \theta(r)\theta(a) = \theta(r) \cdot 0 = 0$. This proves that Ker θ is an ideal of R.
    (c) This is a direct consequence of the last part of Theorem 21.1; simply change to additive notation.   ■

Example 38.3.   The kernel of the homomorphism $\theta : \mathbb{Z} \to \mathbb{Z}_n$ in Example 38.1 is the set (ideal) consisting of all multiples of the integer n. The kernel of the homomorphism $\pi_1 : R \times S \to R$ in Example 38.2 is $\{(0, s) : s \in S\}$; this ideal of $R \times S$ is isomorphic of S (Problem 38.8).   ■

The ideals in the first part of the preceding example, the kernels of the homomorphisms $\theta : \mathbb{Z} \to \mathbb{Z}_n$, belong to an important general class defined as follows. Let R be a commutative ring with unity e, and let $a \in R$; then (a) will denote the set of all multiples of a by elements of R:

$$(a) = \{ra : r \in R\}.$$

We shall verify that (a) is an ideal of R. First, (a) is a subgroup of the additive group of R:

(i)   If $r, s \in R$, so that $ra, sa \in (a)$, then $ra + sa = (r + s)a \in (a)$.

(ii) $0 = 0a \in (a)$.

(iii)   The negative of an element ra in (a) is $(-r)a$, and it is also in (a).

Second, if $ra \in (a)$ and $s \in R$, then $s(ra) = (sr)a \in (a)$. Thus (a) is an ideal of R, as claimed.

Ideals of the form (a) are called *principal ideals*. The ideal (a) is the smallest ideal of R containing a. Every ideal of $\mathbb{Z}$ is a principal ideal (Problem 38.17). We shall prove in Theorem 40.3 that if F is a field, then every ideal of the polynomial ring $F[x]$ is a principal ideal. Problem 38.18 gives an example of an ideal that is not principal.

Some rings, such as $\mathbb{Z}$, have many ideals. At the other extreme are rings having no ideals except the two obvious ones, (0) and the ring itself. The next example shows that any field has this property.

Example 38.4.   If F is a field, then F has no ideals other than (0) and F.   ∎

PROOF.   Assume that I is an ideal of F and that $I \neq (0)$; we shall prove that $I = F$. Let $a \in I$, $a \neq 0$. Then a has an inverse in F because F is a field. If e is the unity of F, then $e = a \cdot a^{-1} \in I$ because I is an ideal. But now if r is any element of F, then $r = e \cdot r \in I$, again because I is an ideal. This proves that $I = F$.   ∎

Problem 38.15 gives a partial converse to the statement proved in Example 38.4.

PROBLEMS

Which of the mappings in Problems 38.1–38.6 are ring homomorphisms? Determine the kernel of each mapping that is a homomorphism.

38.1. $\theta : \mathbb{Z} \to \mathbb{Z}$ by $\theta(a) = 3a$

38.2. $\theta : \mathbb{Z} \to \mathbb{Z}$ by $\theta(a) = a^2$

38.3. $\theta : \mathbb{Z}_6 \to \mathbb{Z}_3$ by $\theta([a]) = [a]$

38.4. $\theta : \mathbb{C} \to \mathbb{R}$ by $\theta(z) = |z|$

38.5. $\theta : \mathbb{C} \to \mathbb{C}$ by $\theta(z) = iz$

38.6. $\theta : M(2, \mathbb{Z}) \to M(2, \mathbb{Z})$ by $\theta\left(\begin{bmatrix} a & b \\ c & d \end{bmatrix}\right) = \begin{bmatrix} a & c \\ b & d \end{bmatrix}$.

38.7. Verify that if R and S are rings, and $\theta : R \to S$ is defined by $\theta(r) = 0_s$ for each $r \in R$, then θ is a homomorphism.

38.8. (a) Prove that the mapping $\pi_1 : R \times S \to R$ defined in Example 38.2 is a (ring) homomorphism.

(b) Prove that $\operatorname{Ker} \pi_1 \approx S$. (See Example 38.3.)

38.9. Prove that a homomorphic image of a commutative ring is commutative.

38.10. Prove or disprove that if a ring R has a unity, then every homomorphic image of R has a unity.

38.11. (a) Determine the smallest subgroup containing $\frac{1}{2}$ in the additive group of $\mathbb{Q}$.

(b) Determine the smallest subgroup containing $\frac{1}{2}$ in the multiplicative group of $\mathbb{Q}$.

(c) Determine the smallest subring of $\mathbb{Q}$ containing $\frac{1}{2}$.

(d) Determine the smallest ideal of $\mathbb{Q}$ containing $\frac{1}{2}$

(e) Determine the smallest subfield of $\mathbb{Q}$ containing $\frac{1}{2}$.

38.12. Every subring of $\mathbb{Z}$ is an ideal of $\mathbb{Z}$. Why?

38.13. Prove that the constant polynomials in $\mathbb{Z}[x]$ form a subring that is not an ideal.

38.14. Prove that if R is a commutative ring with unity, and $a \in R$, then (a) is the smallest ideal of R containing a. (Compare Theorem 15.2.)

38.15. Prove that if R is a commutative ring with unity, and R has no ideals other than (0) and R, then R is a field. (*Suggestion*: Use Problem 38.14 to show that each nonzero element has a multiplicative inverse.)

38.16. Determine all of the ideals of $\mathbb{Z}_{12}$.

38.17. Prove that if I is an ideal of $\mathbb{Z}$, then either $I = (0)$ or $I = (n)$, where n is the least positive integer in I. (See Problem 15.29.)

38.18. (An ideal that is not a principal ideal.)
  (a) Let I denote the set of all polynomials in $\mathbb{Z}[x]$ that have an even number as the constant term. Prove that I is an ideal of $\mathbb{Z}[x]$.
  (b) Prove that I is not a principal ideal of $\mathbb{Z}[x]$. (For the concept of principal ideal for noncommutative rings, see a book on general ring theory.)

38.19. Prove or disprove that if $\theta : R \to S$ is a ring homomorphism, then the image of θ is an ideal of S. (*Suggestion*: See Problem 38.13.)

38.20. Prove that if F is a field, R is a ring, and $\theta : F \to R$ is a ring homomorphism, then either θ is one-to-one or $\theta(a) = 0$ for all $a \in F$.

38.21. Prove that if R is a ring, $\alpha : \mathbb{Q} \to R$ and $\beta : \mathbb{Q} \to R$ are ring homomorphisms, and $\alpha(a) = \beta(a)$ for each $a \in \mathbb{Z}$, then $\alpha = \beta$ [that is, $\alpha(a) = \beta(a)$ for each $a \in \mathbb{Q}$].

38.22. The *center* of a ring R is the subring defined by $\{c \in R : cr = rc \text{ for every } r \in R\}$. (See Problems 25.22, 25.23, and 25.24.) Prove that the center of $M(2, \mathbb{Z})$ consists of all matrices of the form

$$\begin{bmatrix} a & 0 \\ 0 & a \end{bmatrix}, \quad a \in \mathbb{Z}.$$

Verify that this is not an ideal of $M(2, \mathbb{Z})$. (Therefore the center of a ring need not be an ideal.)

38.23. A subring I of a ring R is a *left ideal* of R if $ra \in I$ for all $r \in R$ and all $a \in I$. A *right ideal* is defined similarly.
  (a) Verify that the set of all matrices of the form

$$\begin{bmatrix} a & 0 \\ b & 0 \end{bmatrix}, \quad (a, b \in \mathbb{Z})$$

  is a left but not a right ideal of $M(2, \mathbb{Z})$.
  (b) Find a right ideal of $M(2, \mathbb{Z})$ that is not a left ideal.
  (c) Verify that if R is a ring and $a \in R$, then $\{r \in R : ra = 0\}$ is a left ideal of R. Determine this left ideal for $R = M(2, \mathbb{Z})$ and

$$a = \begin{bmatrix} 0 & 1 \\ 0 & 0 \end{bmatrix}.$$

  (d) Verity that if R is a ring and $a \in R$, then $\{r \in R : ar = 0\}$ is a right ideal of R. Determine this right ideal for R and a as in part (c).

38.24. An element a in a commutative ring is said to be *nilpotent* if $a^n = 0$ for some positive integer n (which may depend on a). Prove that the set of all nilpotent elements in a commutative ring R is an ideal of R.

SECTION 39 QUOTIENT RINGS

In this section we discuss quotient rings, the ring analogues of quotient groups, and use them to prove that every ideal of a ring is a kernel of some homomorphism. We also prove the ring version of the Fundamental Homomorphism Theorem.

If I is an ideal of a ring R, then I is a subgroup of the additive group of R, and it is even normal. (The additive group of R is Abelian, and every subgroup of an Abelian group is normal.) Therefore, we can talk about the quotient group R/I. Because the elements of R/I are the cosets of I formed relative to addition, these elements will be written in the form $I + r$ $(r \in R)$. The next theorem shows that R/I can be made into a ring in a very natural way. The construction is merely a generalization of that used for $\mathbb{Z}_n$ in Example 24.2: there $\mathbb{Z}_n$ can be thought of as $\mathbb{Z}/(n)$ for (n) the ideal consisting of all integral multiples on n.

Theorem 39.1. *Let I be an ideal of a ring R, and let R/I denote the set of all right cosets of I considered as a subgroup of the additive group of R. For $I + a \in R/I$ and $I + b \in R/I$, let*

$$(I + a) + (I + b) = I + (a + b)$$

and

$$(I + a)(I + b) = I + (ab)$$

With these operations R/I is a ring, called the quotient ring *of R by I.*

PROOF. Because R/I is a group with respect to addition (by Theorem 22.1 in additive notation) it suffices to verify the properties that involve multiplication. The first step in this is to verify that the multiplication on R/I is well-defined. Thus assume that $I + a_1 = I + a_2$ and $I + b_1 = I + b_2$; we must show that $I + a_1 b_1 = I + a_2 b_2$. From $I + a_1 = I + a_2$ we have $a_1 = n_1 + a_2$ for some $n_1 \in I$. Also, from $I + b_1 = I + b_2$ we have $b_1 = n_2 + b_2$ for some $n_2 \in I$. This implies that

$$a_1 b_1 = (n_1 + a_2)(n_2 + b_2) = n_1 n_2 + n_1 b_2 + a_2 n_2 + a_2 b_2.$$

But $n_1 n_2 + n_1 b_2 + a_2 n_2 \in I$ because $n_1, n_2 \in I$ and I is an ideal of R. Therefore $a_1 b_1$ has the form $a_1 b_1 = n_3 + a_2 b_2$ with $n_3 \in I$, so that $I + a_1 b_1 = I + a_2 b_2$, as required.

Here is a verification of one of the distributive laws. Assume that $a, b, c \in R$. Then

$$(I + a)[(I + b) + (I + c)] = (I + a)[I + (b + c)]$$
$$= I + a(b + c) = I + (ab + ac)$$
$$= (I + ab) + (I + ac)$$
$$= (I + a)(I + b) + (I + a)(I + c).$$

The verification of associativity of multiplication, as well as the other distributive law, will be left as an exercise (Problem 39.2). ∎

If R is commutative, and I is any ideal in R, then R/I will be commutative (Problem 39.3). If R has a unity e, then $I + e$ will be a unity for R/I (Problem 39.4). On the other hand, R/I will not necessarily be an integral domain when R is an integral domain. For instance, $\mathbb{Z}$ is an integral domain, but $\mathbb{Z}_6$ is not (Example 25.2). More will be said about the properties of particular quotient rings in the next section.

Theorem 39.2.   *If R is a ring and I is an ideal of R, then the mapping $\eta : R \rightarrow R/I$ defined by*

$$\eta(a) = I + a \quad \text{for each} \quad a \in R$$

is a homomorphism of R onto R/I, and $\text{Ker } \eta = I$.

PROOF.   This proof is similar to that of Theorem 22.2 (Problem 39.5).   ∎

As in the case of groups, the mapping $\eta : R \rightarrow R/I$ in Theorem 39.2 is called the *natural homomorphism* of R onto R/I. Notice that Theorem 39.2 shows that every ideal is a kernel. The next theorem shows that every homomorphic image of a ring R is isomorphic to a quotient ring of R.

Fundamental Homomorphism Theorem for Rings.   *Let R and S be rings, and let $\theta : R \rightarrow S$ be a homomorphism from R onto S with $\text{Ker } \theta = I$. Then the mapping $\phi : R/I \rightarrow S$ defined by*

$$\phi(I + a) = \theta(a) \quad \text{for each} \quad I + a \in R/I$$

is an isomorphism of R/I onto S. Therefore

$$R/I \approx S.$$

PROOF.   If the Fundamental Homomorphism Theorem for groups, and its proof (Section 23), are changed into additive notation, then all that remains to be shown is that ϕ preserves multiplication. To do this, let $I + a \in R/I$ and $I + b \in R/I$. Then

$$\phi((I + a)(I + b)) = \phi(I + ab) = \theta(ab) = \theta(a)\theta(b) = \phi(I + a)\phi(I + b),$$

as required.   ∎

PROBLEMS

39.1. Define $\theta : \mathbb{Z}_{12} \rightarrow \mathbb{Z}_4$ by $\theta([a]_{12}) = [a]_4$ for each $[a]_{12} \in \mathbb{Z}_{12}$.
   (a) Verify that θ is well-defined.
   (b) Verify that θ is a ring homomorphism.
   (c) Use the Fundamental Homomorphism Theorem for Rings to explain why $\mathbb{Z}_{12}/([4]) \approx \mathbb{Z}_4$.
   (d) Construct Cayley tables for the ring operations on $\mathbb{Z}_{12}/([4])$. (Compare Problem 22.3.)

39.2. (a) Prove that multiplication is associative in every quotient ring R/I.
   (b) Verify the distributive law

$$[(I + a) + (I + b)](I + c) = (I + a)(I + c) + (I + b)(I + c)$$

   for every quotient ring. (See the end of the proof of Theorem 39.1.)

39.3. Prove that if R is commutative and I is an ideal of R, then R/I is commutative.

39.4. Prove that if R has a unity e, then $I + e$ is a unity for R/I.

39.5. Prove Theorem 39.2.

39.6. Prove that if I is an ideal of ring R, then R/I is commutative iff $ab - ba \in I$ for all $a, b \in R$.

39.7. An ideal P of a commutative ring R is a *prime* ideal if $P \neq R$ and, for all $a, b \in R$, $ab \in P$ implies that $a \in P$ or $b \in P$. Prove that if n is a positive integer, then (n) is a prime ideal of $\mathbb{Z}$ iff n is a prime. (For the concept of prime ideal in noncommutative rings, see a book on general ring theory.)

39.8. Assume that R is a commutative ring with unity and $P \neq R$ is an ideal of R. Prove that P is a prime ideal iff R/P is an integral domain. (Prime ideals are defined in Problem 39.7.)

39.9. (First Isomorphism Theorem for Rings) If I and J are ideals of a ring R, then $I + J$ is defined to be $\{a + b : a \in I \text{ and } b \in J\}$. Verify that $I + J$ is an ideal of R, J is an ideal of $I + J$, $I \cap J$ is an ideal of I, and

$$\frac{I}{I \cap J} \approx \frac{I + J}{J}.$$

(*Suggestion*: See Problem 23.20.)

39.10. (Second Isomorphism Theorem for Rings) Prove that if I and J are ideals of a ring R, with $J \subseteq I$, then I/J is an ideal of R/J, and $(R/J)/(I/J) \approx R/I$. (*Suggestion*: See Problem 23.21.)

39.11. Prove that if I is a subring of a ring R, and the two operations in Theorem 39.1 are well-defined on the set R/I of all right cosets of I in R, then I is an ideal of R. (Compare Problem 22.15.)

39.12. An ideal M of a ring R is a *maximal ideal* if $M \neq R$ and there is no ideal I of R such that $M \subsetneqq I \subsetneqq R$. Prove that if R is a commutative ring with unity, and I is an ideal of R, then I is a maximal ideal iff R/I is a field.

SECTION 40   QUOTIENT RINGS OF $F[X]$

We know that a polynomial over a field F need not have a root in F. In Chapter VII it was shown how this problem could be overcome in special cases by appropriate field extensions: the real numbers produced a root for $x^2 - 2 \in \mathbb{Q}[x]$, and the complex numbers produced a root for $x^2 + 1 \in \mathbb{R}[x]$. We shall see that by using quotient rings the problem can be solved in general. Specifically, we shall see that if $f(x)$ is any nonconstant polynomial over any field F, then $f(x)$ has a root in some extension of F; and that extension can be chosen to be (isomorphic to) a quotient ring of $F[x]$. The proof of this is given in the next chapter, and makes use of the facts about quotient rings developed in this section.

Theorem 40.1.   *Assume that F is a field and that $p(x) \in F[x]$. Then $F[x]/(p(x))$ is a field iff $p(x)$ is irreducible over F.*

PROOF.   Let I denote the principal ideal $(p(x))$ throughout the proof. Suppose first that $p(x)$ is reducible over F, say $p(x) = a(x)b(x)$ with both $a(x)$ and $b(x)$ of degree less than that of $p(x)$. We shall prove that in this case $F[x]/I$ is not a field. The degree of any nonzero polynomial in I must be at least as great as $\deg p(x)$; thus $a(x) \notin I$ and $b(x) \notin I$. Therefore, $I + a(x)$ and $I + b(x)$ are both nonzero elements of $F[x]/I$. But

$$(I + a(x))(I + b(x)) = I + a(x)b(x) = I + p(x) = I,$$

the zero element of $F[x]/I$. We conclude that $F[x]/I$ has divisors of zero so that $F[x]/I$ is not a field (it is not even an integral domain). This proves that if $F[x]/I$ is a field, then $p(x)$ must be irreducible.

Suppose now that $p(x)$ is irreducible. Problem 40.1 asks you to show that $F[x]/I$ is commutative and that $I + e$ is a unity for $F[x]/I$ (where e is the unity of F). Thus it suffices to prove that each nonzero element of $F[x]/I$ has a multiplicative inverse in $F[x]/I$. Assume that $I + f(x)$ is nonzero. Then $f(x) \notin I$, which means that $f(x)$ is not a multiple of $p(x)$ in $F[x]$. Because $p(x)$ is irreducible, this implies that $p(x)$ and $f(x)$ have greatest common divisor e (Problem 40.2). Therefore, by Theorem 36.2, $e = p(x)u(x) + f(x)v(x)$ for some $u(x), v(x) \in F[x]$. It follows that $e - f(x)v(x) = p(x)u(x) \in I$, and hence $I + e = I + f(x)v(x) = (I + f(x))(I + v(x))$. This shows that $I + v(x)$ is a multiplicative inverse of $I + f(x)$. ∎

The following theorem is helpful when working with quotient rings $F[x]/(p(x))$, whether $p(x)$ is irreducible or not.

Theorem 40.2. *Assume that F is a field, $p(x) = a_0 + a_1 x + \cdots + a_n x^n$ is a polynomial of degree n over F, and I is the ideal $(p(x))$ of $F[x]$. Then each element of $F[x]/I$ can be expressed uniquely in the form*

$$I + (b_0 + b_1 x + \cdots + b_{n-1} x^{n-1}), \quad \text{with} \quad b_0, b_1, \ldots, b_{n-1} \in F. \quad (40.1)$$

Moreover, $\{I + b : b \in F\}$ is a subfield of $F[x]/I$ isomorphic to F.

PROOF. If $I + f(x) \in F[x]/I$, then by the Division Algorithm $f(x) = p(x)q(x) + r(x)$ for some $q(x), r(x) \in F[x]$ with $r(x) = 0$ or $\deg r(x) < \deg p(x)$. Since $f(x) - r(x) = p(x)q(x) \in I$, we have $I + f(x) = I + r(x)$; therefore, each element of $F[x]/I$ can be expressed in at least one way in the form (40.1). On the other hand, if

$$I + (b_0 + b_1 x + \cdots + b_{n-1} x^{n-1}) = I + (c_0 + c_1 x + \cdots + c_{n-1} x^{n-1}),$$

then

$$(b_0 - c_0) + (b_1 - c_1)x + \cdots + (b_{n-1} - c_{n-1})x^{n-1} \in I,$$

so that $p(x)$ divides

$$(b_0 - c_0) + (b_1 - c_1)x + \cdots + (b_{n-1} - c_{n-1})x^{n-1}.$$

This implies that

$$(b_0 - c_0) + (b_1 - c_1)x + \cdots + (b_{n-1} - c_{n-1})x^{n-1} = 0,$$

because $\deg p(x) = n > n - 1$. Therefore

$$b_0 = c_0, b_1 = c_1, \ldots, b_{n-1} = c_{n-1}.$$

This improves uniqueness.

It is now easy to verify that $b \mapsto I + b$ is a one-to-one ring homomorphism, and this will prove the last part of the theorem (Problem 40.3). ∎

Notice that the proof shows that if $f(x) \in F[x]$, and if $f(x) = p(x)q(x) + r(x)$ with $r(x) = 0$ or $\deg r(x) < \deg p(x)$, then $I + f(x) = I + r(x)$. This is important when computing in $F[x]/(p(x))$, as we shall see in the following example and again in Chapter X.

It allows us to represent elements of $F[x]/(p(x))$ by polynomials of degree less than $\deg p(x)$, just as the elements in $\mathbb{Z}_n$ can be represented by the integers $0, 1, \ldots, n - 1$.

Example 40.1.   We now show how the preceding ideas can be used to construct the field of complex numbers from the field of real numbers. This is a special case of what is to come in the next chapter. Let $p(x) = 1 + x^2$, which is irreducible over $\mathbb{R}$. Also let $I = (1 + x^2)$. By Theorem 40.1, $\mathbb{R}[x]/I$ is a field. In fact, $\mathbb{R}[x]/I \approx \mathbb{C}$, as we now verify.

Each element of $\mathbb{R}[x]/I$ can be written uniquely as $I + (a + bx)$ with $a, b \in \mathbb{R}$, by Theorem 40.2. Define $\theta : \mathbb{R}[x]/I \to \mathbb{C}$ by

$$\theta(I + (a + bx)) = a + bi.$$

That θ is one-to-one and onto, and preserves addition, is left as an exercise (Problem 40.4). To verify that θ preserves multiplication, we first write

$$\theta[I + (a + bx))(I + (c + dx))] = \theta[I + (ac + (ad + bc)x + bdx^2)]$$

To continue, we must determine $u, v \in \mathbb{R}$ such that

$$I + (ac + (ad + bc)x + bdx^2) = I + (u + vx).$$

This is done by using the remark following the proof of Theorem 40.2: divide $ac + (ad + bc)x + bdx^2$ by $1 + x^2$; the remainder will be $u + vx$. The result is

$$ac + (ad + bc)x + bdx^2 = (1 + x^2)bd + (ac - bd) + (ad + bc)x,$$

so that $u + vx = (ac - bd) + (ad + bc)x$. Therefore

$$\begin{aligned}
\theta[(I + (a + bx))(I + (c + dx))] &= \theta[I + (ac - bd) + (ad + bc)x)] \\
&= (ac - bd) + (ad + bc)i \\
&= (a + bi)(c + di) \\
&= \theta(I + (a + bx))\theta(I + (c + dx))
\end{aligned}$$

This proves that θ preserves multiplication. Thus $\mathbb{R}[x]/I \approx \mathbb{C}$, as claimed.   ∎

With the notation $a + bi$ $(a, b \in \mathbb{R})$ for the elements of $\mathbb{C}$, the number i is a root of $1 + x^2$. To verify the corresponding statement with $\mathbb{R}[x]/I$ in place of $\mathbb{C}$, we must check that $1 + z^2 = 0$ for some element z of $\mathbb{R}[x]/I$. Because of the isomorphism $b \mapsto I + b$ from Theorem 40.2, in this equation $(1 + z^2 = 0)$ 1 should be interpreted as $I + 1$, and 0 should be interpreted as $I + 0$. Because $\theta(I + x) = i$, one root must be $z = I + x \in \mathbb{R}[x]/I$. To check this directly, write

$$\begin{aligned}
(I + 1) + (I + x)^2 &= (I + 1) + (I + x^2) \\
&= I + (1 + x^2) \\
&= I + 0.
\end{aligned}$$

The next theorem shows that Theorem 40.2 covers all quotient rings of $F[x]$.

Theorem 40.3.   *If F is a field, then every ideal of the polynomial ring $F[x]$ is a principal ideal.*

PROOF.   Let I denote an ideal of $F[x]$. If $I = \{0\}$, then I is the principal ideal (0). Assume $I \neq \{0\}$, and let $g(x)$ denote any polynomial of least degree among the nonzero polynomials in I. We shall show that $I = (g(x))$; certainly $I \supseteq (g(x))$. Let $f(x) \in I$. By the Division Algorithm there are polynomials $q(x), r(x) \in F[x]$ such that

$$f(x) = g(x)q(x) + r(x), \quad \text{with} \quad r(x) = 0 \quad \text{or} \quad \deg r(x) < \deg g(x).$$

Then $f(x) \in I$ and $g(x)q(x) \in I$ and so $r(x) = f(x) - g(x)q(x) \in I$. Thus $r(x) = 0$, for otherwise $r(x) \in I$ and $\deg r(x) < \deg g(x)$, contradicting the choice of $g(x)$ as a polynomial of least degree in I. Therefore $f(x) = g(x)q(x) \in (g(x))$, so that $I = (g(x))$. ∎

Notice that the proof of Theorem 40.3 shows that $I = (g(x))$ for $g(x)$ any polynomial of least degree among the nonzero polynomials in I.

PROBLEMS

40.1. Prove that if F is a field with unity e, and I is an ideal of $F[x]$, then $F[x]/I$ is commutative with unity $I + e$.

40.2. Prove that if F is a field and $p(x), f(x) \in F[x]$, with $p(x)$ irreducible and $p(x) \nmid f(x)$, then $p(x)$ and $f(x)$ have greatest common divisor e, where e is the unity of F (Compare Problem 12.15.)

40.3. Verify that the mapping $b \mapsto I + b$ in the proof of Theorem 40.2 is a one-to-one ring homomorphism.

40.4. Prove that the mapping θ in Example 40.1 is one-to-one and onto, and preserves addition.

40.5. Prove that if F is a subfield of a field E, and $c \in E$, then $\theta : F[x] \to E$ defined by $\theta(f(x)) = f(c)$ is a homomorphism.

If θ is any homomorphism of the type in Problem 40.5, then $\mathrm{Ker}\, \theta = (b(x))$ for some $b(x) \in F[x]$, because $F[x]$ is a principal ideal domain. Determine such a polynomial $b(x)$ for each of the following examples. (Suggestion: Use the Factor Theorem.)

40.6. $\theta : \mathbb{Q}[x] \to \mathbb{Q}$ by $\theta(f(x)) = f(0)$ **40.7.** $\theta : \mathbb{Q}[x] \to \mathbb{Q}$ by $\theta(f(x)) = f(3)$

40.8. $\theta : \mathbb{Q}[x] \to \mathbb{R}$ by $\theta(f(x)) = f(\sqrt{2})$ **40.9.** $\theta : \mathbb{R}[x] \to \mathbb{C}$ by $\theta(f(x)) = f(i)$

40.10. $\theta : \mathbb{R}[x] \to \mathbb{C}$ by $\theta(f(x)) = f(-i)$

40.11. Suppose that F is a field and $f(x), g(x) \in F[x]$. Prove that $(f(x)) = (g(x))$ iff $f(x)$ and $g(x)$ are associates. (Associates are defined in Section 36.)

40.12. Prove that if F is a field and I is an ideal of $F[x]$, then there is a unique monic polynomial $m(x) \in F[x]$ such that $I = (m(x))$.

40.13. (a) Prove or disprove that if $(f(x)) = (g(x))$, then $\deg f(x) = \deg g(x)$.
      (b) Prove or disprove that if $\deg f(x) = \deg g(x)$, then $(f(x)) = (g(x))$.

40.14. True or false: If $f(x) \in (g(x))$ and $\deg f(x) = \deg g(x)$, then $(f(x)) = (g(x))$.

40.15. The proof of Theorem 40.3 uses the Least Integer Principle. Where?

40.16. Determine all of the prime ideals of $F[x]$, where F is a field. (See Problems 39.7 and 39.8.)

SECTION 41   FACTORIZATION AND IDEALS

The theme in Chapter VIII was factorization; the theme in this chapter has been the mutually equivalent ideas of ring homomorphism, ideal, and quotient ring. In this section we shall see that these apparently unrelated themes are, in fact, not unrelated. Specifically, Theorem 41.2 will show that information about the ideals in a ring can often tell us when there is unique factorization in the ring. Then, by looking at some penetrating discoveries from nineteenth-century number theory, we shall see that in many rings factorization can most appropriately be studied by considering "products of ideals" in the ring, rather than just products of elements of the ring. This section is designed to show the relation between some general ideas connecting number theory and rings—details and proofs will be omitted or left to the problems. You should be familiar with Sections 37 and 38, and Theorem 40.3. We begin with a key definition.

Definition.   An integral domain in which every ideal is a principal ideal is called a *principal ideal domain.*

Examples of principal ideal domains include the ring of integers (Problem 38.17) and the ring $F[x]$ of polynomials over any field F (Theorem 40.3). Recall that both $\mathbb{Z}$ and $F[x]$ are Euclidean domains; therefore, the fact that they are also principal ideal domains is a special case of the following theorem.

Theorem 41.1.   *Every Euclidean domain is a principal ideal domain.*

The proof of Theorem 41.1 is similar to the proof of Theorem 40.3, and will be left as an exercise (Problem 41.1). The converse of Theorem 41.1 is false; that is, not every principal ideal domain is a Euclidean domain. An example is given by the ring of all complex numbers of the form $a + b(1 + \sqrt{-19})/2$ for $a, b \in \mathbb{Z}$. (See [3] for a discussion of this.)

The following theorem gives a direct link between factorization and ideals.

Theorem 41.2.   *Every principal ideal domain is a unique factorization domain.*

A proof of Theorem 41.2 can be constructed by working through Problems 41.2 through 41.8; they provide a thorough test of your grasp of the ideas in the last few chapters. The converse of Theorem 41.2, like the converse of Theorem 41.1, is false, For example, $\mathbb{Z}[x]$ is a unique factorization domain, but it is not a principal ideal domain: the set of all polynomials that have an even number as constant term forms an ideal that is not principal (Problem 38.18).

Here is a summary of the relationships between several important classes of integral domains.

$$\begin{array}{ccccccc} \text{Euclidean} & & \text{principal ideal} & & \text{unique factorization} & & \text{integral} \\ \text{domains} & \subset & \text{domains} & \subset & \text{domains} & \subset & \text{domains} \end{array} \quad (41.1)$$

Each class is contained in, but different from, the class that follows it. Problem 41.9 asks you to place a number of specific examples in the smallest possible class of this sequence.

The extension of factorization theorems to rings other than the integers first arose in the last century, from two different sources. The first was in work on biquadratic reciprocity, and the second was in work on Fermat's Last Theorem. Here briefly, is what was involved.

If the congruence $x^2 \equiv a(\bmod m)$ has a solution, then a is said to be a *quadratic residue* of m; if there is no solution, then a is said to be a *quadratic nonresidue* of m. (For example, 1 and 4 are quadratic residues of 5; 2 and 3 are quadratic nonresidues of 5.) A. M. Legendre (1752–1833) introduced the following convenient notation for working with quadratic residues: if p is an odd prime and $p \nmid a$, then the symbol (a/p) is defined by

$$(a/p) = \begin{cases} 1 & \text{if } a \text{ is a quadratic residue of } p \\ -1 & \text{if } a \text{ is a quadratic nonresidue of } p. \end{cases}$$

(Thus $(1/5) = (4/5) = 1$ and $(2/5) = 3/5 = -1$.)

The most famous theorem about quadratic residues is the *law of quadratic reciprocity*: If p and q are distinct odd primes, then

$$(p/q)(q/p) = (-1)^{(1/4)(p-1)(q-1)}.$$

Another way to say this is as follows: If either $p \equiv 1 \ (\bmod 4)$ or $q \equiv 1(\bmod 4)$, then p is a quadratic residue of q iff q is a quadratic residue of p; if both $p \equiv 3(\bmod 4)$ and $q \equiv 3(\bmod 4)$, then p is a quadratic residue of q iff q is a quadratic nonresidue of p. The law of quadratic reciprocity was stated by Euler and Legendre, but Gauss gave the first complete proof. Gauss then searched for an equally comprehensive law for biquadratic residues: a is a *biquadratic residue* of m if $x^4 \equiv a(\bmod m)$ has a solution. He found such a law: the *law of biquadratic reciprocity*. The detailed statement of this law is not important here. What is important is that in order to solve this problem about biquadratic residues, Gauss stepped outside of $\mathbb{Z}$—the ring in which the problem was posed—and worked in the larger ring $\mathbb{Z}[i]$, now known as the *ring of Gaussian integers* (Example 37.2). In the process he proved that unique factorization holds in $\mathbb{Z}[i]$. Gauss also used similar ideas to develop a theory of cubic reciprocity (replacing x^4 by x^3).

Now we turn to the connection between factorization and Fermat's Last Theorem. There are infinitely many triples (x, y, z) of integers, called *Pythagorean triples*, such that $x^2 + y^2 = z^2$. (Only solutions with nonzero integers are of interest, of course. See Problem 41.11.) Many such triples were known to the Babylonians as early as 1600 B.C., and they were also studied by the Greek mathematician Diophantus in his book *Arithmetica* (c. A.D. 250). In 1637, in a marginal note in a copy of *Arithmetica*, Fermat wrote that there are no solutions in positive integers of $x^n + y^n = z^n$ if $n > 2$. This became known as Fermat's Last Theorem, even though Fermat gave no proof. In fact, it is very unlikely that Fermat had a proof.

By the 1840s, mathematicians who worked on Fermat's Last Theorem were concentrating on those cases where n is an odd prime, and several realized the usefulness of factoring the left side of $x^p + y^p = z^p$ as

$$x^p + y^p = (x + y)(x + uy) \cdots (x + u^{p-1}y),$$

where u is an imaginary pth root of unity. This revealed the problem as one of factorization: Are there nonzero integers x, y, and z such that

$$z^p = (x + y)(x + uy) \cdots (x + u^{p-1}y)?$$

Because u is imaginary, this took them outside of $\mathbb{Z}$. Specifically, they were faced with questions about factorization in the ring $\mathbb{Z}[u]$, where u is a solution of

$$u^{p-1} + u^{p-2} + \cdots + u + 1 = 0. \tag{41.2}$$

At least one erroneous proof of Fermat's Last Theorem rested on the mistaken assumption that unique factorization holds in $\mathbb{Z}[u]$. Although unique factorization holds in $\mathbb{Z}[u]$ for many values of p [with u a solution of (41.2)], it does not hold for all values. The smallest value for which unique factorization fails is $p = 23$. Although the ring $\mathbb{Z}\left[\sqrt{-5}\right]$ is not of the type being considered here, the failure of unique factorization in it shows what can happen (Example 37.1).

The German mathematician Ernst Kummer (1810–1893) was able to overcome the failure of unique factorization in part by inventing new "ideal numbers." By using these ideal numbers, along with the elements in the original ring $\mathbb{Z}[u]$, Kummer was able to prove the impossibility of $x^p + y^p = z^p$ for many values of p, thereby adding considerably to what was known about Fermat's Last Theorem. It was nearly 150 years later, in 1994, that Fermat's Last Theorem was finally proved in all cases by the English-born mathematician Andrew Wiles. The book [5] by Simon Singh listed at the end of this chapter gives a fascinating account of the history of Fermat's Last Theorem and of the work of Andrew Wiles.

From the modern point of view, the proper setting for questions about factorization in the rings considered by Gauss and Kummer is algebraic number theory. This theory was created by Richard Dedekind (1831–1916) in the 1870s. An *algebraic number* is any complex number that is a solution of an equation

$$a_n z^n + \cdots + a_1 z + a_0 = 0, \tag{41.3}$$

where the coefficients are in $\mathbb{Z}$. It can be shown that the set of all algebraic numbers forms a field. An *algebraic integer* is any solution of an equation of the form (41.3) where the coefficients are in $\mathbb{Z}$ and $a_n = 1$. The set of all algebraic integers forms an integral domain. This integral domain of algebraic integers contains many other integral domains, and some of these, such as $\mathbb{Z}[i]$, are unique factorization domains, while others, such as $\mathbb{Z}\left[\sqrt{-5}\right]$, are not.

One of the basic problems of algebraic number theory is to determine just which rings of algebraic integers are unique factorization domains. Theorem 41.2 gives a partial answer: if every ideal is principal, then factorization is unique. By replacing products of numbers by products of ideals, Dedekind was able to construct a theory of unique factorization for all rings of algebraic numbers. If I and J are ideals of a ring R, then the product IJ is defined to be the ideal generated by all products ab for $a \in I$, $b \in J$. An ideal $I \neq R$ is said to be *prime* if $ab \in I$ implies that $a \in I$ or $b \in I$. The fundamental theorem of ideal theory states that every nonzero ideal in a ring of algebraic integers can be factored uniquely as a product of prime ideals. In particular, if a denotes any algebraic integer, than the principal ideal (a) can be factored uniquely as a product of prime ideals. With these ideas it is possible to develop theorems for the set of ideals in a general ring of algebraic integers that are similar to the theorems for the set of elements in every unique factorization domain.

The preceding discussion shows part of what grew from the problems studied by Gauss and Kummer. These ideas form the basis for much of modern algebraic number theory, which has come to be seen as the appropriate place to study many of the more difficult questions about ordinary integers. These ideas also provided many of the problems that have shaped modern ring theory, including the abstract theory of ideals; this part of algebra owes a great deal to Emmy Noether (1882–1935) and her students.

PROBLEMS

41.1. Prove Theorem 41.1.

41.2. Assume that D is an integral domain and that $a, b \in D$.
  (a) Prove that $a \mid b$ iff $(b) \subseteq (a)$.
  (b) Prove that a and b are associates iff $(b) = (a)$.

41.3. If a and b are elements of a principal ideal domain D, not both zero, then a and b have a greatest common divisor in D. Prove this by justifying each of the following statements.
  (a) If I is defined by $I = \{ax + by : x, y \in D\}$, then I is an ideal of D.
  (b) $I = (d)$ for some $d \in D$.
  (c) An element d, chosen as in part (b), is a greatest common divisor of a and b.

41.4. Prove that if a and b are elements of a principal ideal domain D with unity e, and a and b have greatest common divisor e, then $e = ar + bs$ for some $r, s \in D$. (See Problem 41.3.)

41.5. Prove that if a, b, and p are nonzero elements of a principal ideal domain D, and p is irreducible and $p \mid ab$, then $p \mid a$ or $p \mid b$. (See Problem 41.4.)

41.6. Prove that if D is a principal ideal domain, $a_1, a_2, \dots, a_n, p \in D$, and p is irreducible and $p \mid a_1 a_2 \cdots a_n$, then $p \mid a_i$ for some i. (See Problem 41.5.)

41.7. If D is a principal ideal domain, $a \in D$, and a is not a unit of D, then a can be written as a product of irreducible elements of D. Give an indirect proof of this by justifying each of the following statements.
  (a) Assume that a cannot be written as a product of irreducible elements. Then a is not irreducible, and so $a = a_1 b_1$ where neither a_1 nor b_1 is a unit.
  (b) Either a_1 or b_1 cannot be written as a product of irreducible elements; assume that a_1 cannot be written as a product of irreducible elements.
  (c) Since $a_1 \mid a$, Problem 41.2 implies that $(a) \subseteq (a_1)$.
  (d) Since b_1 is not a unit, Problem 41.2 implies that $(a) \neq (a_1)$. Therefore $(a) \subset (a_1)$.
  (e) Repeat parts (a) through (d) with a_1 in place of a to deduce the existence of an element $a_2 \in D$ such that $(a) \subset (a_1) \subset (a_2)$.
  (f) This can be done repeatedly, yielding a sequence

$$(a) \subset (a_1) \subset (a_2) \subset \cdots$$

  of ideals of D. Let $I = \{x \in D : x \in (a_i) \text{ for some } a_i\}$ and prove that I is an ideal of D.
  (g) Because I is an ideal, $I = (c)$ for some $c \in I$. Therefore $c \in (a_k)$ for some k, and then $I = (c) \subseteq (a_k)$. This contradicts the fact that all of the inclusions $(a_i) \subset (a_{i+1})$ in part (f) are strict inequalities.
  (h) This completes this proof.

41.8. Use Problems 41.6 and 41.7 to prove Theorem 41.2.

41.9. Consider the four classes of rings in (41.1). Determine the smallest class to which each of the following rings belongs.
  (a) $\mathbb{Z}$ (b) $\mathbb{Z}_5$ (c) $\mathbb{Q}$ (d) $\mathbb{Z}_5[x]$
  (e) $\mathbb{Z}[i]$ (f) $\mathbb{Z}[\sqrt{-5}]$ (g) $\mathbb{Z}[x]$ (h) $\mathbb{Q}[x]$

41.10. (a) Which integers $k \in \{1, 2, \dots, 10\}$ are quadratic residues of 11?
  (b) Is 11 a quadratic residue of 7?
  (c) Verify the law of quadratic reciprocity for $p = 7$ and $q = 11$.

41.11. Prove that if m and n are positive integers with $m > n$, and $x = m^2 - n^2$, $y = 2mn$, $z = m^2 + n^2$, then (x, y, z) is a Pythagorean triple.

41.12. Prove that a nonzero ideal (a) of a principal ideal domain D is a maximal ideal iff a is an irreducible element of D.

NOTES ON CHAPTER IX

Reference [1] is an excellent source for the detailed history of many of the ideas in Section 41. Reference [4] is an elementary introduction to algebraic number theory.

1. Edwards, H. M., *Fermat's Last Theorem: A Generic Introduction to Algebraic Number Theory*, Springer-Verlag, New York/Berlin, 1977.
2. Ireland, K., and M. Rosen, *A Classical Introduction to Modern Number Theory*, Springer-Verlag, 2nd ed., New York/Berlin, 1991.
3. Motzkin, T., The Euclidean Algorithm, *Bulletin of the American Mathematical Society*, **55** (1949), 1142–1146.
4. Pollard, H., and Harold G. Diamond, *Theory of Algebraic Numbers*, 3rd ed., Dover, 1999.
5. Singh, S., *Fermat's Enigma*, Walker, New York, 1997.

CHAPTER X

FIELD EXTENSIONS

Fields are most conveniently analyzed by studying how they are built up from smaller subfields. Recall that a field E is an extension of a field F if E contains F, or, more generally, a subfield isomorphic to F. We know from Sections 21 and 24 that any field of prime characteristic p is an extension of $\mathbb{Z}_p$, and any field of characteristic 0 is an extension of $\mathbb{Q}$. Therefore, if we were to determine all extensions of $\mathbb{Z}_p$ and $\mathbb{Q}$, we would have determined all fields. Although that is asking too much, it is nonetheless true that a lot *is* known about field extensions. In the first section of this chapter we look at how to construct extensions that will solve a particular kind of problem, namely, that of providing roots for polynomials; the extension of $\mathbb{R}$ to $\mathbb{C}$ to obtain a root for $1 + x^2$ is a special case. We also introduce some basic definitions about field extensions that will be used in Chapters XI and XII. In the second section of the chapter, ideas from the first section are used to explain how to construct all finite fields. This chapter depends heavily on Section 40.

SECTION 42   ADJOINING ROOTS

Let E be an extension field of a field F; for convenience, assume $F \subset E$. Also let S be a subset of E. There is at least one subfield of E containing both F and S, namely, E itself. The intersection of all the subfields of E that contain both F and S is a subfield of E (Problem 42.1); it will be denoted $F(S)$. If $S \subseteq F$, then $F(S) = F$. If $S = \{a_1, a_2, \ldots, a_n\}$, then $F(S)$ will be denoted $F(a_1, a_2, \ldots, a_n)$. For example, $\mathbb{R}(i) = \mathbb{C}$. The field $F(S)$ consists of all the elements of E that can be obtained from F and S by repeated applications of the operations of E — addition, multiplication, and the taking of additive and multiplicative inverses (Problem 42.3).

If $E = F(a)$ for some $a \in F$, then E is said to be a *simple extension* of F. We can classify the simple extensions of F by making use of $F[x]$, the ring of polynomials in the indeterminate x over F, and

$$F[a] = \{a_0 + a_1 a + \cdots + a_n a^n : a_0, a_1, \ldots, a_n \in F\},$$

the ring of all polynomials in a. The difference between $F[x]$ and $F[a]$ is that two polynomials in $F[x]$ are equal only if the coefficients on like powers of x are equal, whereas in some cases two polynomials in $F[a]$ can be equal without the coefficients on like powers of a being equal. For example,

$$1 + 3\sqrt{2} = -1 + 3\sqrt{2} + \sqrt{2}^2 \quad \text{in} \quad \mathbb{Q}[\sqrt{2}],$$

but

$$1 + 3x \neq -1 + 3x + x^2 \quad \text{in} \quad \mathbb{Q}[x].$$

Theorem 42.1.   *If E is a simple extension of F, with $E = F(a)$, then*

$$F[a] \approx F[x]/(p(x)),$$

where $(p(x))$ is the ideal consisting of all $f(x) \in F[x]$ such that $f(a) = 0$. Either $p(x) = 0$ or $p(x)$ is irreducible over F.

PROOF.   Define $\theta\colon F[x] \to F[a]$ by

$$\theta(a_0 + a_1 x + \cdots + a_n x^n) = a_0 + a_1 a + \cdots + a_n a^n.$$

It can be verified that θ is a ring homomorphism (Problem 40.5). Therefore, by the Fundamental Homomorphism Theorem for Rings, $F[x]/I \approx F[a]$, where $I = \operatorname{Ker} \theta$ is an ideal of $F[x]$. Notice that I consists precisely of those polynomials having a as a root, because $f(x) \in I$ iff $\theta(f(x)) = f(a) = 0$. By Theorem 40.3 every ideal of $F[x]$ is a principal ideal, so $I = (p(x))$ for some $p(x) \in F[x]$. Because $F[a] \subseteq E$, and E has no zero divisors, $F[a]$ cannot have zero divisors. Therefore $F[x]/(p(x))$ cannot have zero divisors. It follows that either $p(x) = 0$ or $p(x)$ is irreducible over F. (See the proof of Theorem 40.1.)   ∎

If $p(x) = 0$ in Theorem 42.1, then $F(a)$ is said to be a *simple transcendental extension* of F, and a is said to be *transcendental* over F. If $p(x)$ is irreducible over F in Theorem 42.1, then $F(a)$ is said to be a *simple algebraic extension* of F, and a is said to be *algebraic* over F. The next two theorems say more about these two cases.

Theorem 42.2.   *If a is transcendental over F, then $F(a)$ is isomorphic to the field of quotients of $F[x]$.*

PROOF.   If a is transcendental over F, then $p(x) = 0$ in Theorem 42.1, and $F[x] \approx F[a]$. The field of quotients of $F[x]$ can be thought of as the set of all quotients $f(x)/g(x)$, $g(x) \neq 0$, where $f(x)$ and $g(x)$ have no common factor of positive degree (Problem 42.8). Also, $F(a)$ consists of all "quotients" $f(a)g(a)^{-1}$ with $f(a), g(a) \in F[a]$ and $g(a) \neq 0$. It follows that $F(a)$ is isomorphic to the field of quotients of $F[x]$.   ∎

Example 42.1.   It can be shown—but not easily—that the real number π is not a root of any polynomial with rational coefficients [2]. Therefore, $\mathbb{Q}(\pi)$ is a simple transcendental extension of $\mathbb{Q}$, and π is transcendental over $\mathbb{Q}$. The elements of $\mathbb{Q}(\pi)$ are the equivalence classes of rational expressions

$$\frac{a_0 + a_1 \pi + a_2 \pi^2 \cdots + a_n \pi^n}{b_0 + b_1 \pi + b_2 \pi^2 \cdots + b_m \pi^m}$$

for $a_0, a_1, \ldots, a_n, b_0, b_1, \ldots, b_m \in \mathbb{Q}$.   ∎

Theorem 42.3.   *Assume that F is a field, and that $p(x) \in F[x]$ is irreducible over F. Then $F[x]/(p(x))$ is a field extension of F, and $p(x)$ has a root in $F[x]/(p(x))$.*

PROOF. As before, let $I = (p(x))$. Theorem 40.1 shows that $F[x]/I$ is a field. In Theorem 40.2 we showed that by identifying each element $I + b \in F[x]/I$ with $b \in F$ we obtain a subfield of $F[x]/I$ that is isomorphic to F. Therefore $F[x]/I$ is an extension of F.

Assume that $p(x) = a_0 + a_1 x + \cdots + a_n x^n$, and let α denote the element $I + x \in F[x]/I$. Then

$$
\begin{aligned}
p(\alpha) &= a_0 + a_1(I + x) + \cdots + a_n(I + x)^n \\
&= I + (a_0 + a_1 x + \cdots + a_n x^n) \\
&= I + p(x) \\
&= I.
\end{aligned}
$$

But I is the zero of $F[x]/I$. Thus α is a root of $p(x)$ in $F[x]/I$.   ■

Corollary. *If F is a field, and $f(x)$ is a polynomial of positive degree over F, then $f(x)$ has a root in some extension of F.*

PROOF. If $f(x)$ is irreducible over F, then Theorem 42.3 applies directly. Otherwise $f(x)$ has some irreducible factor $p(x) \in F[x]$, and then $f(x) = p(x)q(x)$ for some $q(x) \in F[x]$. Apply Theorem 42.3 to $p(x)$. This gives a root α for $p(x)$ in $F[x]/(p(x))$, and this α is also a root for $f(x)$ because $f(\alpha) = p(\alpha)q(\alpha) = 0 \cdot q(\alpha) = 0$.   ■

In the next section it will be useful to have the following remarks based on Theorem 40.2 and the proof of Theorem 42.3. If $p(x)$, I, and α are as in Theorem 42.3 and its proof, then Theorem 40.2 shows that each element of $F[x]/I$ can be expressed uniquely in the form

$$
I + (b_0 + b_1 x + b_2 x^2 + \cdots + b_{n-1} x^{n-1}) \quad \text{with} \quad b_0, b_1, \ldots, b_{n-1} \in F.
$$

Because of the way the operations are defined in $F[x]/I$, and because $\alpha = I + x$, this means that each element can be expressed uniquely in the form

$$
b_0 + b_1 \alpha + b_2 \alpha^2 + \cdots + b_{n-1} \alpha^{n-1} \quad \text{with} \quad b_0, b_1, \ldots, b_{n-1} \in F \qquad (42.1)
$$

Making further use of the way the operations are defined in $F[x]/I$, we have the following corollary.

Corollary. *Assume F and $p(x)$ as in Theorem 42.3. Then $F[x]/(p(x))$ contains a root α of $p(x)$, and each element of $F[x]/(p(x))$ can be expressed uniquely in the form (42.1). Elements of the form (42.1) are added and subtracted using the usual addition and subtraction of polynomials. To multiply elements $f(\alpha)$ and $g(\alpha)$ of the form (42.1), multiply them as polynomials and divide the result by $p(\alpha)$; the remainder will equal $f(\alpha)g(\alpha)$.*

An important fact about Theorem 42.3 is that it shows how to construct a field having a root for $p(x)$ beginning with only F and $p(x)$; we need not have an extension before we begin. Example 40.1 illustrated this for the case of the polynomial $x^2 + 1$, irreducible over $\mathbb{R}$. Here is another example.

Example 42.2. The polynomial $x^2 - 2$ is irreducible over $\mathbb{Q}$. By Theorem 42.3, $\mathbb{Q}[x]/(x^2 - 2)$ is an extension of $\mathbb{Q}$, and it contains a root for $x^2 - 2$. The point being stressed here is that this requires no previous knowledge of $\mathbb{R}$, which we know to contain the root $\sqrt{2}$ of $x^2 - 2$.

The last corollary tells us that each element of $\mathbb{Q}[x]/(x^2 - 2)$ can be written uniquely in the form

$$a + b\alpha, \quad a, b \in \mathbb{Q}.$$

Here is the calculation of a typical product in $\mathbb{Q}[x]/(x^2 - 2)$.

$$(1 - 2\alpha)(2 + \alpha) = 2 - 3\alpha - 2\alpha^2 = -2 - 3\alpha, \tag{42.2}$$

because

$$\frac{2 - 3\alpha - 2\alpha^2}{\alpha^2 - 2} = -2 + \frac{-2 - 3\alpha}{\alpha^2 - 2}.$$

The mapping $\theta : \mathbb{Q}[x]/(x^2 - 2) \to \mathbb{Q}(\sqrt{2})$ defined by

$$\theta(a + b\alpha) = a + b\sqrt{2}$$

is an isomorphism. A direct proof of this is similar to that in Example 40.1, and is left as an exercise (Problem 42.9). The calculation in $\mathbb{Q}(\sqrt{2})$ corresponding to that in (42.2) would be

$$(1 - 2\sqrt{2})(2 + \sqrt{2}) = 2 - 3\sqrt{2} - 2(\sqrt{2})^2 = -2 - 3\sqrt{2}. \qquad \blacksquare$$

We close with two more important ideas relating to field extensions. If $f(x) \in F[x]$, then an extension field E of F such that $E = F(a_1, a_2, \ldots, a_n)$ and $f(x) = a(x - a_1)(x - a_2) \cdots (x - a_n) \in F[x]$ is called a *splitting* (or *root*) field of $f(x)$ over F. In other words, a splitting field of $f(x)$ is just large enough to contain F and the roots of $f(x)$. It can be proved that such a field does exist (Problem 42.10). Moreover, it is uniquely determined by F and $f(x)$, in the sense that if E_1 and E_2 are splitting fields of $f(x)$ over F, then $E_1 \approx E_2$. The splitting field of $x^2 - 2$ over $\mathbb{Q}$ is $\mathbb{Q}(\sqrt{2})$. If F is a field, and P is the intersection of all the subfields of F, then P is a subfield of F. If F has characteristic 0, then $P \approx \mathbb{Q}$. If F has prime characteristic p, then $P \approx \mathbb{Z}_p$. (See Problem 42.11.) In either case P is called the *prime subfield* of F.

PROBLEMS

Assume that F is a field throughout these problems.

42.1. Prove that if E is a field and T is a subset of E, then the intersection of all the subfields of E that contain T is a subfield of E. (See Problem 26.20.) [If F is a subfield of E and S is a subset of E, and $T = F \cup S$, then the subfield obtained here is $F(S)$.]

42.2. Prove that if $a + bi$ is imaginary, then $\mathbb{R}(a + bi) = \mathbb{C}$.

42.3. Prove the last statement in the first paragraph of this section.

42.4. Let $\alpha \in \mathbb{Q}[x]/(x^2 - 7)$ be a root of the irreducible polynomial $x^2 - 7 \in \mathbb{Q}[x]$. Express each of the following elements in the form $a + b\alpha$ with $a, b \in \mathbb{Q}$. (See Example 42.2.)
   (a) α^3 (b) $(1 - \alpha)(2 + \alpha)$ (c) $(1 + \alpha)^2$ (d) $(1 + \alpha)^{-1}$

42.5. Show that every element of $\mathbb{Q}(\sqrt{5})$ can be written in the form $a + b\sqrt{5}$, with $a, b \in \mathbb{Q}$.

42.6. Show that every element of $\mathbb{Q}(\sqrt[3]{5})$ can be written in the form $a + b\sqrt[3]{5} + c\sqrt[3]{25}$, with $a, b, c \in \mathbb{Q}$.

42.7. Prove that $\mathbb{Q}(\sqrt{2}) \neq \mathbb{Q}(\sqrt{3})$.

42.8. Prove that the set of all quotients $f(x)/g(x)$ with $f(x), g(x) \in F[x]$, and $g(x) \neq 0$, is a field with respect to the usual equality of fractions and the usual operations from elementary algebra. Also explain why this field is isomorphic to the field of quotients of $F[x]$.

42.9. Prove that the mapping θ in Example 42.2 is an isomorphism.

42.10. Prove that there is a splitting field for each $p(x) \in F[x]$. (*Suggestion:* Use the first corollary of Theorem 42.3.)

42.11. Prove that the prime subfield of F is isomorphic to $\mathbb{Q}$ if F has characteristic 0, and is isomorphic to $\mathbb{Z}_p$ if F has characteristic p.

SECTION 43   FINITE FIELDS

In Section 26 we proved that $\mathbb{Z}_n$ is a field iff n is a prime. This established the existence of a field of order p for each prime p, but it left open the question of whether there are fields of other finite orders. We settled this in a special case with Example 26.1, which gave Cayley tables defining operations for a field of order 4. It can be proved that there is a finite field of order q iff q is a power of a prime. We prove half of this in Theorem 43.1: the order of a finite field must be a power of a prime. We then construct fields of order 8 and 9; these constructions make use of theorems that we have proved about polynomials and field extensions. We close with some remarks about what is needed regarding polynomials and field extensions to prove the existence of a field of *each* prime power order. The proof of Theorem 43.1 makes use of the idea of the dimension of a vector space. This is reviewed in Appendix D; if it is not familiar, simply pass on to the examples that follow Theorem 43.1.

Assume that E is a field extension of F. We can think of E as a vector space over F in the following way: the vectors are the elements of E; the scalars are the elements of F; and the "product" of a scalar $a \in F$ and a vector $b \in E$ is simply the product ab in the field E. Problem 43.1 asks for a proof that this does yield a vector space. The dimension of this vector space is called the *degree* of E over F, and is denoted $[E : F]$. For example, $[\mathbb{C} : \mathbb{R}] = 2$, because $\{1, i\}$ is a basis for $\mathbb{C}$ over $\mathbb{R}$: each element of $\mathbb{C}$ can be written uniquely as $a \cdot 1 + b \cdot i$ with $a, b \in \mathbb{R}$.

Theorem 43.1.   *If F is a finite field, then the order of F is p^n, where p is the (prime) characteristic of F and n is a positive integer.*

PROOF.   The characteristic of F cannot be 0, so it must be a prime, which we denote by p. The prime subfield of F can be identified with $\mathbb{Z}_p$, and F will be a vector space over this prime subfield. If $[F : \mathbb{Z}_p] = n$, with $\{b_1, b_2, \ldots, b_n\}$ a basis for F over $\mathbb{Z}_p$, then each element of F can be written uniquely as a linear combination

$$a_1 b_1 + a_2 b_2 + \cdots + a_n b_n. \tag{43.1}$$

with $a_1, a_2, \ldots, a_n \in \mathbb{Z}_p$. The choices for $a_1, a_2, \ldots, a_n$ can be made independently, so that there are p^n total linear combinations (43.1). Thus the order of F is p^n.   ∎

To construct finite fields we make use of the ideas in Section 42. If $p(x) \in \mathbb{Z}_p[x]$, and $p(x)$ is irreducible and of degree n over $\mathbb{Z}_p$, then $\mathbb{Z}_p[x]/(p(x))$ will be a field by Theorem 40.1. Its order will be p^n, because each element of the field can be expressed uniquely in the form (42.1). Following are two examples. To simplify notation in these examples, *we write a in place of [a] for the elements of Z_n*; addition and multiplication of coefficients is performed modulo n, of course. Also, we write elements of $\mathbb{Z}_n[x]/(p(x))$ in the form $b_0 + b_1\alpha + \cdots + b_{n-1}\alpha^{n-1}$, as in (42.1).

Example 43.1.   The polynomial $1 + x + x^2 \in \mathbb{Z}_2[x]$ is irreducible over $\mathbb{Z}_2$ (Problem 43.2). Therefore $\mathbb{Z}_2[x]/(1 + x + x^2)$ is a field of order $2^2 = 4$. Tables 43.1 and 43.2 are the Cayley tables for the field operations. They are followed by several examples of calculations of entries in the tables.

Table 43.1

+	0	1	α	$1+\alpha$
0	0	1	α	$1+\alpha$
1	1	0	$1+\alpha$	α
α	α	$1+\alpha$	0	1
$1+\alpha$	$1+\alpha$	α	1	0

Table 43.2

$\cdot$	0	1	α	$1+\alpha$
0	0	0	0	0
1	0	1	α	$1+\alpha$
α	0	α	$1+\alpha$	1
$1+\alpha$	0	$1+\alpha$	1	α

Remember that computations with the coefficients are to be performed using the rules of $\mathbb{Z}_2$. Thus, for example, $\alpha + (1 + \alpha) = 1 + 2\alpha = 1 + 0\alpha = 1$. Also remember the second corollary of Theorem 43.3: if $f(\alpha) = (1 + \alpha + \alpha^2)q(\alpha) + r(\alpha)$, then $f(\alpha) = r(\alpha)$. Hence $\alpha(1 + \alpha) = \alpha + \alpha^2 = 1$ in the table, because

$$\alpha + \alpha^2 = (1 + \alpha + \alpha^2) \cdot 1 + 1 \quad \text{in} \quad \mathbb{Z}_2[\alpha].$$

Also $(1 + \alpha)(1 + \alpha) = 1 + 2\alpha + \alpha^2 = 1 + \alpha^2 = \alpha$, because

$$1 + \alpha^2 = (1 + \alpha + \alpha^2) \cdot 1 + \alpha \quad \text{in} \quad \mathbb{Z}_2[\alpha]. \qquad \blacksquare$$

Example 43.2.   The polynomial $1 + x^2$ is irreducible over $\mathbb{Z}_3$ (Problem 43.3). Therefore $\mathbb{Z}_3[x]/(1 + x^2)$ is a field of order $3^2 = 9$. Tables 43.3 and 43.4 are the Cayley tables for the operations. Here are several examples of the necessary calculations.

The elements are added as any polynomials, with the coefficients reduced by the rules in $\mathbb{Z}_3$. For example,

$$(1 + 2\alpha) + (2 + 2\alpha) = 3 + 4\alpha = 0 + \alpha = \alpha.$$

Table 43.3

+	0	1	2	α	2α	$1+\alpha$	$1+2\alpha$	$2+\alpha$	$2+2\alpha$
0	0	1	2	α	2α	$1+\alpha$	$1+2\alpha$	$2+\alpha$	$2+2\alpha$
1	1	2	0	$1+\alpha$	$1+2\alpha$	$2+\alpha$	$2+2\alpha$	α	2α
2	2	0	1	$2+\alpha$	$2+2\alpha$	α	2α	$1+\alpha$	$1+2\alpha$
α	α	$1+\alpha$	$2+\alpha$	2α	0	$1+2\alpha$	1	$2+2\alpha$	2
2α	2α	$1+2\alpha$	$2+2\alpha$	0	α	1	$1+\alpha$	2	$2+\alpha$
$1+\alpha$	$1+\alpha$	$2+\alpha$	α	$1+2\alpha$	1	$2+2\alpha$	2	2α	0
$1+2\alpha$	$1+2\alpha$	$2+2\alpha$	2α	1	$1+\alpha$	2	$2+\alpha$	0	α
$2+\alpha$	$2+\alpha$	α	$1+\alpha$	$2+2\alpha$	2	2α	0	$1+2\alpha$	1
$2+2\alpha$	$2+2\alpha$	2α	$1+2\alpha$	2	$2+\alpha$	0	α	1	$1+\alpha$

Table 43.4

$\cdot$	0	1	2	α	2α	$1+\alpha$	$1+2\alpha$	$2+\alpha$	$2+2\alpha$
0	0	0	0	0	0	0	0	0	0
1	0	1	2	α	2α	$1+\alpha$	$1+2\alpha$	$2+\alpha$	$2+2\alpha$
2	0	2	1	2α	α	$2+2\alpha$	$2+\alpha$	$1+2\alpha$	$1+\alpha$
α	0	α	2α	2	1	$2+\alpha$	$1+\alpha$	$2+2\alpha$	$1+2\alpha$
2α	0	2α	α	1	2	$1+2\alpha$	$2+2\alpha$	$1+\alpha$	$2+\alpha$
$1+\alpha$	0	$1+\alpha$	$2+2\alpha$	$2+\alpha$	$1+2\alpha$	2α	2	α	α
$1+2\alpha$	0	$1+2\alpha$	$2+\alpha$	$1+\alpha$	$2+2\alpha$	2	α	2α	1
$2+\alpha$	0	$2+\alpha$	$1+2\alpha$	$2+2\alpha$	$1+\alpha$	1	2α	α	2
$2+2\alpha$	0	$2+2\alpha$	$1+\alpha$	$1+2\alpha$	$2+\alpha$	α	1	2	2α

Simplifications of products are made as in the second corollary of Theorem 42.3. Thus

$$(1+2\alpha)(2+2\alpha) = 2 + 6\alpha + 4\alpha^2 = 2 + \alpha^2 \quad \text{in} \quad \mathbb{Z}_3[\alpha],$$

and

$$2 + \alpha^2 = \left(1 + \alpha^2\right) \cdot 1 + 1 \quad \text{in} \quad \mathbb{Z}_3[\alpha],$$

so that

$$2 + \alpha^2 = 1, \quad \text{and} \quad (1+2\alpha)(2+2\alpha) = 1$$

in the table.   ∎

Assume now that F is a field of order p^n. Because F has field characteristic p, $pa = 0$ for each $a \in F$, so each nonzero element of F has order p in the additive group of F. Our remarks about a finite field as a vector space over $\mathbb{Z}_p$ can be used to show that, *as additive groups, $F \approx \mathbb{Z}_p \times \mathbb{Z}_p \times \cdots \times \mathbb{Z}_p$* ($n$ copies of $\mathbb{Z}_p$; see Problem 43.4). The multiplicative group of F is of order $p^n - 1$; therefore, by a corollary of Lagrange's theorem, $a^{p^n - 1} = 1$ for each $a \in F$. (As in Example 43.1 and 43.2 we are using 1 for the unity of F.) It can be proved that the multiplicative group of F will have at least one element of order $p^n - 1$. This means that *the multiplicative group of a finite field is cyclic.* In Example 43.1, either α or $1 + \alpha$ is a generator for the multiplicative group of F. In Example 43.2, either $1 + \alpha$ or $1 + 2\alpha$ or $2 + \alpha$ or $2 + 2\alpha$ is a generator.

Because $a^{p^n-1} = 1$ for each $a \in F$, it follows that each $a \in F$ is a root of $x^{p^n} - x \in \mathbb{Z}_p[x]$. It can be shown that F is a splitting field for $x^{p^n} - x$ over $\mathbb{Z}_p$. By the remarks at the end of Section 42 we know that this splitting field is essentially unique. It has order p^n and is called the *Galois field* of order p^n; it is commonly denoted by $GF(p^n)$. By the existence and uniqueness of splitting fields we can say that *there is a field of each prime power order p^n, and all fields of order p^n are isomorphic.*

In practice, Galois fields are constructed from irreducible polynomials of degree n, as in our two examples. Such polynomials always exist, although there is no general rule that produces them.

PROBLEMS

43.1. Prove that if E is a field extension of F, then E is a vector space over F with respect to the operations described in the second paragraph of this section.

43.2. Verify that $1 + x + x^2 \in \mathbb{Z}_2[x]$ is irreducible over $\mathbb{Z}_2$. (*Suggestion*: Use the Factor Theorem, from Section 35.)

43.3. Verify that $1 + x^2 \in \mathbb{Z}_3[x]$ is irreducible over $\mathbb{Z}_3$. (*Suggestion*: Use the Factor Theorem, from Section 35.)

43.4. Assume that F is a finite field of order p^n and let $\{b_1, b_2, \ldots, b_n\}$ be a basis for F as a vector space over $\mathbb{Z}_p$. If $a \in F$, then a can be written uniquely in the form of Equation (43.1); define $\theta : F \to \mathbb{Z}_p \times \mathbb{Z}_p \times \cdots \times \mathbb{Z}_p$ by

$$\theta(a) = (a_1, a_2, \ldots, a_n).$$

Prove that θ is an isomorphism of additive groups.

43.5. The first step in the proof of Theorem 43.1 states that the characteristic of a finite field cannot be 0. Give a reason.

43.6. (a) Verify that $1 + x + x^3 \in \mathbb{Z}_2[x]$ is irreducible over $\mathbb{Z}_2$.
(b) Construct addition and multiplication tables for the field $\mathbb{Z}_2[x]/(1 + x + x^3)$.

43.7. (a) Verify that $1 + x + x^4 \in \mathbb{Z}_2[x]$ is irreducible over $\mathbb{Z}_2$.
(b) Find a generator for the (cyclic) multiplicative group of the field $\mathbb{Z}_2[x]/(1 + x + x^4)$.

43.8. Construct addition and multiplication tables for the ring $\mathbb{Z}_2[x]/(x^2)$. Give two proofs that the ring is not a field, one based on Theorem 40.1 and another based directly on the tables that you construct.

NOTES ON CHAPTER X

General references for the topics in this chapter are the same as those listed in the notes at the end of Chapter XI. The second book below contains proofs that π and e are transcendental.

1. Lidl, R., and Harald Niederreiter, *Finite Fields*, Encyclopedia of Mathematics and its Applications, Vol. **20**, 2nd ed., Cambridge University Press, England, 1997.
2. Niven, I., *Irrational Numbers* (Carus Monograph No. **11**, The Mathematical Association of America, 1961.

CHAPTER XI

POLYNOMIAL EQUATIONS

The desire to answer questions about roots of polynomials was the most important factor in the early history of modern algebra. This chapter contains some of the basic facts and history concerning these questions.

SECTION 44 ROOTS OF A POLYNOMIAL

Recall that an element c of a field F is a root of a polynomial $f(x) \in F[x]$ if $f(c) = 0$. By the Factor Theorem (Section 35), $f(c) = 0$ iff $x - c$ is a factor of $f(x)$. If $(x - c)^m$ divides $f(x)$, but no higher power of $x - c$ divides $f(x)$, then c is called a root of *multiplicity m*. When we count the number of roots of a polynomial, each root of multiplicity m is counted m times. For example, $x^3 - x^2 - x + 1 = (x - 1)^2(x + 1)$ has 1 as a root of multiplicity two, and -1 as a root of multiplicity one; it has no other root. Thus we say that this polynomial has three roots.

Theorem 44.1. *A polynomial $f(x)$ of degree $n \geq 1$ over a field F has at most n roots in F.*

PROOF. The proof will be by induction on n. If $n = 1$, then $f(x) = a_0 + a_1 x$ with $a_1 \neq 0$, and the only root is $-a_1^{-1} a_0$. Thus assume that $n > 1$, and assume the theorem true for polynomials of degree less than n. If $f(x)$ has no root, we are through. If c is a root, then by the Factor Theorem $f(x) = (x - c) f_1(x)$ for some $f_1(x) \in F[x]$, and $\deg f_1(x) = n - 1$. By the induction hypothesis $f_1(x)$ has at most $n - 1$ roots in F. It will follow that $f(x)$ has at most n roots in F if $f(x)$ has no roots in F except c and the roots of $f_1(x)$. But this is so because if $a \in F$, then $f(a) = (a - c) f_1(a)$, so that $f(a) = 0$ only if $a - c = 0$ or $f_1(a) = 0$ (because F has no divisors of zero). Thus $f(a) = 0$ only if $a = c$ or a is a root of $f_1(x)$. ∎

We have seen that a polynomial over a field may have no roots in that field. For example, $x^2 - 2$ and $x^2 + 1$ have no roots in the field of rationals. However, the Fundamental Theorem of Algebra (Section 32) ensures that each polynomial over the complex numbers has at least one complex root. In fact, we can prove more, as in the following corollary.

Corollary. Each polynomial of degree $n \geq 1$ over the field $\mathbb{C}$ of complex numbers has n roots in $\mathbb{C}$.

PROOF. We shall use induction on n. Assume that $f(x) = a_0 + a_1 x + \cdots + a_n x^n \in \mathbb{C}[x]$. If $n = 1$, then $f(x)$ has one root, namely, $-a_1^{-1} a_0$. Assume the corollary to be true for all polynomials of degree less than n. If $n > 1$, then $f(x)$ has a root, say c, by the Fundamental Theorem of Algebra. Thus, by the Factor Theorem, $f(x) = (x - c)f_1(x)$ for some $f_1(x) \in \mathbb{C}[x]$, and deg $f_1(x) = n - 1$. By the induction hypothesis, $f_1(x)$ has $n - 1$ roots in $\mathbb{C}$. Each of these is clearly a root of $f(x)$, as is c, and hence $f(x)$ has *at least* n roots in $\mathbb{C}$.

By Theorem 44.1, $f(x)$ has at most n roots in $\mathbb{C}$. Thus $f(x)$ has exactly n roots in $\mathbb{C}$. ∎

PROBLEMS

44.1. Construct a polynomial over $\mathbb{R}$ having 2 as a root of multiplicity one and 3 as a root of multiplicity two.

44.2. Construct a polynomial over $\mathbb{C}$ having i as a root of multiplicity two and $-i$ as a root of multiplicity one.

44.3. There are eight polynomials of degree three over $\mathbb{Z}_2$. For each one, find the number of roots over $\mathbb{Z}_2$.

44.4. There are nine monic polynomials of degree two over $\mathbb{Z}_3$. For each one, find the number of roots over $\mathbb{Z}_3$.

44.5. Construct a polynomial over $\mathbb{Z}_5$ having 3 (that is, [3]) as a root of multiplicity two and 1 as a root of multiplicity one.

44.6. Prove that $x^2 - 1 \in \mathbb{Z}_{12}[x]$ has four roots in $\mathbb{Z}_{12}$. Does this contradict Theorem 44.1? Why?

44.7. Prove that if Q denotes the division ring of quarternions introduced in Problem 32.19, then $x^2 + 1 \in Q[x]$ has infinitely many roots in Q. (Here 1 denotes the unity of Q, that is, the 2×2 identity matrix.)

44.8. Assume that $a_0, a_1, \ldots, a_n$ are distinct elements of a field F, that $b_0, b_1, \ldots, b_n \in F$, and that $f(x), g(x) \in F[x]$ are of degree n or less and satisfy

$$f(a_j) = b_j = g(a_j) \quad \text{for } 0 \le j \le n.$$

Prove that $f(x) = g(x)$ in $F[x]$. (Compare Problem 35.19.)

44.9. Assume that F is a field, $f(x) \in F[x]$, and E is an extension of F. Prove that an element c of E is a multiple root of $f(x)$ iff $f(c) = f'(c) = 0$. *Multiple root* means root of multiplicity greater than one. The polynomial $f'(x)$ is the formal derivative of $f(x)$, defined in Problem 34.13. [*Suggestion:* Use the Division Algorithm to write $f(x) = (x - c)^2 q(x) + r(x)$, where $r(x)$ is linear. Verify that $f(c) = r(c)$ and $f'(c) = r'(c)$. Next, explain why $f(x) = (x - c)^2 q(x) + (x - c)f'(c) + f(c)$. Finally, use the latter equation to examine when $(x - c)^2 \mid f(x)$.]

SECTION 45   RATIONAL ROOTS. CONJUGATE ROOTS

The first theorem in this section gives information about the *rational roots* of a polynomial with *integral coefficients*; the second gives information about the *complex roots* of a polynomial with *real coefficients*.

Theorem 45.1.   *Let $f(x) = a_0 + a_1x + \cdots + a_nx^n$ be a polynomial with integral coefficients. If r/s is a rational root of $f(x)$, and $(r, s) = 1$, then $r \mid a_0$ and $s \mid a_n$.*

Example 45.1.   The rational roots of $f(x) = 2 - 3x - 8x^2 + 12x^3$ will be represented by fractions with numerators chosen from the divisors of 2, and denominators chosen from the divisors of 12. The list of all possibilities is ± 1, ± 2, $\pm \frac{1}{2}$, $\pm \frac{1}{3}$, $\pm \frac{2}{3}$, $\pm \frac{1}{4}$, $\pm \frac{1}{6}$, $\pm \frac{1}{12}$. It can be verified that $\frac{1}{2}$, $-\frac{1}{2}$, and $\frac{2}{3}$ are roots of $f(x)$. Theorem 45.1 gives the same possibilities for the rational roots of $g(x) = 12x^3 - 3x + 2$ as for the rational roots of $f(x)$. But $g(x)$ has no rational root (Problem 45.4).   ∎

PROOF OF THEOREM 45.1.   If r/s is a root of $f(x)$, then $a_0 + a_1(r/s) + \cdots + a_n(r/s)^n = 0$. Therefore $a_0s^n + a_1rs^{n-1} + \cdots + a_nr^n = 0$. Solving this first for a_0s^n, and then for a_nr^n, we get

$$a_0s^n = -[a_1rs^{n-1} + \cdots + a_nr^n] \tag{45.1}$$

and

$$a_nr^n = -[a_0s^n + \cdots + a_{n-1}r^{n-1}s]. \tag{45.2}$$

Because r is a divisor of the right side of (45.1), $r \mid a_0s^n$. But $(r, s) = 1$, so $(r, s^n) = 1$, and therefore $r \mid a_0$ by Lemma 13.1. Similarly, because s is a divisor of the right side of (45.2), $s \mid a_nr^n$. This implies that $s \mid a_n$ because $(s, r^n) = 1$, again by Lemma 13.1.   ∎

Corollary.   *A rational root of a monic polynomial $a_0 + a_1x + \cdots + x^n$ with integral coefficients must be an integer and a divisor of a_0.*

PROOF.   Apply Theorem 45.1 with $a_n = 1$.   ∎

Before stating the next theorem, we first recall from elementary algebra that each quadratic equation $ax^2 + bx + c = 0$ has two solutions, given by

$$x = \frac{-b \pm \sqrt{b^2 - 4ac}}{2a}. \tag{45.3}$$

The number $b^2 - 4ac$ is called the *discriminant* of $ax^2 + bx + c$. Assume that a, b and c are real. Then the two solutions in (45.3) will be real and equal if $b^2 - 4ac = 0$, real and unequal if $b^2 - 4ac > 0$, and imaginary and unequal if $b^2 - 4ac < 0$. In the latter case the two solutions will be complex conjugates of each other (that is, of the form $u + vi$ and $u - vi$, where $u, v \in \mathbb{R}$). Notice that the conjugate of a real number u is u itself. Thus we can say in any case that if a complex number $u + vi$ is a solution of a quadratic equation with real coefficients, then its conjugate $u - vi$ is also a solution. This is a special case of the next theorem.

Theorem 45.2.   *If $a + bi$ is a root of a polynomial $f(x)$ with real coefficients, then its complex conjugate $a - bi$ is also a root of $f(x)$.*

PROOF.   We assume that $b \neq 0$; otherwise the theorem is trivially true. In light of the Factor Theorem, we know that $x - (a + bi)$ is a factor of $f(x)$, and it suffices to establish that $x - (a - bi)$ is also a factor of $f(x)$. We shall do the latter by showing that

$$[x - (a + bi)][x - (a - bi)] = [(x - a) - bi][(x - a) + bi] = (x - a)^2 + b^2$$

is a factor of $f(x)$.

Since $(x - a)^2 + b^2 = x^2 - 2ax + (a^2 + b^2)$ is a polynomial with real coefficients, we can divide it into $f(x)$, applying the Division Algorithm for polynomials over $\mathbb{R}$. This yields unique $q(x), r(x) \in \mathbb{R}[x]$ such that

$$f(x) = [(x - a)^2 + b^2]q(x) + r(x), \quad \text{with} \quad r(x) = 0 \quad \text{or} \quad \deg r(x) < 2.$$

If $r(x) \neq 0$, then $r(x)$ must have the form $cx + d$. Then for $x = a + bi$ we have

$$0 = f(a + bi) = r(a + bi) = c(a + bi) + d = (ca + d) + cbi. \qquad (45.4)$$

(Here we have used the fact that $(x - a)^2 + b^2 = 0$ for $x = a + bi$ because $(x - a)^2 + b^2 = [x - (a + bi)][x - (a - bi)]$.) But (45.4), in turn, implies that $ca + d = 0$ and $cb = 0$. Because $b \neq 0$, we conclude that $c = 0$; and then $d = 0$. Therefore $r(x) = 0$, and $[x - (a + bi)][x - (a - bi)]$ is a factor of $f(x)$. ∎

Corollary.   *A polynomial over the field $\mathbb{R}$ of real numbers if irreducible over $\mathbb{R}$ iff it is linear (first degree), or quadratic (second degree) with a negative discriminant.*

PROOF.   Any linear polynomial is irreducible, and any quadratic polynomial over $\mathbb{R}$ with a negative discriminant is irreducible over $\mathbb{R}$ because its roots are imaginary. It now suffices to prove the converse.

Assume that $f(x) \in \mathbb{R}[x]$ and that $f(x)$ is irreducible and nonlinear; we shall prove that $f(x)$ must be quadratic with a negative discriminant.

We can think of $f(x)$ as a polynomial over $\mathbb{C}$; as such, it has a root, say $a + bi$, in $\mathbb{C}$. If $b = 0$, then $f(a) = 0$ so that $x - a$ is a factor of $f(x)$, contradicting that $f(x)$ is nonlinear and irreducible over $\mathbb{R}$. Thus $b \neq 0$. The proof of the preceding theorem shows that $(x - a)^2 + b^2$, which is in $\mathbb{R}[x]$, must be a factor of $f(x)$. Since $f(x)$ is irreducible, it follows that $f(x)$ must be an element of $\mathbb{R}$ times $(x - a)^2 + b^2$, and thus, in particular, $f(x)$ must be quadratic. The discriminant of $f(x)$ must be negative in this case because its roots are imaginary ($b \neq 0$). ∎

PROBLEMS

45.1. Prove that if $f(x) \in \mathbb{R}[x]$ has an imaginary root of multiplicity two, then $\deg f(x) \geq 4$.

45.2. Give an example of a quadratic polynomial over $\mathbb{C}$ that has an imaginary root of multiplicity two. (Compare Problem 45.1.)

45.3. Give an example of a polynomial over $\mathbb{Z}_2$ that is irreducible and of degree 3. (Compare the corollary of Theorem 45.2.)

45.4. Prove that $12x^3 - 3x + 2$ has no rational root.

45.5. Find all rational roots of each of the following polynomials over $\mathbb{Q}$.
   (a) $4x^3 - 7x - 3$ (b) $2 - 11x + 17x^2 - 6x^3$ (c) $2x^3 - x^2 + 8x - 4$

45.6. Write each of the following polynomials over $\mathbb{Q}$ as a product of factors that are irreducible over $\mathbb{Q}$.

(a) $x^3 - x^2 - 5x + 5$ (b) $3x^3 - 2x^2 + 3x - 2$ (c) $x^3 - 2x^2 + 2x$

45.7. (a) to (c). Repeat Problem 45.6 using factors that are irreducible over $\mathbb{R}$.

45.8. (a) to (c) Repeat Problem 45.6 using factors that are irreducible over $\mathbb{C}$.

SECTION 46   INTRODUCTION TO GALOIS THEORY

The challenge of solving polynomial equations

$$a_n x^n + \cdots + a_1 x + a_0 = 0 \tag{46.1}$$

has been one of the most important in the history of algebra. The next three paragraphs give an overview of this subject. Some of the details are then considered in this and the next section.

Methods for solving first- and second-degree equations go back to the early Egyptians and Babylonians. In the sixteenth century, Italian mathematicians (del Ferro, Tartaglia, and Ferrari) succeeded in solving cubic (third-degree) and quartic (fourth-degree) equations. (For cubic equations, see Problem 46.17.) Their solutions give formulas for writing the roots in terms of the coefficients $a_0, a_1, \ldots, a_n$, much as the quadratic formula (45.3) for the solutions of second-degree equations. It then became a challenge to do the same for equations of degree higher than four. Specifically, the problem, as it eventually came to be interpreted, was to show that it is possible to express the roots of an equation (46.1) in terms of the coefficients $a_0, a_1, \ldots, a_n$ using only addition, subtraction, multiplication, division, and the extraction of roots, each applied only finitely many times. When the roots can be so expressed, it is said that the equation is *solvable by radicals*.

In 1770–1771 Lagrange took the first steps in settling this problem by introducing methods for studying polynomial equations that in reality involved group theory. Lagrange sensed that the key to understanding these equations and their roots was related to the effect, on the original equations and on certain related equations, of permutations of the roots of the equation.

Early in the nineteenth century Paulo Ruffini (Italian) and N. H. Abel (Norwegian), drawing on the ideas introduced by Lagrange, showed that, in fact, there are equations of each degree higher than four that are not solvable by radicals. The complete answer to the question of which equations are solvable by radicals was finally given by Évariste Galois, another French mathematician, in the mid-nineteenth century. Galois was able to show that a group can be associated with each polynomial equation (46.1) in such a way that the property of solvability by radicals is directly related to a corresponding property of the group. This group property is called *solvability:* A group G is *solvable* if it contains a finite sequence of subgroups

$$G = G_0 \rhd G_1 \rhd G_2 \rhd \cdots \rhd G_n = \{e\},$$

with each subgroup normal in the preceding, as indicated, such that each of the successive quotient groups G_{i-1}/G_i is Abelian. What Galois proved is that an equation is solvable by radicals iff its associated group (now called its *Galois group*) is solvable. The point of the earlier work by Ruffini and Abel was that for each degree higher than four there exist polynomial equations whose Galois groups are not solvable.

The work by Galois became the basis for what is now known as *Galois theory*. Today this theory has to do with the study of groups of automorphisms of field extensions; theorems concerning solvability by radicals are merely a special part of the general theory.

The remainder of this chapter will give a sketch of the parts of Galois theory relating to solvability by radicals. For simplicity, *all fields are assumed to be subfields of the field of complex numbers.*

We begin by recalling some facts from Section 42. If F is a field, and $a \in \mathbb{C}$, then $F(a)$ denotes the smallest subfield of $\mathbb{C}$ that contains both F and a. Also, if F is a field and S is a subset of $\mathbb{C}$, then $F(S)$ denotes the smallest subfield of $\mathbb{C}$ that contains both F and S. If $f(x) \in F[x]$, then an extension field E of F such that $E = F(a_1, \ldots, a_n)$ and $f(x) = a(x - a_1) \cdots (x - a_n) \in E[x]$ is called a *splitting* (or *root*) *field* of $f(x)$ over F. In other words, a splitting field of $f(x)$ is just large enough to contain F and the roots of $f(x)$. It can be proved that such a field exists, and that it is uniquely determined by F and $f(x)$.

The first key to what follows is that of a field *automorphism*, which is an isomorphism of the field onto itself.

Example 46.1. If $\sigma : \mathbb{C} \to \mathbb{C}$ is defined by $\sigma(a + bi) = a - bi$ for all $a, b \in \mathbb{R}$, then σ is an automorphism of $\mathbb{C}$ (Problem 46.1). ∎

Example 46.2. If E is an extension of $\mathbb{Q}$, and σ is an automorphism of E, then $\sigma(a) = a$ for all $a \in \mathbb{Q}$. Here is an outline of the proof.

PROOF. First, $\sigma(1) = 1$ (Problem 27.1). Therefore, $\sigma(2) = \sigma(1 + 1) = \sigma(1) + \sigma(1) = 2$. We can extend this by mathematical induction to prove that $\sigma(n) = n$ for every positive integer n (Problem 46.7). But then for $m \in \mathbb{Z}$ and $m < 0$, we have $-m > 0$ and $\sigma(-m) = -m$, so $\sigma(m) = \sigma(-(-m)) = -\sigma(-m) = -(-m) = m$. Thus $\sigma(r) = r$ for all $r \in \mathbb{Z}$.

Also, if $s \in \mathbb{Z}$ and $s \neq 0$, then $\sigma(1/s) = 1/\sigma(s)$ by Problem 27.9. Thus if $r, s \in \mathbb{Z}$ with $s \neq 0$, then $\sigma(r/s) = \sigma(r)\sigma(1/s) = r/s$. This completes the proof. ∎

Example 46.3. Every element of $\mathbb{Q}(\sqrt{2})$ can be written uniquely in the form $a + b\sqrt{2}$, with $a, b \in \mathbb{Q}$ (Section 42). If α is an automorphism of $\mathbb{Q}(\sqrt{2})$, then $\alpha(a) = a$ and $\alpha(b) = b$ by Example 46.2, so

$$\alpha(a + b\sqrt{2}) = \alpha(a) + \alpha(b)\alpha(\sqrt{2}) = a + b\alpha(\sqrt{2}).$$

However, $[\alpha(\sqrt{2})]^2 = \alpha(\sqrt{2}^2) = \alpha(2) = 2$, so $\alpha(\sqrt{2})$ satisfies $x^2 = 2$ and $\alpha(\sqrt{2}) = \pm\sqrt{2}$. This means there are only two automorphisms of $\mathbb{Q}(\sqrt{2})$, the identity ι and the mapping α defined by $\alpha(a + b\sqrt{2}) = a - b\sqrt{2}$. ∎

The set $\{\iota, \alpha\}$ of all automorphisms of $\mathbb{Q}(\sqrt{2})$ in Example 46.3 forms a group with respect to composition. More generally, if E is any field, then the set of all automorphisms of E forms a group with respect to composition (Problem 46.2).

If E is an extension of a field F, then $\mathrm{Gal}(E/F)$ will denote the set of all automorphisms σ of E such that $\sigma(a) = a$ for all $a \in F$. Problem 46.8 asks you to prove that $\mathrm{Gal}(E/F)$ is a subgroup of $\mathrm{Aut}(E)$. This group is called the *Galois group* of E over F.

If H is any subgroup of $\mathrm{Gal}(E/F)$, then E_H will denote the set of all $x \in E$ such that $\sigma(x) = x$ for all $\sigma \in H$. Problem 46.9 asks you to prove that E_H is a subfield of E. This field is called the *fixed field* of H.

Note the similarities: every element of the group $\mathrm{Gal}(E/F)$ fixes every element of F, while every element of the field E_H is fixed by every element of H. The following theorem summarizes the last few paragraphs.

Theorem 46.1. *If E is a field, then the set of all automorphisms of E forms a group $\mathrm{Aut}(E)$ with respect to composition. If F is a subfield of E, then*

$$\mathrm{Gal}(E/F) = \{\sigma \in \mathrm{Aut}(E) : \sigma(x) = x \quad \text{for all } x \in F\}$$

is a subgroup of $\mathrm{Aut}(E)$, called the Galois group of E over F; and if H is a subgroup of $\mathrm{Gal}(E/F)$, then

$$E_H = \{x \in E : \sigma(x) = x \quad \text{for all} \quad \sigma \in \mathrm{Gal}(E/F)\}$$

is a subfield of E, called the fixed field of H.

The key to Galois theory is a natural connection between the subgroups of $\mathrm{Gal}(E/F)$ and the subfields of E that contain F, especially in the case where E is the splitting field of a polynomial $f(x)$ over F. Before describing this connection, in the Fundamental Theorem of Galois Theory, we recall from Section 43 that $[E : F]$ denotes the *degree* of E over F, that is, the dimension of E as a vector space over F. Also, recall that all fields are assumed to be subfields of the field of complete numbers. As you read the theorem, it may help to refer to the following relations, noting that the larger the subfield, the smaller the corresponding subgroup, and conversely.

$$H_1 \subseteq H_2 \quad \text{implies} \quad E_{H_1} \supseteq E_{H_2} \tag{46.2}$$

$$K_1 \subseteq K_2 \quad \text{implies} \quad \mathrm{Gal}(E/K_1) \supseteq \mathrm{Gal}(E/K_2) \tag{46.3}$$

Fundamental Theorem of Galois Theory. *Assume that E is the splitting field of a polynomial $f(x)$ over a field F (a subfield of $\mathbb{C}$). Consider the correspondence defined by*

$$K \to \mathrm{Gal}(E/K) \tag{46.4}$$

for each subfield K of E such that K contains F. The correspondence (46.4) is one-to-one between the set of all subfields of E that contain F and the set of all subgroups $\mathrm{Gal}(E/F)$. Moreover, for such K,

$$[E : K] = |\mathrm{Gal}(E/K)|, \tag{46.5}$$

$$K = E_H \quad \text{for} \quad H = \mathrm{Gal}(E/K), \tag{46.6}$$

and K is a splitting field for some polynomial over F iff $\mathrm{Gal}(E/K)$ is a normal subgroup of $\mathrm{Gal}(E/F)$, in which case $\mathrm{Gal}(K/F) \approx \mathrm{Gal}(E/F)/\mathrm{Gal}(E/K)$.

The problems ask for proofs of parts of the Fundamental Theorem. Full proofs can be found in many of the references listed at the end of this chapter. We'll now look at several examples, and then turn to applications in Section 47.

Example 46.4.   The splitting field of the polynomial $(x^2 - 3)(x^2 - 5)$ over $\mathbb{Q}$ is

$$Q(\sqrt{3}, \sqrt{5}) = \{a + b\sqrt{3} + c\sqrt{5} + d\sqrt{3}\sqrt{5} : a, b, c, d \in \mathbb{Q}\}.$$

If α and β are defined by

$$\alpha(a + b\sqrt{3}) = a - b\sqrt{3}, \quad \alpha(c + d\sqrt{5}) = c + d\sqrt{5}$$
$$\beta(a + b\sqrt{3}) = a + b\sqrt{3}, \quad \beta(c + d\sqrt{5}) = c - d\sqrt{5}$$

for all $a, b, c, d \in \mathbb{Q}$, and γ is defined by $\gamma = \beta\alpha$, then

$$\mathrm{Gal}(\mathbb{Q}(\sqrt{3}, \sqrt{5})/\mathbb{Q}) = \{\iota, \alpha, \beta, \gamma\}.$$

(Problem 27.23 shows that $\sigma(a + b\sqrt{3}) = a + b\sqrt{5}$ does not define an isomorphism.) Problem 46.15 asks you to verify the following details, where $E = \mathbb{Q}(\sqrt{2}, \sqrt{3})$.

(i) $\mathrm{Gal}(E/\mathbb{Q}) \approx \mathbb{Z}_2 \times \mathbb{Z}_2$

(ii)   The lattice of subgroups of $\mathrm{Gal}(E/\mathbb{Q})$ and the lattice of subfields of $\mathbb{Q}(\sqrt{2}, \sqrt{3})$ are as shown in Figure 46.1. (The integers on the lines in the figure show the indexes for subgroups and the degrees for subfields.)

(iii) $E_{\{\iota,\alpha\}} = \mathbb{Q}(\sqrt{5})$, $E_{\{\iota,\beta\}} = \mathbb{Q}(\sqrt{3})$, $E_{\{\iota,\gamma\}} = \mathbb{Q}(\sqrt{3}\sqrt{5})$

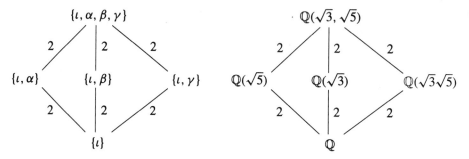

Figure 46.1   ∎

Example 46.5.   The splitting field of $x^3 - 2$ over $\mathbb{Q}$ is $Q(\omega, \sqrt[3]{2})$, where ω is the primitive cube root $-1/2 + i\sqrt{3}/2$ of 2 (Section 33). If α and β are defined by

$$\alpha(\sqrt[3]{2}) = \sqrt[3]{2} \quad \text{and} \quad \alpha(\omega) = \omega^2$$
$$\beta(\sqrt[3]{2}) = \omega\sqrt[3]{2} \quad \text{and} \quad \beta(\omega) = \omega$$

then α and β determine automorphisms of $\mathbb{Q}(\omega, \sqrt[3]{2})$ and

$$\mathrm{Gal}(\mathbb{Q}(\omega, \sqrt[3]{2})/\mathbb{Q}) = \{\iota, \alpha, \beta, \beta^2, \alpha\beta, \alpha\beta^2\}.$$

Figure 46.2 shows the lattice of subgroups of the Galois group and the lattice of subfields of $\mathbb{Q}(\omega, \sqrt[3]{2})$. As in Figure 46.1, the integers on the lines show the indexes for subgroups and the degrees for subfields. Also, corresponding subgroups and subfields are in corresponding positions. For example, $\mathrm{Gal}(\mathbb{Q}(\omega, \sqrt[3]{2})/\mathbb{Q}(\sqrt[3]{2}))$ is $\{\iota, \alpha\}$.

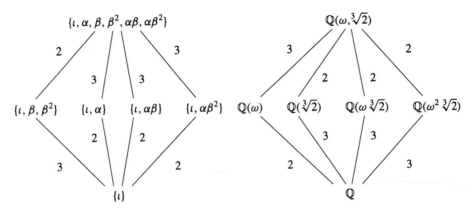

Figure 46.2

∎

PROBLEMS

46.1. Prove that the mapping σ in Example 46.1 is an isomorphism.

46.2. Prove that the set $\text{Aut}(E)$ of all automorphisms of a field E is a group.

46.3. Prove that $\mathbb{Q}(\sqrt{2}, i) = \mathbb{Q}(\sqrt{2} + \sqrt{2}\,i)$.

46.4. Prove that $x^2 - 3$ and $x^2 - 2x - 2$ have the same Galois group over $\mathbb{Q}$. (*Suggestion:* It suffices to show that they have the same splitting field.)

46.5. Prove the statement in (46.2).

46.6. Prove the statement in (46.3).

46.7. Prove that in Example 46.2, $\sigma(n) = n$ for every positive integer n.

46.8. Prove that $\text{Gal}(E/F)$ is a subgroup of $\text{Aut}(E)$.

46.9. Prove that E_H, as defined in Theorem 46.1, is a subfield of E.

46.10. In Example 46.5, verify that $\text{Gal}(\mathbb{Q}(\omega, \sqrt[3]{2})/\mathbb{Q}(\sqrt[3]{2}))$ is $\{\iota, \alpha\}$.

46.11. In Example 46.5, what is the fixed field of $\langle \alpha \rangle$?

46.12. Verify that $\text{Gal}(\mathbb{Q}(\sqrt{5})/\mathbb{Q}) \approx \mathbb{Z}_2$.

46.13. Verify that $\text{Gal}(\mathbb{Q}(\sqrt[3]{5})/\mathbb{Q})$ has order one.

46.14. (a) Show that the splitting field of $x^4 - 5$ over $\mathbb{Q}$ is $Q(\sqrt[4]{5}, i)$.
(b) Show that $[\mathbb{Q}(\sqrt[4]{5}, i) : \mathbb{Q}] = 8$.
(c) What is the order of the Galois group of $x^4 - 5$ over $\mathbb{Q}$?

46.15. Verify statements (i), (ii), and (iii) in Example 46.4.

46.16. Prove that if a is a primitive nth root of unity, then the splitting field of $x^n - 1$ over $\mathbb{Q}$ is $\mathbb{Q}(a)$. (See Problem 33.21.)

46.17. Any cubic equation can be written in the form $x^3 + bx^2 + cx + d = 0$ by dividing by the leading coefficient. The following steps show how to solve the equation by radicals.
(a) Substitute $x = y - b/3$ to get $y^3 + py + q = 0$ for appropriate p and q. Find p and q.
(b) Show that substitution of $y = z - p/(3z)$ leads to $z^3 - p^3/(27z^3) + q = 0$.

(c) Multiply through by z^3 to get a quadratic equation in z^3. Show that a solution for z^3 is

$$z^3 = -\frac{q}{2} + \sqrt{\frac{q^2}{4} + \frac{p^3}{27}}$$

(d) The cube roots of z^3 give z. From this we can get y and then x. Carry out the process outlined here to solve the equation $x^3 + 4x^2 + 4x + 3 = 0$.

(e) Solve the cubic equation in part (d) by using Theorem 45.1 to find a rational root as a first step.

SECTION 47   SOLVABILITY BY RADICALS

A polynomial equation $f(x) = 0$, with $f(x) \in F[x]$, is said to be *solvable by radicals* over F if the splitting field E of $f(x)$ over F can be embedded in a field D such that

$$F = F_0 \subset F_1 \subset \cdots \subset F_k = D,$$

where $F_j = F_{j-1}(a_{j-1})$ for $a \leq j \leq k$ with each a_{j-1} a root of an equation

$$x^{n_j} - c_{j-1} = 0 \quad \text{with} \quad c_{j-1} \in F_{j-1}.$$

If E is the splitting field of a polynomial $f(x)$ over F, then $\mathrm{Gal}(E/F)$ is called the *Galois group* of $f(x)$ over F. Following is the basic theorem on solvability by radicals. A proof can be found in any of the references listed at the end of this chapter. To provide an overview with the fewest possible distractions, we are considering only subfields of $\mathbb{C}$, but the references are less restrictive.

Theorem 47.1.   *An equation $f(x) = 0$ is solvable by radicals over F iff the Galois group of $f(x)$ over F is a solvable group.*

Theorem 47.1 will be used in Example 47.1 to exhibit a polynomial that is not solvable by radicals. But first we need several preliminary results.

Theorem 47.2.   *If $\sigma \in \mathrm{Gal}(E/F)$ and a is a root of $f(x) \in F[x]$, then $\sigma(a)$ is also a root of $f(x)$.*

Problem 47.1 asks for a proof of Theorem 47.2. Problem 47.2 asks for a proof of Theorem 47.3.

Theorem 47.3.   *Assume that E is a splitting field for $f(x)$ over F, with*

$$f(x) = c(x - a_1)(x - a_2) \cdots (x - a_n)$$

where $c \in F$, and the roots $a_1, a_2, \ldots, a_n$ are distinct elements of E. Then $\mathrm{Gal}(E/F)$ is isomorphic to a subgroup of S_n, the symmetric group of degree n.

Theorem 47.4.   *If $n \geq 5$, then the symmetric group S_n is not solvable.*

PROOF.   The proof will be by contradiction. Thus assume that $n \geq 5$ and that S_n is solvable. By the definition of solvability (Section 46), S_n has a finite sequence of subgroups such that

$$S = G_0 \triangleright G_1 \triangleright G_2 \triangleright \cdots \triangleright G_m = \{e\}$$

and G_{i-1}/G_i is Abelian for all i.

We'll first show that G_1 must contain every 3-cycle of S_n. Let (rst) denote any 3-cycle of S_n. Because $n \geq 5$, we can choose $u, v \in \{1, 2, \ldots, n\}$ such that u and v are different from r, s, and t. Let $a = (tus)$ and $b = (srv)$. By Problem 23.14, $aba^{-1}b^{-1} \in G_1$ since S_n/G_1 is Abelian. But $aba^{-1}b^{-1} = (tus)(srv)(tsu)(svr) = (rst)$. Thus $(rst) \in G_1$, as claimed.

If we repeat the argument just given with G_1 in place of S_n and G_2 in place of G_1, we see that G_2 contains all 3-cycles of S_n. We can continue in this way to deduce that $G_m = \{e\}$ contains all 3-cycles of S_n, which is a contradiction. Thus, as claimed, S_n is not solvable.   ∎

Example 47.1.   Consider the polynomial $f(x) = 4x^5 - 20x + 5$. It will be shown that the Galois group $f(x)$ is isomorphic to S_5. Because S_5 is not solvable by Theorem 47.4, it will follow from Theorem 47.1 that $f(x)$ is not solvable by radicals.

First we'll show that $f(x)$ has five distinct roots, three real and two imaginary. By Eisenstein's irreducibility criterion (Problem 36.22) with $p = 5$, $f(x)$ is irreducible. The derivative of $f(x)$ is $f'(x) = 20x^4 - 20 = 20(x^4 - 1) = 20(x + 1)(x - 1)(x^2 + 1)$, and $f''(x) = 80x^3$. Thus $f'(x) = 0$ iff $x = -1$ or $x = 1$, and $f''(-1) < 0$ while $f''(1) > 0$. Thus $f(x)$ has a maximum at $x = -1$, a minimum at $x = 1$, and a point of inflection at $x = 0$. It follows that $f(x)$ has three real zeros, one less than -1, one between -1 and 1, and one greater than 1. The other two roots must be imaginary conjugates (Sections 44 and 45).

Assume that the roots are $a_1, a_2, \ldots, a_5$. By Theorem 47.3 the Galois group G of $f(x)$ over $\mathbb{Q}$ is isomorphic to a subgroup of S_5. We'll show that, in fact, G is isomorphic to S_n.

Because a_1 is a root of the irreducible polynomial $f(x)$ of degree 5 over $\mathbb{Q}$, the second corollary of Theorem 42.3 implies that $[\mathbb{Q}(A_1) : \mathbb{Q}] = 5$. Therefore, by the Fundamental Theorem of Galois Theory, $|G|$ is divisible by 5. Thus G has an element of order 5 (Problems 53.15 and 14.34). Such an element of order 5 in G corresponds to a 5-cycle (a permutation of $a_1, a_2, \ldots, a_5$) in S_5. Also, G contains the automorphism that results from interchanging the two imaginary (conjugate) roots of $f(x)$, as in Example 46.1; this corresponds to a 2-cycle in S_5. Since G is isomorphic to a subgroup of S_5 containing both a 5-cycle and a 2-cycle, this subgroup is, in fact, S_5 (Problem 47.6). Thus the Galois group of $f(x)$ is not solvable, so $f(x)$ is not solvable by radicals.   ∎

PROBLEMS

47.1. Prove Theorem 47.2.

47.2. Prove Theorem 47.3.

47.3. Prove that the group S_3 is solvable.

47.4. Prove that if groups A and B are solvable, then so is $A \times B$.

47.5. Prove that the dihedral group D_4 (Example 8.1) is solvable. (The rotations form a subgroup.)

47.6. Prove that if $\alpha = (1\ 2\ 3\ 4\ 5)$, $\beta = (1\ 2)$, and $G = \langle \alpha, \beta \rangle$, then $G = S_5$. (*Suggestion:* Show that $(1\ 2)$, $(2\ 3)$, $(3\ 4)$, $(4\ 5)$ are in G by considering $\alpha^k \beta \alpha^{-k}$ for $0 \le k \le 3$. Then show that $(1\ 3)$, $(1\ 4)$, $(1\ 5)$ are in G by considering $(1\ 2)(2\ 3)(1\ 2)$, and so on. Finally, use Problem 6.9)

47.7. Show that $2x^5 - 5x^4 + 5 = 0$ is not solvable by radicals.

47.8. Find the Galois group of $x^3 - 5$. Solve the equation by radicals (over $\mathbb{Q}$).

47.9. Find the Galois group of $x^4 - 9x^2 + 14$. Is $x^4 - 9x^2 + 14 = 0$ solvable by radicals?

NOTES ON CHAPTER XI

References [1] through [6] are general works on modern algebra. Each of these references contains more extensive discussions of fields and polynomial equations than this book.

The early editions of [6], which were written in the 1930s, have had a major influence on most other textbooks in modern algebra. Reference [1] has been widely used since the appearance of its first edition in 1941. References [5] and [6] are graduate-level textbooks; reference [3] is written at a level between those books and this one. Reference [2] is a well-known series treating algebra as well as other parts of mathematics; although originally published in French, some volumes are available in English translation.

1.   Birkhoff, G., and S. MacLane, *A Survey of Modern Algebra*, A. K. Peters, 1997.
2.   Bourbaki, N., *Éléments de mathématique*, Hermann, Paris.
3.   Herstein, I. N., *Topics in Algebra*, 2nd ed., Wiley, New York, 1975.
4.   Hungerford, T. W., *Algebra*, 2nd ed., Harcourt Brace, 1990.
5.   Jacobson, N., *Basic Algebra I*, 2nd ed., Freeman, San Francisco, 1985.
6.   Van der Waerden, B. I., *Algebra*, 9th ed., Springer Verlag, New York, 1994.

The following books are specialized, treating fields and polynomial equations.

7.   Artin, E., *Galois Theory*, Dover, 1998.
8.   Gaal, L., *Classical Galois Theory with Examples*, 5th ed., American Mathematical Society, 1998.
9.   Rotman, J., *Galois Theory*, 2nd ed., Springer Verlag, 1998.
10.   Stewart, I., *Galois Theory*, 2nd ed., Chapman & Hall, London, 1990.

CHAPTER XII

GEOMETRIC CONSTRUCTIONS

In the last chapter we saw that in trying to solve certain polynomial equations by radicals, mathematicians in the seventeenth and eighteenth centuries had attempted to do the impossible. The story in this chapter is similar: in working to carry out certain geometric constructions using only an unmarked straightedge and collapsible compass, mathematicians in ancient Greece also had tried to do the impossible. What makes this story especially interesting is that although the constructions were to be geometric, the proofs of their impossibility involve algebra. In fact, the key algebraic concepts are the same as those used to analyze solvability by radicals.

After showing why geometric construction problems are equivalent to problems in algebra, we determine which of these algebraic problems have solutions. We then show how these results apply to the constructions originally attempted by the Greeks.

SECTION 48   THREE FAMOUS PROBLEMS

In the fifth century B.C., early in the history of Greek geometry, three problems began to attract increasing attention.

    I.   The duplication of the cube.
   II.   The trisection of an arbitrary angle.
  III.   The quadrature of the circle.

Each involved the construction of one geometrical segment from another, using only an (unmarked) straightedge and a (collapsible) compass. With the first the problem was to construct the edge of a cube having twice the volume of a given cube; with the second the problem was to show that any angle could be trisected; and with the third the problem was to construct the side of a square having the same area as a circle of given radius.

It must be stressed that these problems are concerned only with the question of whether the constructions can, in theory, be carried out in a finite number of steps using only a straightedge and compass. With the straightedge we can draw the line through two given points, and with the compass we can draw the circle through a given point with a given radius. For practical purposes there is no reason to restrict the tools to a straightedge and compass, and the constructions can be carried out to any desired degree of accuracy by other means.

As a first step in analyzing the three problems, let us rephrase each of them using numbers. In I, if the edge of the given cube is taken as the unit of length, and the edge of the required cube is denoted by x, then the volumes of the two cubes are 1 cubic unit and x^3 cubic units, respectively. Thus I can be rephrased as follows:

> **I.′**   *Given a segment of length* 1, *construct a segment of length* x *with* $x^3 = 2$.

Ultimately, we shall show that the construction II is impossible in general by showing that an angle of $60°$ cannot be trisected. (Some angles can, in fact, be trisected. But the problem in II is whether they all can be trisected, and one example will suffice to prove otherwise.) Thus we restrict attention now to an angle of $60°$. It is easy to show that an angle can be trisected iff a segment the length of its cosine can be constructed from a segment of unit length (Problem 48.1; here and elsewhere we can assume a segment of unit length as given). It can be shown with elementary trigonometry (Problem 48.2) that if A is any angle, then

$$\cos A = 4 \cos^3\left(\frac{A}{3}\right) - 3\cos\left(\frac{A}{3}\right). \tag{48.1}$$

If we take $A = 60°$, and let $x = \cos(A/3)$, this simplifies to

$$8x^3 - 6x - 1 = 0.$$

Thus the problem of trisecting a $60°$ angle can be rephrased like this:

> **II.′**   *Given a segment of length* 1, *construct a segment of length* x *with* $8x^3 - 6x - 1 = 0$.

For III, we can take the radius of the given circle as the unit of length. Then the area of the circle is π square units, and the problem becomes that of constructing a segment of length x such that $x^2 = \pi$. So III can be rephrased as follows:

> **III.′**   *Given a segment of length* 1, *construct a segment of length* x *with* $x^2 = \pi$.

With I′, II′, and III′, each of the original problems becomes a problem of whether a certain multiple of a given length can be constructed. The key in each case is in the following fact, which will be proved in the next section.

> *A necessary and sufficient condition that a segment of length x can be constructed with a straightedge and compass, beginning with a segment of length* 1, *is that x can be obtained from* 1 *by a finite number of rational operations and square roots.*

(The rational operations referred to here are addition, subtraction, multiplication, and division.) After we have proved this fact, the only remaining questions will be whether the solutions of the equations in I′, II′, and III′ satisfy the specified condition; those questions will be considered in Section 50.

Now assume that a unit of length has been given. Just what this length is can vary from problem to problem, but, whatever it is, we shall regard it as fixed until the end of the next section. We can construct perpendicular lines with a straightedge and compass, so we do that, and then mark off the given unit of length on each of the perpendicular lines (axes), to establish coordinates for points in the plane of the axes. Following the usual conventions, we label the axes with x and y, directed as shown in Figure 48.1.

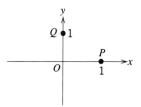

Figure 48.1

The points O, P, and Q will be called *constructible points*, as will any other points that can be obtained by starting with the points O, P, and Q and making repeated use of a straightedge and compass. To determine which points are constructible, we must introduce two further notions. A *constructible line* is a line (in the xy-plane) passing through two constructible points. A *constructible circle* is a circle (in the xy-plane) having its center at a constructible point and its radius equal to the distance between two constructible points. Thus a point other than O or P or Q is constructible if it is a point of intersection of two constructible lines, of two constructible circles, or of a constructible line and a constructible circle. Finally, a *constructible number* is a number that is either the distance between two constructible points or the negative of such a distance. (This definition depends on the chosen unit of length, of course.)

Example 48.1.   The unit circle is constructible: its center is at point O and its radius is OP. The line bisecting the first and third quadrants is constructible, because the lines OP and OQ are constructible, and the bisector of an angle between two constructible lines is constructible (Problem 48.5). Therefore, point R (Figure 48.2) is constructible, being the intersection of a constructible line and a constructible circle. Also, $\sqrt{2}$ is a constructible number, because it is the distance between P and Q.

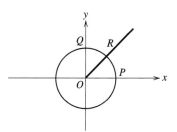

Figure 48.2   ■

Example 48.2.   A point is constructible iff each of its coordinates is a constructible number. Because, given constructible numbers a and b, the points $(a, 0)$ and $(0, b)$ are constructible in one step; and then the point (a, b) is constructible as a point of intersection of two circles, one with center at $(a, 0)$ and radius $|b|$, the other with center at $(0, b)$ and radius $|a|$. Conversely, given a constructible point S with coordinates (a, b), we can construct lines through S perpendicular to each coordinate axis; the points of intersection of these lines with the coordinate axes have coordinates $(a, 0)$ and $(0, b)$, so that a and b are constructible.   ■

We are primarily interested in constructible numbers, but we must get at those by working with constructible points, lines, and circles. Let $\mathbb{K}$ denote the set of all constructible numbers. This is a subset of $\mathbb{R}$. In fact, by the following theorem, $\mathbb{K}$ is a sub*field* of $\mathbb{R}$.

Theorem 48.1.   *The set $\mathbb{K}$ of constructible numbers is a field.*

PROOF.   We are assuming that 1, a unit of length, has been given. Thus it suffices to prove that if $a \in \mathbb{K}$ and $b \in \mathbb{K}$, then $a + b \in \mathbb{K}$, $a - b \in \mathbb{K}$, $ab \in \mathbb{K}$, and, if $b \neq 0$ as well, then $a/b \in \mathbb{K}$. The proof for a/b will be given here; the other parts will be left to the problems.

Assume that a and b are constructible numbers, with $b \neq 0$. It will suffice to prove that $a/b \in \mathbb{K}$ for $a > 0$ and $b > 0$. As before, assume OP of unit length (Figure 48.3). The point S, b units from O on the x-axis, is constructible; and the point T, a units from O on the line $x = y$, is constructible (Problem 48.6). The line through P parallel to ST is constructible (Problem 48.7); let this line intersect OT at U. The point U is constructible. Also, because triangles OPU and OST are similar, $OU/OP = OT/OS$. Therefore, the distance between O and U is a/b, so that $a/b \in \mathbb{K}$.

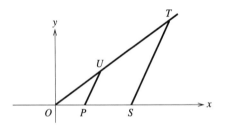

Figure 48.3   ■

Being a subfield of $\mathbb{R}$, the field $\mathbb{K}$ must contain all rational numbers. In the next section we shall determine what other numbers are in $\mathbb{K}$.

PROBLEMS

All constructions are to involve only an (unmarked) straightedge and a compass.

48.1. Show how to construct an angle given a segment the length of its cosine and a segment of unit length.

48.2. Derive the trigonometric identity (48.1).

48.3. Show how to construct the perpendicular bisector of a given segment.

48.4. Show how to construct a perpendicular to a given line at a given point on the line.

48.5. Show how to construct the bisector of a given angle.

48.6. Explain why the point T is constructible in the proof of Theorem 48.1.

48.7. Explain why the line through P parallel to ST is constructible in the proof of Theorem 48.1.

48.8. Prove that if a and b are constructible numbers, then so are $a + b$, $a - b$, and ab.

48.9. Find the cubic equation (48.1) for the trisection of a $90°$ (right) angle by letting $A = 90°$ and $x = \cos(A/3)$. Show how to construct a segment of length x. (By Problem 48.1, this shows that a $90°$ angle can be trisected.)

SECTION 49 CONSTRUCTIBLE NUMBERS

The next lemma is used in proving Theorem 49.1, which characterizes the constructible numbers.

Lemma 49.1. *Let F be a subfield of $\mathbb{R}$.*

    (a) *If a line contains two points whose coordinates are in F, then the line has an equation $ax + by + c = 0$ for some $a, b, c \in F$.*

    (b) *If a circle has its radius in F and its center at a point whose coordinates are in F, then the circle has an equation $x^2 + y^2 + dx + ey + f = 0$ for some $d, e, f \in F$.*

    PROOF.   (a) If $x_1 \neq x_2$, an equation for the line through (x_1, y_1) and (x_2, y_2) is

$$\frac{y - y_1}{x - x_1} = \frac{y_2 - y_1}{x_2 - x_1}$$

This can be put in the form $ax + by + c = 0$ with each of the coefficients in F because each is a rational combination of $x_1, y_1, x_2,$ and y_2. If $x_1 = x_2$, the line through (x_1, y_1) and (x_2, y_2) is $x - x_1 = 0$, which also has the form $ax + by + c = 0$ with $a, b, c \in F$.

    (b) Similar to part (a), beginning with

$$(x - x_1)^2 + (y - y_1)^2 = r^2$$

with $x_1, y_1, r \in F$.   ■

    If F is a subfield of $\mathbb{R}$, and $a \in \mathbb{R}$, then $F(a)$ is the smallest subfield of $\mathbb{R}$ containing both F and a (see Section 42). For example, $\mathbb{Q}(\sqrt{2})$ consists of all real numbers $c + d\sqrt{2}$ with $c, d \in \mathbb{Q}$. Notice that each number in $\mathbb{Q}(\sqrt{2})$ is constructible, because $\mathbb{Q} \subset \mathbb{K}$ and $\sqrt{2} \in \mathbb{K}$ (Example 48.1). Each number in $\mathbb{Q}(\sqrt{2})(\sqrt{3}) = \mathbb{Q}(\sqrt{2}, \sqrt{3})$ is also constructible, because $\sqrt{3} \in \mathbb{K}$ (see Problem 49.2). We now prove that all constructible numbers belong to fields that are built up in this way.

Theorem 49.1. *A real number r is constructible iff there is a finite sequence $\mathbb{Q} = F_0 \subset F_1 \subset \cdots \subset F_k$ of subfields of $\mathbb{R}$ with $r \in F_k$, where $F_j = F_{j-1}(\sqrt{a_{j-1}})$ and a_{j-1} is a positive number in F_{j-1} for $1 \leq j \leq k$.*

    PROOF.   Assume first that r is constructible. Then r can be constructed in a finite number of steps beginning with 1, which is in $\mathbb{Q}$. In light of Lemma 49.1, it suffices to prove that if F is any subfield of $\mathbb{R}$, and a real number r is a coordinate of an intersection point of lines or circles having equations with coefficients in F, then $r \in F(\sqrt{a})$ for some positive real number $a \in F$.

It is easy to see that in solving two simultaneous linear equations $a_1 x + b_1 y + c_1 = 0$ and $a_2 x + b_2 y + c_2 = 0$, with coefficients in F, we shall not be led outside F. Thus, if r is a coordinate of the intersection of the corresponding lines, then $r \in F$, which is sufficient.

If y is eliminated from $ax + by + c = 0$ and $x^2 + y^2 + dx + ey + f = 0$, the result is a quadratic equation $Ax^2 + Bx + C = 0$ with $A, B, C \in F$. Therefore, the x-coordinates of the intersection of the corresponding line and circle are given by

$$x = \frac{-B \pm \sqrt{B^2 - 4AC}}{2A}.$$

If these are real numbers, then $x \in F(\sqrt{a})$ for $a = B^2 - 4AC > 0$, as required. The case of the y-coordinate is similar.

The coordinates of the intersection points of two circles, $x^2 + y^2 + d_1 x + e_1 y + f_1 = 0$ and $x^2 + y^2 + d_2 x + e_2 y + f_2 = 0$, can be obtained by first subtracting, thereby getting $(d_1 - d_2)x + (e_1 - e_2)y + (f_1 - f_2) = 0$, and then using this equation with that of the first circle. The result is similar to that in the case of a line and a circle.

Now assume that r is an element of a field F_k as described in the theorem. We know that each number in $\mathbb{Q}$ is constructible, and so in order to prove that r is constructible it suffices to prove that if each number in a subfield F of $\mathbb{R}$ is constructible, and if a is a positive number in F, then each number in $F(\sqrt{a})$ is constructible. Because each number in $F(\sqrt{a})$ is of the form $c + d\sqrt{a}$ for $c, d \in F$, and because products and sums of constructible numbers are constructible, it suffices to prove that $\sqrt{a}$ is constructible. But it is an elementary construction problem to show that $\sqrt{a}$ is constructible from 1 and a (Problem 49.3). ∎

PROBLEMS

49.1. Carefully fill in the details in the proof of Lemma 49.1.

49.2. Prove that $\sqrt{3} \in \mathbb{K}$ without using Theorem 49.1.

49.3. Prove that $\sqrt{a}$ is constructible from 1 and a.

49.4. Prove that $\sqrt[4]{2} \in \mathbb{K}$.

SECTION 50   IMPOSSIBLE CONSTRUCTIONS

We now use the characterization of constructible numbers given in Theorem 49.1 to prove the impossibility of duplication of the cube and trisection of a $60°$ angle. We also indicate, without proof, why quadrature of the circle is impossible. In settling the first two problems we need the following theorem.

Theorem 50.1.   *If a cubic equation with rational coefficients has no rational root, then it has no constructible root.*

The following two lemmas are used in the proof of the theorem.

Lemma 50.1.   *If the roots of $ax^3 + bx^2 + cx + d$ are r_1, r_2, and r_3, then $r_1 + r_2 + r_3 = -b/a$.*

PROOF.   The polynomials $ax^3 + bx^2 + cx + d$ and $a(x - r_1)(x - r_2)(x - r_3) = ax^3 - a(r_1 + r_2 + r_3)x^2 + a(r_1 r_2 + r_1 r_3 + r_2 r_3)x - a r_1 r_2 r_3$ have the same roots and the same leading coefficients, and therefore the coefficients on all like power of x must be equal (Problem 50.1). Thus, in particular, $b = -a(r_1 + r_2 + r_3)$.   ∎

Lemma 50.2.   *If F is a subfield of $\mathbb{R}$, and a cubic polynomial $ax^3 + bx^2 + cx + d \in F[x]$ has a root $p + q\sqrt{r}$, where $p, q, r \in F, r > 0$, and $\sqrt{r} \notin F$, then $p - q\sqrt{r}$ is also a root.*

PROOF.   The proof is similar to that of Theorem 45.2 and is left as an exercise (Problem 50.2).   ∎

PROOF OF THEOREM 50.1.   Assume that $ax^3 + bx^2 + cx + d \in \mathbb{Q}[x]$ has no rational root, but that it does have a root in $\mathbb{K}$. We shall show that this leads to a contradiction. Let $u \in \mathbb{K}$ be a root of the given cubic, and assume u chosen so that the corresponding field F_k in Theorem 49.1 satisfies the condition that k is minimal with respect to the condition of containing such a root. Then $k \neq 0$ because $u \notin \mathbb{Q} = F_0$. Therefore, $u = p + q\sqrt{r}$ for some $p, q, r \in F_{k-1}$. By Lemma 50.2, $p - q\sqrt{r}$ is also a root of the given cubic. By Lemma 50.1, the sum of the roots of the cubic is $-b/a$, so that if v is the third root, then $-b/a = v + (p + q\sqrt{r}) + (p - q\sqrt{r})$, and $v = -b/a - 2p$. But $a, b, p \in F_{k-1}$, and hence $v \in F_{k-1}$, contradicting the minimality of k.   ∎

IMPOSSIBILITY OF THE DUPLICATION OF THE CUBE

From I′ in Section 48, the question is whether $x^3 - 2$ has a constructible root. Theorem 45.1 can be used to show that $x^3 - 2$ has no rational root (Problem 50.3), and therefore, by Theorem 50.1, $x^3 - 2$ has no constructible root.

IMPOSSIBILITY OF THE TRISECTION OF AN ARBITRARY ANGLE

From II′ in Section 48, if a 60° angle could be trisected, then $8x^3 - 6x - 1$ would have a constructible root. But Theorem 45.1 can be used to show that $8x^3 - 6x - 1$ has no rational root (Problem 50.3), and therefore, again by Theorem 50.1, $8x^3 - 6x - 1$ has no constructible root.

DISCUSSION OF THE QUADRATURE OF THE CIRCLE

It can be proved from Theorem 49.1 that any constructible number is a root of a polynomial with integral coefficients; any root of such a polynomial is called an *algebraic number*. Thus from III′ in Section 48 the impossibility of the quadrature of the circle would follow if it were shown that $\sqrt{\pi}$ is not algebraic. This was done by the German mathematician C. L. F. Lindemann, in 1882. The proof requires methods that are not algebraic, however, and thus it will not be given here. (See reference [2] of Chapter X.)

PROBLEMS

50.1. Prove that if two polynomials over $\mathbb{C}$ have the same degree, the same roots, and the same leading coefficients, then the coefficients on all like powers of x must be equal.

50.2. Prove Lemma 50.2.

50.3. Prove that neither $x^3 - 2$ nor $8x^3 - 6x - 1$ has a rational root.

50.4. Prove the following generalization of Lemma 50.1: If the roots of $a_n x^n + \cdots + a_1 x + a_0$ are $r_1, r_2, \ldots, r_n$, then

$$r_1 + r_2 + \cdots + r_n = -a_{n-1}/a_n$$

and

$$r_1 r_2 \cdots r_n = (-1)^n a_0/a_n.$$

NOTES ON CHAPTER XII

1. Klein, F., *Famous Problems of Elementary Geometry*, Chelsea, New York, 1955.
2. Dudley, U., *The Trisectors*, Mathematical Association of America, 1994.

CHAPTER XIII

APPLICATIONS OF PERMUTATION GROUPS

This chapter introduces the use of permutation groups for the solution of counting problems. This is one of the applications of modern algebra to the branch of mathematics known as *combinatorics* or *combinatorial analysis*. Section 51 introduces some basic terminology, which is then used in Section 52 in the proof and application of Burnside's Counting Theorem. In Section 53 the ideas in Section 51 are applied to prove a theorem within group theory itself.

SECTION 51   GROUPS ACTING ON SETS

We begin by recalling the group of symmetries of a square, from Example 8.1. For convenience, here are the definitions of the elements and also the accompanying figure, Figure 51.1.

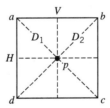

Figure 51.1

Group of symmetries of a square (Figure 51.1)

$\mu_1 =$ identity permutation
$\mu_2 =$ rotation $90°$ clockwise around p
$\mu_3 =$ rotation $180°$ clockwise around p
$\mu_4 =$ rotation $270°$ clockwise around p
$\mu_5 =$ reflection through H
$\mu_6 =$ reflection through V
$\mu_7 =$ reflection through D_1
$\mu_8 =$ reflection through D_2.

Denote the group by G. The elements of G are isometries of the plane; therefore, they are permutations of the set of all points of the plane. But these isometries also induce permutations of the set of vertices of the square in a natural way. If we assign to each isometry the corresponding permutation of $\{a, b, c, d\}$, then we have a mapping from G to $\mathrm{Sym}\{a, b, c, d\}$. The mapping with the induced permutations written in cycle notation is

$$
\begin{aligned}
\mu_1 &\mapsto (a)(b)(c)(d) & \mu_5 &\mapsto (a \quad d)(b \quad c) \\
\mu_2 &\mapsto (a \quad b \quad c \quad d) & \mu_6 &\mapsto (a \quad b)(c \quad d) \\
\mu_3 &\mapsto (a \quad c)(b \quad d) & \mu_7 &\mapsto (a)(c)(b \quad d) \\
\mu_4 &\mapsto (a \quad d \quad c \quad b) & \mu_8 &\mapsto (a \quad c)(b)(d)
\end{aligned}
$$

Because the operations on both G and $\mathrm{Sym}\{a, b, c, d\}$ are composition, this mapping is a homomorphism.

The isometries in G also induce permutations of the set of diagonals $\{D_1, D_2\}$ in a natural way. For instance, $\mu_3(D_1) = D_1$ and $\mu_3(D_2) = D_2$ (μ_3 interchanges the ends of each of the diagonals, but that is not important here). In this case we can assign to each isometry the corresponding permutation of $\{D_1, D_2\}$. This gives a mapping from G to $\mathrm{Sym}\{D_1, D_2\}$. The induced permutations are

$$
\begin{aligned}
\mu_1 &\mapsto (D_1)(D_2) & \mu_5 &\mapsto (D_1 \quad D_2) \\
\mu_2 &\mapsto (D_1 \quad D_2) & \mu_6 &\mapsto (D_1 \quad D_2) \\
\mu_3 &\mapsto (D_1)(D_2) & \mu_7 &\mapsto (D_1)(D_2) \\
\mu_4 &\mapsto (D_1 \quad D_2) & \mu_8 &\mapsto (D_1)(D_2)
\end{aligned}
$$

Again, the mapping is a homomorphism.

These two examples are special cases of the following generalization of the notion of a permutation group on a set. (Here, and throughout this chapter, the notation $\beta \circ \alpha$ for composition will be shortened to $\beta\alpha$.)

Definition.   A group G *acts on a set* S if to each $g \in G$ there is assigned a permutation π_g of S in such a way that $\pi_{ab} = \pi_a \pi_b$ for all $a, b \in G$.

In other words, G acts on S if there is a homomorphism $g \mapsto \pi_g$ of G into $\mathrm{Sym}(S)$. [If $g \in G$, so that $\pi_g \in \mathrm{Sym}(S)$, and $s \in S$, then $\pi_g(s) \in S$. We are writing π_g rather than $\pi(g)$ because $\pi_g(s)$ seems preferable to $\pi(g)(s)$.] Any subgroup of $\mathrm{Sym}(S)$ acts on S. The idea of a group's acting on S is more general than that of a permutation group on S because, in the former case, unequal group elements can give rise to equal permutations; that is, the mapping $g \mapsto \pi_g$ need not be one-to-one. If the mapping $g \mapsto \pi_g$ is one-to-one, then G is said to act *faithfully* on S. In the first example above, G acts faithfully—unequal group elements give rise to unequal permutations of $\{a, b, c, d\}$. In the second example, G does not act faithfully—considered only as permutations of $\{D_1, D_2\}$, $\mu_1 = \mu_3 = \mu_7 = \mu_8$ and $\mu_2 = \mu_4 = \mu_5 = \mu_6$.

Because $g \mapsto \pi_g$ is a homomorphism, necessarily $\pi_e = \iota_s$ and $\pi_{g^{-1}} = \pi_g^{-1}$ for all $g \in G$. This is used in the proof of the following theorem.

Theorem 51.1.   *Assume that G is a group acting on a set S. Define the relation $\sim$ on S by*

$$ s_1 \sim s_2 \quad \textit{iff} \quad \pi_g(s_1) = s_2 \quad \textit{for some} \quad g \in G. $$

Then $\sim$ is an equivalence relation on S.

PROOF.   *Reflexive*: If $s \in S$, then $\pi_e(s) = \iota_s(s) = s$ so that $s \sim s$.
Symmetric: If $s_1 \sim s_2$ with $\pi_g(s_1) = s_2$, then $\pi_{g^{-1}}(s_2) = \pi_g^{-1}(s_2) = s_1$ so $s_2 \sim s_1$.
Transitive: If $s_1 \sim s_2$ and $s_2 \sim s_3$, with $\pi_g(s_1) = s_2$ and $\pi_h(s_2) = s_3$, then $\pi_{hg}(s_1) = \pi_h \pi_g(s_1) = \pi_h(s_2) = s_3$ so that $s_1 \sim s_3$.   ∎

The equivalence classes relative to $\sim$ in Theorem 51.1 are called *orbits*. The term *G-orbit* can be used if it is necessary to specify the group. Also, $\sim_G$ can be written in place of $\sim$.

Example 51.1.   Let G denote the group of all rotations of a plane about a point p in the plane (Example 5.7). Then G acts on the set of points in the plane, and the orbits are the circles with centers at p.   ∎

Example 51.2.   If S is any nonempty set, then Sym(S) acts on S, and there is only one orbit, namely S.   ∎

Example 51.3.   The subgroup $\{\mu_1, \mu_3\}$ of the symmetry group of a square acts on $\{a, b, c, d\}$ (Figure 51.1). The orbits are $\{a, c\}$ and $\{b, d\}$.   ∎

For G acting on S and $s \in S$, let Orb(s) denote the orbit of s:

$$\text{Orb}(s) = \{\pi_g(s): g \in G\}.$$

Then $|\text{Orb}(s)|$ is the number of elements in the orbit of s. Let G_s denote the set of all $g \in G$ such that $\pi_g(s) = s$. Problem 51.8 asks for verification that G_s is a subgroup of G.

Example 51.4.   The group $G = \{(1), (1 \quad 2), (3 \quad 4), (1 \quad 2)(3 \quad 4)\}$ acts on $\{1, 2, 3, 4\}$. In this case

$$\text{Orb}(1) = \text{Orb}(2) = \{1, 2\}, \qquad \text{Orb}(3) = \text{Orb}(4) = \{3, 4\},$$
$$G_1 = G_2 = \{(1), (3 \quad 4)\}, \qquad G_3 = G_4 = \{(1), (1 \quad 2)\}.$$

The following theorem, which gives a formula for the number of elements in an orbit, is the key to the connection between group theory and combinatorics.   ∎

Theorem 51.2.   *If a finite group G acts on a set S, and $s \in S$, then*

$$|\text{Orb}(s)| = [G : G_s] = \frac{|G|}{|G_s|}.$$

PROOF.   To compute $|\text{Orb}(s)|$, we must compute the number of distinct elements $\pi_g(s)$ for $g \in G$. To do that, we examine when $\pi_g(s) = \pi_h(s)$ for $g, h \in G$.

$$\pi_g(s) = \pi_h(s) \quad \text{iff} \quad (\pi_h^{-1} \pi_g)(s) = s \quad \text{iff} \quad \pi_h^{-1}(\pi_g(s)) = s$$
$$\text{iff} \quad \pi_{h^{-1}g}(s) = s \quad \text{iff} \quad h^{-1}g \in G_s \quad \text{iff} \quad gG_s = hG_s.$$

The equivalence $\pi_g(s) = \pi_h(s)$ iff $gG_s = hG_s$ shows that there is a one-to-one correspondence between the set of elements in Orb(s) and the set of all left cosets of G_s in G. (Use Lemma 16.1 with left cosets in place of right cosets.) Therefore, $|\text{Orb}(s)| = [G : G_s]$.
The equation $[G : G_s] = |G| / |G_s|$ follows from Lagrange's Theorem.   ∎

PROBLEMS

51.1. Write the permutation induced on $\{H, V\}$ by each isometry (symmetry) of the square.

51.2. Let E_1, E_2, E_3, and E_4 denote, respectively, the edges ab, bc, cd, and da of the square in Figure 51.1. Write the permutation induced on $\{E_1, E_2, E_3, E_4\}$ by each isometry (symmetry) of the square. [Example: $\mu_5 \mapsto (E_1 E_3)$.] Does the symmetry group of the square act faithfully on $\{E_1, E_2, E_3, E_4\}$?

51.3. Determine the orbits for each of the following subgroups of the symmetry group of the square, acting on the set of diagonals, $\{D_1, D_2\}$. (Compare Example 51.3.)
(a) $\langle \mu_2 \rangle$ (b) $\langle \mu_5 \rangle$ (c) $\langle \mu_6 \rangle$

51.4. (a) to (c). Repeat problem 51.3, except replace $\{D_1, D_2\}$ by $\{a, b, c, d\}$.

51.5. The group $G = \langle (1 \quad 2 \quad 3)(4 \quad 5) \rangle$ is of order 6 and acts on $\{1, 2, 3, 4, 5\}$.
(a) Determine Orb(k) for $1 \leq k \leq 5$.
(b) Determine G_k for $1 \leq k \leq 5$.
(c) Use parts (a) and (b) to verify that $|\text{Orb}(k)| = |G| / |G_k|$ for $1 \leq k \leq 5$. (Compare Theorem 51.2.)

51.6. The group $G = \langle (1 \quad 2 \quad 3 \quad 4)(5 \quad 6) \rangle$ is of order 4 and acts on $\{1, 2, 3, 4, 5, 6\}$.
(a) Determine Orb(k) for $1 \leq k \leq 6$.
(b) Determine G_k for $1 \leq k \leq 6$.
(c) Use parts (a) and (b) to verify that $|\text{Orb}(k)| = |G| / |G_k|$ for $1 \leq k \leq 6$. (Compare Theorem 51.2.)

51.7. The paragraph preceding Theorem 51.1 states that because $g \mapsto \pi_g$ is a homomorphism if G acts on a set, necessarily $\pi_e = \iota_s$ and $\pi_{g^{-1}} = \pi_g^{-1}$. Give a precise reference to justify this. (See Section 21.)

51.8. Prove that if G acts on S by $g \mapsto \pi_g$, and $s \in S$ and

$$G_s = \{g \in G : \pi_g(s) = s\},$$

then G_s is a subgroup of G.

51.9. Assume that G is a group, and for each $a \in G$ define $\pi_a : G \to G$ by $\pi_a(x) = axa^{-1}$ for each $x \in G$.
(a) Verify that with this definition G acts on G. (The orbits in this case are known as the *conjugate classes* of G; elements $s, t \in G$ are *conjugate* if $t = asa^{-1}$ for some $a \in G$.)
(b) For $G = S_3$, determine Orb(s) for each $s \in G$.
(c) Verify that G is Abelian iff $|\text{Orb}(s)| = 1$ for each $s \in G$.
(d) Verify that, in general, $|\text{Orb}(s)| = 1$ iff $sa = as$ for each $a \in G$.
(e) Verify that the set of all elements s satisfying the two equivalent conditions in part (d) is a subgroup of G. This subgroup is called the *center* of G; denote it by $Z(G)$.
(f) Verify that $Z(S_3) = \{(1)\}$.
(g) Verify that the center of the symmetry group of a square is $\{\mu_1, \mu_3\}$.

51.10. Let G be a group, and let Aut(G) denote the automorphism group of G (Problem 19.25).
(a) Verify that if π_a is defined as in Problem 51.9, then $\pi_a \in \text{Aut}(G)$. (An automorphism of the form π_a is called an *inner automorphism* of G.)
(b) Define $\theta : G \to \text{Aut}(G)$ by $\theta(a) = \pi_a$ for each $a \in G$. Verify that θ is a homomorphism.
(c) With θ as in part (b), $\theta(G)$, being the image of a homomorphism, must be a subgroup of Aut(G). Denote this subgroup by Inn(G) (it is called the *group of inner automorphisms* of G). Verify that $|\text{Inn}(G)| = 1$ iff G is Abelian.
(d) Explain why $G/Z(G) \approx \text{Inn}(G)$. [See Problem 51.9(e) for the definition of $Z(G)$.]

51.11. Verify that $\mathbb{Z}_3$ has an automorphism that is not an inner automorphism. [Compare Problem 51.10(c).]

51.12. Let S denote the collection of all subgroups of a finite group G. For $a \in G$ and $H \in S$, let $\pi_a(H) = aHa^{-1}$.
  (a) Verify that with this definition G acts on S. (Each subgroup aHa^{-1} is called a *conjugate* of H.)
  (b) For $G = S_3$, determine $\text{Orb}(\langle(1 \quad 2)\rangle)$.
  (c) For $G = S_3$, determine $G_{\langle(1 \quad 2)\rangle}$.
  (d) Use parts (b) and (c) to verify that
  $$|\text{Orb}(\langle(1 \quad 2)\rangle)| = \frac{|G|}{|G_{\langle(1 \quad 2)\rangle}|}$$
  (Compare Theorem 51.2).
  (e) For $H \in S$, the *normalizer* of H in G is defined by $N_G(H) = \{a \in G : aHa^{-1} = H\}$. Using results from this section, explain why $N_G(H)$ is a subgroup of G, and the number of conjugates of H is $[G : N_G(H)]$.

51.13. Prove that if G acts on S and $\pi_a(s) = t$, then $G_s = a^{-1}G_t(a)$.

51.14. If G acts on S and H is a subgroup of G, then H acts on S.
  (a) Verify that $s \sim_H t$ implies $s \sim_G t$.
  (b) Give an example (specific S, G, H, s, t) showing that $s \sim_G t$ does not imply $s \sim_H t$.

51.15. Assume that H acts on S and that $\theta : G \to H$ is a homomorphism from G onto H.
  (a) Verify that G acts on S if π_a is defined to be $\pi_{\theta(a)}$ for each $a \in G$.
  (b) Verify that if $s, t \in S$, then $s \sim_H t$ iff $s \sim_G t$.

51.16. Explain Cayley's Theorem (Section 20) in terms of a group acting on a set. Is the action faithful?

51.17. (Generalization of Cayley's Theorem) Let G be a group, H a subgroup of G, and S the set of left cosets of H in G. For $a \in G$, define $\pi_a : S \to S$ by $\pi_a(gH) = (ag)H$.
  (a) Verify that in this way G acts on S, that is, $a \mapsto \pi_a$ defines a homomorphism from G into $\text{Sym}(S)$.
  (b) Prove that $\text{Ker } \pi = \cap\{gHg^{-1} : g \in G\}$.
  (c) Explain and prove the statement "$\text{Ker } \pi$ is the largest normal subgroup of G that is contained in H."

51.18. Prove that if a finite group G contains a subgroup $H \neq G$ such that $|G| \nmid [G : H]!$, then H contains a nontrivial normal subgroup of G. (Use Problem 51.17, Lagrange's Theorem, and facts about homomorphisms.)

51.19. (a) Prove that if a finite group G contains a subgroup $H \neq G$ such that $|G| \nmid [G : H]!$, then G is not simple. (Use Problem 51.18. Simple groups, and their importance, were discussed in Section 23.)
  (b) Assume that each group of order 100 contains a subgroup of order 25 (this will be proved in Section 53). Prove that there is no simple group of order 100.

SECTION 52   BURNSIDE'S COUNTING THEOREM

One of the central problems of combinatorics is to compute the number of distinguishable ways in which something can be done. A simple example is to compute the number of distinguishable ways the three edges of an equilateral triangle can be painted so that one edge is red (R), one is white (W), and one is blue (B). The six possibilities are shown in the top row of Figure 52.1.

If we permit rotation of the triangle in the plane, then the first three possibilities become indistinguishable—they collapse into the first possibility in the second row of Figure 52.1. The last three possibilities in the top row collapse into the second possibility in the second row.

If we permit reflections through lines, as well as rotation in the plane, then the only possibility is that shown in the third row of Figure 52.1.

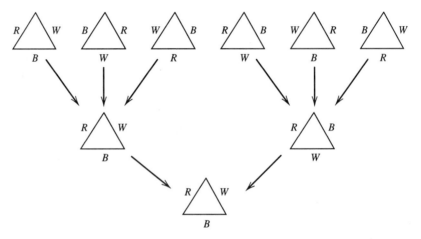

Figure 52.1

In passing from one row to the next in the example, we have treated different ways of painting the triangle as being equivalent, and we have shown one representative from each equivalence class. The problem at each step is to compute the number of equivalence classes. In the terminology used most often in combinatorics, possibilities in the same equivalence class are *indistinguishable*, while possibilities in different equivalence classes are *distinguishable*. So the problem of computing the number of distinguishable ways in which something can be done is the same as that of computing the number of equivalence classes under an appropriate equivalence relation. The link between group theory and combinatorics rests on the fact that the equivalence classes are very often orbits under the action of an appropriate group. In the example, the appropriate group for passing to the second row of Figure 52.1 is the group of rotations of the triangle; the appropriate group for passing to the third row contains these three rotations and also three reflections, each through a line connecting a vertex with the midpoint of the opposite side (this group is D_3 in the notation of Section 54).

To solve a combinatorics problem with these ideas, the first step is to get clearly in mind the total set of possibilities (without regard to equivalence). Next, decide the condition under which possibilities are to be considered equivalent (indistinguishable). This will mean membership in a common orbit under the action of some group; so in effect this means to choose an appropriate group. Finally, compute the number of orbits relative to this group; this will be the number of distinguishable possibilities. This method derives its power from Burnside's Counting Theorem, proved below, which gives a way to compute the number of orbits relative to a group action.

In considering a group acting on a set, it is often less cumbersome to use the same symbol for a group element and the permutation assigned to it, that is, to write g for π_g; we do so in this section. (If this troubles you, do Problem 52.10). *All groups and sets in this section are assumed to be finite.*

For a group G acting on a set S, we shall use $\psi(g)$ to denote the number of elements of S that are invariant under $g \in G$:

$$\psi(g) = |\{s \in S: g(s) = s\}|.$$

Burnside's Counting Theorem.   *If a group G acts on a set S, then the number of orbits is*

$$\frac{1}{|G|} \sum_{g \in G} \psi(g).$$

Example 52.1.   Consider the group $\{(1), (1 \quad 2), (3 \quad 4), (1 \quad 2)(3 \quad 4)\}$ acting on $\{1, 2, 3, 4\}$. Then, for example, the permutation $(1 \quad 2)$ leaves only 3 and 4 invariant, so that $\psi((1 \quad 2)) = 2$. Here are all values $\psi(g)$:

$$\psi((1)) = 4 \qquad \psi((3 \quad 4)) = 2$$
$$\psi((1 \quad 2)) = 2 \quad \psi((1 \quad 2)(3 \quad 4)) = 0$$

The formula in Burnside's Theorem gives $\left(\frac{1}{4}\right)(4 + 2 + 2 + 0) = 2$. The two orbits are $\{1, 2\}$ and $\{3, 4\}$.   ∎

Other examples will follow the proof of the theorem, which will depend on the following lemma.

Lemma 52.1.   *If a group G acts on a set S, $s, t \in S$, and s and t are in the same orbit, then $|G_s| = |G_t|$.*

PROOF.   If s and t are in the same orbit, then $|\text{Orb}(s)| = |\text{Orb}(t)|$, so Theorem 51.2 implies

$$|G_s| = |G| / |\text{Orb}(s)| = |G| / |\text{Orb}(t)| = |G_t|.$$ ∎

PROOF OF BURNSIDE'S THEOREM.   Let n denote the number of pairs (g, s) such that $g(s) = s$ $(g \in G, s \in S)$. The number of s appearing with a given g is $\psi(g)$, so that

$$n = \sum_{g \in G} \psi(g). \tag{52.1}$$

The number of g appearing with a given s is $|G_s|$, so that

$$n = \sum_{s \in S} |G_s|. \tag{52.2}$$

If we select an orbit $\text{Orb}(t)$, sum $|G_s|$ over all $s \in \text{Orb}(t)$, and use first Lemma 52.1 and then Theorem 51.2, we get

$$\sum_{s \in \text{Orb}(t)} |G_s| = |\text{Orb}(t)| \cdot |G_t| = |G|. \tag{52.3}$$

If we add together all sums like those in (52.3), one such sum for each orbit, we get the sum in (52.2). Therefore,

$$n = \sum_{s \in S} |G_s| = \text{(number of orbits)} \cdot |G|.$$

Solve this for the number of orbits, and then use (52.1):

$$\text{number of orbits} = \frac{n}{|G|} = \frac{1}{|G|} \sum_{g \in G} \psi(g). \qquad \blacksquare$$

Example 52.2. In how many distinguishable ways can the four edges of a square be painted with four different colors if there is no restriction on the number of times each color can be used, and two ways are considered indistinguishable if one can be obtained from the other by an isometry in the group of symmetries of the square? (This would be the case for a square that could be either rotated in the plane or turned over; the latter corresponds to reflection through a line in the plane.)

The appropriate set S in this case is the set of $4^4 = 256$ ways of painting the edges without regard to equivalence. The group is the one described at the beginning of Section 51. If ρ is a group element, then $\psi(\rho) = 4^k$, where k is the number of independent choices to be made in painting the edges so as to have invariance under ρ.

$\psi(\mu_1) = 4^4$ (always $\psi(\iota) = |S|$ if ι is the identity—four choices)

$\psi(\mu_2) = 4$ (for invariance under μ_2, all edges must be of the same color—one choice)

$\psi(\mu_3) = 4^2$ (pairs of opposite edges must be of the same color—two choices)

$\psi(\mu_4) = 4$ (like μ_2—one choice)

$\psi(\mu_5) = 4^3$ (*ab* and *dc* must be of the same color; *ad* and *bc* are independent—three choices)

$\psi(\mu_6) = 4^3$ (like μ_5—three choices)

$\psi(\mu_7) = 4^2$ (*ab* and *ad* must be of the same color; *cb* and *cd* must be of the same color—two choices)

$\psi(\mu_8) = 4^2$ (like μ_7 two choices).

Therefore, by Burnside's Theorem, the number of distinguishable ways is

$$\left(\tfrac{1}{8}\right)(4^4 + 4 + 4^2 + 4 + 4^3 + 4^3 + 4^2 + 4^2) = 55. \qquad \blacksquare$$

Example 52.3. In how many distinguishable ways can the six faces of a cube be painted with six different colors if each face is to be a different color and two ways are considered indistinguishable if one can be obtained from the other by rotation of the cube?

The answer here is the number of orbits when the group G of all rotations of a cube acts on the set S of all cubes painted with the six different colors. The group G contains 24 elements, which can be classified as follows (the references are to Figure 52.2):

1. The identity.
2. Three 180° rotations around lines (such as ij) joining the centers of opposite faces.
3. Six 90° rotations around lines (such as ij) joining the centers of opposite faces.
4. Six 180° rotations around lines (such as kl) joining the midpoints of opposite edges.
5. Eight 120° rotations around lines (such as ag) joining opposite vertices.

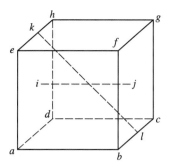

Figure 52.2

The number of elements of S is $6! = 720$ (the number of ways of painting the cube without regard to equivalence). If ι denotes the identity of G, then $\psi(\iota) = 720$. If $\rho \neq \iota$, then $\psi(\rho) = 0$, because any way of painting the cube with all faces different colors will be carried into a different way under the action of ρ. Therefore, by Burnside's Theorem, the number of orbits is $\left(\frac{1}{24}\right)(720) = 30$.

If we identify the six colors here with the integers from 1 to 6, we can interpret this answer as saying that there are 30 distinguishable ways of assigning the numbers to the faces of a die. (One of these ways is standard.)   ∎

The book *Combinatorial Enumeration of Groups, Graphs, and Chemical Compounds*, by G. Pólya and R. C. Read (Springer-Verlag, New York, 1987), discusses the history and applications of the part of combinatorial analysis that has been built on Burnside's Theorem.

PROBLEMS

52.1. Use Burnside's Counting Theorem to compute the number of orbits for the group $\langle (1 \quad 2 \quad 3)(4 \quad 5) \rangle$ acting on $\{1, 2, 3, 4, 5\}$. What are the orbits?

52.2. Use Burnside's Counting Theorem to compute the number of orbits for the group $\langle (1 \quad 2 \quad 3 \quad 4)(5 \quad 6) \rangle$ acting on $\{1, 2, 3, 4, 5, 6\}$. What are the orbits?

52.3. Consider the problem of painting the edges of a square so that one is red, one is white, one is blue, and one is yellow.
   (a) In how many distinguishable ways can this be done if the edges of the square are distinguishable? (This is the analogue of the case of a triangle represented by the first row in Figure 52.1.)
   (b) Repeat (a), except count different ways as being indistinguishable if one can be obtained from the other by rotation of the square in the plane. (Compare the second row in Figure 52.1.)
   (c) Repeat (b), except permit reflections through lines as well as rotation in the plane.

52.4. State and solve the problem that results from replacing a square by a regular pentagon in Problem 52.3 (with some appropriate fifth color). (The group will have order 10. It is D_5 in the notation of Section 54.)

52.5. A bead is placed at each of the four vertices of a square, and each bead is to be painted either red (R) or blue (B). Under equivalence relative to the group of rotations of the square, there are six distinguishable patterns:

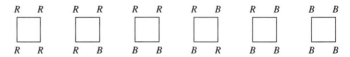

Verify that Burnside's Counting Theorem gives the correct number of distinguishable patterns.

52.6. A bead is placed at each of the six vertices of a regular hexagon, and each bead is to be painted either red or blue. How many distinguishable patterns are there under equivalence relative to the group of rotations of the hexagon? (Compare Problem 52.5.)

52.7. State and solve the problem that results from replacing a square by a regular pentagon in Example 52.2. Compare Problem 52.4.

52.8. Repeat Problem 52.7 with a regular hexagon in place of a regular pentagon.

52.9. Repeat Problem 52.4 with a regular hexagon in place of a regular pentagon.

52.10. Rewrite Burnside's Counting Theorem and Lemma 52.1, and their proofs, using the notation π_g in place of g wherever appropriate.

52.11. In how many distinguishable ways can the three edges of a triangle be painted with the three colors red, white, and blue if there is no restriction on the number of times each color can be used? State clearly how this problem is like, and also different from, the example at the first of this section, and also Example 52.2.

52.12. In how many distinguishable ways can the four faces of a regular tetrahedron be painted with four different colors if each face is to be a different color and two ways are considered indistinguishable if one can be obtained from the other by rotation of the tetrahedron? (The group of rotations in this case has order 12. In addition to the identity, there are eight 120 rotations around lines such as ae in the following figure, and three 180 rotations around lines such as fg.)

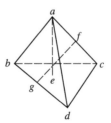

52.13. A bead is placed at each of the eight vertices of a cube, and each bead is to be painted either red or blue. Under equivalence relative to the group of rotations of the cube, how many distinguishable patterns are there?

52.14. Repeat Problem 52.13 with a regular tetrahedron in place of a cube. (The group is described in Problem 52.12.)

SECTION 53 SYLOW'S THEOREM

Sylow's Theorem. *If G is a group of order $p^k m$, where p is a prime and $p \nmid m$, then G has a subgroup of order p^k.*

This theorem should be viewed as a partial converse of Lagrange's Theorem. That theorem tells us that if H is a subgroup of G, then $|H|$ divides $|G|$. However, as we learned in Section 17, there is no guarantee of a subgroup of each possible order dividing the order of a group. Sylow's Theorem provides such a guarantee for the highest power of each prime dividing the order. In fact, the ideas we shall use in proving Sylow's Theorem can also be used to prove the existence of a subgroup of each prime power order—not just the highest—dividing the order of a group (Problem 53.15).

This theorem was first proved by Sylow in 1872. The proof we give here is due to the German mathematician Helmut Wielandt. It is a splendid application of the idea of a group acting on a set.

In the proof of the theorem, S denotes the set of all p^k-element subsets of G, and G acts on this set in a natural way. Before the proof we need some preliminary results about the set S.

The number of r-element subsets of an n-element set is the binomial coefficient

$$\binom{n}{r} = \frac{n!}{r!\,(n-r)!} = \frac{n\,(n-1)\cdots(n-r+1)}{r!}. \tag{53.1}$$

Here n can be any positive integer, and $0 \le r \le n$. You can construct (or review) a proof of this by doing Problem 53.8. We shall need to consider this in the special case $n = p^k m$, $r = p^k$.

Lemma 53.1. *If p is a prime and $p \nmid m$, then*

$$p \nmid \binom{p^k m}{p^k}.$$

PROOF.

$$\binom{p^k m}{p^k} = \frac{p^k m (p^k m - 1) \cdots (p^k m - j) \cdots (p^k m - p^k + 1)}{p^k (p^k - 1) \cdots (p^k - j) \cdots (p^k - p^k + 1)}$$

Except for the factor m, which is not divisible by p, this fraction is equal to a product of fractions

$$\frac{p^k m - j}{p^k - j},$$

with $0 < j \le p^k - 1$. If these fractions are reduced to lowest terms, then none has a numerator divisible by p, because for $0 < j \le p^k - 1$ the highest power of p dividing $p^k - j$ is the same as the highest power of p dividing $p^k m - j$, both being equal to the highest power of p dividing j (Problem 53.9). ∎

PROOF OF SYLOW'S THEOREM. Let S denote the set of all p^k-element subsets of G. By the lemma, $p \nmid |S|$. Also, G acts on S as follows: for $a \in G$ and $T \in S$,

$$\pi_a(T) = aT = \{at : t \in T\}.$$

Because $|S|$ is the sum of the numbers $|\text{Orb}(T)|$ for the different G-orbits, we must have $p \nmid |\text{Orb}(T)|$ for some $T \in S$, for otherwise we would have $p \mid |S|$. Let $\{T_1, \ldots, T_u\}$ be an orbit with $p \nmid u$, and let $H = \{g \in G : gT_1 = T_1\}$. Then H is a subgroup—in fact, in the notation of Section 51, $H = G_{T_1}$. By Theorem 51.2, $|G| = u \cdot |H|$. Since $p^k \mid |G|$, and $p \nmid u$, we must have $p^k \mid |H|$. Therefore $p^k \leq |H|$.

Because $hT_1 = T_1$ for each $h \in H$, we can also think of H as acting on T_1. Let $t \in T_1$. Then H_t, which is $\{h \in H : ht = t\}$, contains only e, so $|H_t| = 1$. By Theorem 51.2, $|H| = |\text{Orb}(t)| \cdot |H_t|$. Therefore $|H| = |\text{Orb}(t)| \leq p^k$.

From $p^k \leq |H| \leq p^k$ we deduce that $|H| = p^k$.   ■

Sylow's Theorem, as stated previously, can be extended to include still more information about subgroups of prime power order. To do this, it will help to introduce the following terminology. Let p be a prime. A subgroup whose order is a power of p is called a *p-subgroup*. A p-subgroup of a finite group G whose order is the highest power of p dividing $|G|$ is called a *Sylow p-subgroup*. If H and K are subgroups of a group G, and $K = aHa^{-1}$ for some $a \in G$, then K is a *conjugate* of H (see Problem 51.12).

Sylow's Theorem (Extended Version).   *Let G be a group of order $p^k m$, where p is a prime and $p \nmid m$. Then*

(a)   *the number n_p of Sylow p-subgroups of G satisfies $n_p \equiv 1 \ (\text{mod } p)$,*

(b)   *$n_p \mid m$, and*

(c)   *any two Sylow p-subgroups are conjugate.*

Although we shall not prove this theorem, an application of it is given in the following example. This example gives some idea of the usefulness of Sylow's Theorem, but only more experience with the theory of finite groups can give a true appreciation of the theorem.

Example 53.1.   Assume that G is a finite group of order $p^k m$, where p is a prime, $p \nmid m, m > 1$, and $k > 0$. It is not hard to prove that if G has only one Sylow p-subgroup, then the subgroup must be normal (Problem 53.10). In particular, if G has only one Sylow p-subgroup, then G is not simple. (Simple groups and their importance are discussed in Section 23.) This fact can be used along with parts (a) and (b) of Sylow's Theorem (the extended version) to show that there are no simple groups of certain orders. For example, suppose that $|G| = 100$. Then $n_5 \equiv 1 \ (\text{mod } 5)$ so that $n_5 = 1$ or 6 or 11 or $\cdots$. But also $n_5 \mid 4$. Therefore $n_5 = 1$, that is, G has only one Sylow 5-subgroup; thus G is not simple. (Problem 51.19 outlines a different proof that there is no simple group of order 100.)   ■

PROBLEMS

53.1. What are the orders of the Sylow p-subgroups of a group of order 180?

53.2. What are the orders of the Sylow p-subgroups of a group of order 700?

53.3. Verify Lemma 53.1 by direct computation for $\binom{12}{4}$.

53.4. Verify Lemma 53.1 by direct computation for $\binom{20}{4}$.

53.5. Find all of the Sylow p-subgroups of S_3 for $p = 2$ and $p = 3$. Verify that the three conditions in Sylow's Theorem (the extended version) are satisfied for both values of p.

53.6. Use Sylow's Theorem (Extended Version) to verify that if $|G| = 24$, then n_2 equals 1 or 3, and n_3 equals 1 or 4. Now find n_2 and n_3 for S_4 by finding all Sylow 2-subgroups and all Sylow 3-subgroups of S_4.

53.7. Explain why every group of order 12 must have a subgroup of every order dividing 12 except, possibly, 6. (Problem 17.28 gives an example of a group of order 12 that has no subgroup of order 6.)

53.8. Assume that n and r are integers such that $0 < r < n$.
   (a) Explain why the number of ways of choosing r elements from an n-element set, taking into account the order in which the elements are chosen, is $n(n - 1) \cdots (n - r + 1)$.
   (b) Let $\binom{n}{r}$ denote the number of ways of choosing r elements from an n-element set, without regard to order. Explain why

$$r! \binom{n}{r} = n(n - 1) \cdots (n - r + 1).$$

   (Notice that each way of choosing r elements without regard to order corresponds to $r!$ ways of choosing r elements taking order into account.)
   (c) Using part (b) and the convention $0! = 1$, verify Equation (53.1).

53.9. Assume that p is a prime.
   (a) Prove that if $0 < j \leq p^k - 1$, then $p^e \mid j$ iff $p^e \mid (p^k - j)$.
   (b) Prove that if $0 < j \leq p^k - 1$ and $p \nmid m$, then

$$p^e \mid j \quad \text{iff} \quad p^e \mid (p^k m - j).$$

53.10. (a) Verify that if H is a subgroup of G, and $a \in G$, then aHa^{-1} is a subgroup of G.
   (b) Prove that if H is a finite subgroup of G, and $a \in G$, then $\left| aHa^{-1} \right| = |H|$. (*Suggestion*: The mapping $h \mapsto aha^{-1}$ is one-to-one.)
   (c) Explain why if H is a Sylow p-subgroup of a finite group, then so is each conjugate of H.
   (d) Prove that if a finite group has only one Sylow p-subgroup for some prime p, then that subgroup must be normal.

53.11. (a) Use Sylow's Theorem (the extended version) to prove that every group of order 6 has only one subgroup of order 3.
   (b) Give an example of a group of order 6 that has more than one subgroup of order 2.
   (c) Give an example of group of order 6 that has only one subgroup of order 2.

53.12. Prove that there is no simple group of order 20. (See Example 53.1.)

53.13. Prove that a group of order 75 must have normal subgroups of orders 3 and 25. (See Example 53.1.)

53.14. Let G be a group of order 56.
   (a) Say as much as you can, based on Sylow's Theorem (the extended version), about the number of Sylow 2-subgroups and the number of Sylow 7-subgroups of G.
   (b) Explain why the total number of nonidentity elements in the Sylow 7-subgroups of G is $6n_7$. (Use Lagrange's Theorem.)
   (c) Using parts (a) and (b), explain why either $n_7 = 1$ or $n_2 = 1$.
   (d) Using (c), explain why there is no simple group of order 56.

53.15. Prove that if p is a prime and G is a finite group such that $p^k \mid |G|$, then G has a subgroup of order p^k, [*Suggestion*: Follow the proof of Sylow's Theorem, but first prove the following generalization of Lemma 53.1: If p is a prime and p^t is the highest power of p dividing m, then p^t is also the highest power of p dividing $\binom{p^k m}{p^k}$.]

CHAPTER XIV

SYMMETRY

This chapter continues the discussion of groups and symmetry that was begun in Section 8. The first section gives complete lists of the finite symmetry groups of plane figures, and the finite rotation groups of three-dimensional figures. The next section gives a complete list of the infinite discrete symmetry groups of two-dimensional figures; these are the symmetry groups of such things as border ornaments and wallpaper patterns. The third section is an introduction to the use of groups in crystallography, one of several important applications of groups to symmetry problems in science. In the last section of the chapter we draw on linear algebra to prove a result known as the *crystallographic restriction*; this is done to illustrate the natural connection between groups, geometry, and linear algebra. The proofs of many of the facts presented in this chapter are too long to be included here. The goal is to summarize some ideas and introduce some others, and to provide more experience with groups and symmetry through the examples and problems. References for proofs and other details can be found in the notes at the end of the chapter. The list of references in [14] gives sources for other applications in physics and chemistry.

SECTION 54   FINITE SYMMETRY GROUPS

The motions (isometries) leaving either a plane figure or a three-dimensional figure invariant form a group, called the *group of symmetries* (or *symmetry group*) of the figure. We saw this for plane figures in Section 8. For applications of group theory to geometry and crystallography it is important to determine just what groups can arise as symmetry groups—it turns out that most groups cannot. In this section we shall get an idea of what the possibilities are among finite groups. We begin with the two-dimensional case.

If a motion preserves orientation, it is called a *proper motion*. (A plane motion preserves orientation if the clockwise direction around a circle before the motion is applied is still the clockwise direction after the motion is applied. Any translation is a proper motion; reflection through any line is not.)

There are three basic types of plane motions: rotations, translations, and glide-reflections. We met rotations, translations, and reflections in Section 8. A *glide-reflection* is a translation in the direction of a line followed by reflection through the line; the set of footprints shown in Figure 54.1 (assumed to extend infinitely far in each direction) is invariant under a glide-reflection consisting of translation through $t/2$ units followed by reflection through the central line (axis). For convenience, the case of reflection without translation is included in the category of glide-reflections.

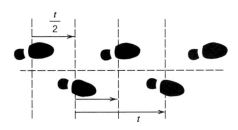

Figure 54.1

It is easy to verify that rotations and translations are proper motions, and that glide-reflections are improper. It is also easy to verify that if two motions are applied in succession, then the result will be proper if both are proper, proper if both are improper, and improper if one is proper and the other is improper. Problems 54.22, 54.23, and 54.24 suggest how to prove the following theorem.

Theorem 54.1. *A motion of the plane is either a rotation, a translation, or a glide-reflection. The symmetry group of a plane figure either contains only proper motions, or its proper motions form a subgroup of index two (in which case half of its motions are proper and half are improper).*

We shall restrict attention hereafter to discrete groups of motions: A group G of plane motions is *discrete* if for each point p in the plane there is a circle with center at p such that for each μ in G either $\mu(p)$ is outside that circle or $\mu(p) = p$. The symmetry group of a square (Example 8.1) is discrete. The symmetry group of a circle is not discrete—any point on the circle can be reached from points arbitrarily close to itself by rotations in the group. A group G of three-dimensional motions is *discrete* if for each point p there is a sphere with center at p such that for each μ in G either $\mu(p)$ is outside that sphere or $\mu(p) = p$. A finite symmetry group is necessarily discrete (Problem 54.12.)

Example 54.1.   The group of Figure 54.2 is cyclic of order 3. It contains the identity, clockwise rotation through $120°$ around p, and clockwise rotation through $240°$ around p. Each is a proper motion. We denote this group by C_3.   ∎

Example 54.2.   The group of Figure 54.3 is of order 6. It contains the group in Example 54.1 as a subgroup, and it also contains three reflections (improper motions), one through each of the three axes intersecting at p. We denote this group by D_3.   ∎

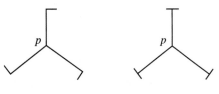

Figure 54.2   **Figure 54.3**

Except for the number of rotations and axes involved, the groups in the two examples just given exhaust the finite plane symmetry groups. To state this precisely, we introduce the following two classes of groups.

A group C_n is cyclic of order n, and consists of clockwise rotations through $k(360/n)°$, $0 \le k < n$, around a fixed point p.

A group D_n has order $2n$, and contains the elements of C_n together with reflections through n axes that intersect at p and divide the plane into $2n$ equal angular regions. The groups D_n are called *dihedral groups*.

Example 54.3. The symmetry group of a square is D_4; the axes of symmetry are the two diagonals and the horizontal and vertical lines through the center of the square (Example 8.1). The symmetry group of an equilateral triangle is D_3. More generally, the symmetry group of a regular n-sided polygon is D_n. Most snowflakes have the symmetry group D_6 (Figure 54.4), although some have the symmetry group D_3. (This phenomenon was studied

Figure 54.4

by Kepler [11].) The symmetry group of a rectangle (Example 8.1) is D_2, and that of a parallelogram (again Example 8.1) is C_2.

In his interesting book *Symmetry* [22], Hermann Weyl credits the discovery of the following theorem, in essence, to Leonardo da Vinci, who wanted to determine the possible ways to attach chapels and niches to a central building without destroying the symmetry of the nucleus. ∎

Theorem 54.2. *A finite symmetry group of a plane figure is either a cyclic group C_n or a dihedral group D_n.*

PROOF. Assume that G is a finite symmetry group of a plane figure, and assume first that G has order n and contains only rotations. Each rotation except the identity can be assumed to be clockwise through a positive angle of less than $360°$. Let α be the one with the smallest angle. Then G contains $\iota = \alpha^0, \alpha, \alpha^2, \alpha^3, \ldots$; in fact, every element of G will appear in this list: For suppose that $\beta \in G$ and $\beta \neq \alpha^k$ for every k. If the angles for α and β are θ_α and θ_β, respectively, then $t\theta_\alpha < \theta_\beta < (t+1)\theta_\alpha$, for some positive integer t. But then $\beta\alpha^{-t} \in G$ and $\beta\alpha^{-t}$ corresponds to a positive clockwise rotation through an angle less than θ_α, a contradiction. Therefore, $G = \{\iota = \alpha^0, \alpha, \ldots, \alpha^{n-1}\}$ with $\alpha^n = \iota$ and $\theta_\alpha = (360/n)°$.

Now assume that G contains a reflection ρ, and let H denote the set of rotations in G. Then H is a subgroup of G and by the first part of the proof we can assume that $H = \{\iota = \alpha^0, \alpha, \ldots, \alpha^{n-1}\}$ for some rotation α. Necessarily, G contains $\iota, \alpha, \ldots, \alpha^{n-1}, \rho, \rho\alpha, \cdots, \rho\alpha^{n-1}$. Each element $\rho\alpha^k$ $(0 \leq k < n)$ is a reflection, and these elements are all distinct by the left cancellation law in G.

Let μ be a reflection in G. Then $\rho\mu$, being the product of two reflections, is a rotation (Problem 54.16). Therefore $\rho\mu = \alpha^k$ for some k so that $\mu = \rho^{-1}\alpha^k = \rho\alpha^k$. It follows from this that $G = \{\iota, \alpha, \ldots, \alpha^{n-1}, \rho, \rho\alpha, \ldots, \rho\alpha^{n-1}\}$. Problem 54.18 is designed to convince you that this makes G a dihedral group. ∎

We turn now to three-dimensional figures. We shall restrict our attention to rotations, and in Theorem 54.3 shall give a complete list of finite rotation groups. (As in the case of plane figures, a finite symmetry group for a three-dimensional figure either contains only rotations or the rotations form a subgroup of index 2.) In describing these groups, an axis of rotation corresponding to rotations of multiples of $(360/m)°$ will be called an *m-fold* axis. Before Theorem 54.3 we give an example of each type of group that will occur.

Example 54.4. An *n-pyramid* $(n \geq 3)$ is a right pyramid whose base is a regular *n*-sided polygon (for $n = 3$ we also require that the edges off the base have a length different from those on the base). A 4-pyramid is shown in Figure 54.5. For completeness we define a *2-pyramid* to be a right "pyramid" whose base has the shape in Figure 54.6. The rotation group of an *n*-pyramid is isomorphic to C_n. The line connecting the center of the base to the vertex off the base is an *n*-fold axis, and the elements of the group are rotations about this axis. ∎

Figure 54.5   **Figure 54.6**

Example 54.5. An *n-prism* $(n \geq 3)$ is a right cylinder whose base is a regular *n*-sided polygon (for $n = 4$ we also require that the height be different from the length of the edges on the base). A *2-prism* is a right cylinder whose base has the shape in Figure 54.6. A

6-prism is shown in Figure 54.7. The rotation group of an n-prism is isomorphic to D_n. The line connecting the centers of the top and bottom bases is an n-fold axis. For n even, the lines connecting midpoints of opposite side-faces are 2-fold axes, and the lines connecting midpoints of opposite side-edges are also 2-fold axes. For n odd, each line connecting the midpoint of a side-face to the midpoint of the opposite side-edge is a 2-fold axis. The elements of the group are the rotations about these different axes.   ∎

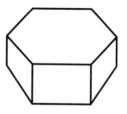

Figure 54.7

The other finite rotation groups are connected with the five regular convex polyhedra, shown in Figure 54.8. These are the only convex solids whose faces are congruent regular polygons and whose vertex angles are all equal. The rotation group of the cube was described in Example 52.3, and that of the tetrahedron in Problem 52.12.

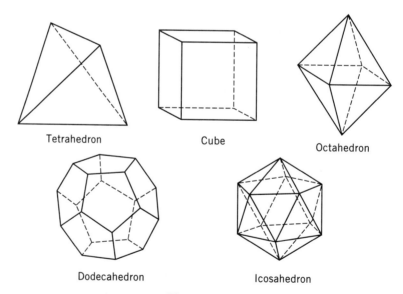

Figure 54.8

Example 54.6.   The rotation group of the tetrahedron has order 12 and is called the *tetrahedral group*. There are four 3-fold axes (one through each vertex), and three 2-fold axes (joining the midpoints of nonintersecting edges).   ∎

Example 54.7.   The rotation group of the cube has order 24. It is the same as the rotation group of the octahedron, and is called the *octahedral group*. For the cube, there are three 4-fold axes (joining centers of opposite faces), six 2-fold axes (joining midpoints of opposite edges), and four 3-fold axes (joining opposite vertices). ∎

Example 54.8.   The rotation group of the icosahedron has order 60. It is the same as the rotation group of the dodecahedron, and is called the *icosahedral group*. For the icosahedron, there are six 5-fold axes (joining pairs of opposite vertices), ten 3-fold axes (joining centers of opposite faces), and fifteen 2-fold axes (joining the midpoints of opposite edges). ∎

Theorem 54.3.   *A finite rotation group of a three-dimensional figure is either a cyclic group, a dihedral group, the tetrahedral group, the octahedral group, or the icosahedral group.*

Proofs of this theorem can be found in references [1], [4], [7], [14], and [23] listed at the end of this chapter. Each of these references also gives a list and derivation of the finite symmetry groups (of three-dimensional figures) that contain improper motions, that is, reflections and reflections combined with rotations.

Reference [5] gives a thorough analysis of Rubik's Cube in terms of group theory.

PROBLEMS

Determine the symmetry group of each of the following figures.

54.1.

54.3.

54.2.

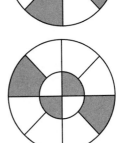

54.4.

What is the symmetry group of the graph (in the xy-plane) of each of the following equations?

54.5. $y = x^2$ **54.6.** $y = x^3$ **54.7.** $3x^2 + 4y^2 = 12$ **54.8.** $xy = 1$

Determine the symmetry group of each of the following figures.

54.9.   **54.10.**

54.11. The groups C_2 and D_1 are isomorphic, but they are not the same as groups of motions. Show this by drawing one figure with C_2 as the symmetry group and another figure with D_1 as the symmetry group.

54.12. Explain why a finite symmetry group must be discrete.

54.13. True or false: Every subgroup of a discrete group is discrete.

54.14. Prove that the dihedral group D_3 is isomorphic to the symmetric group S_3.

54.15. Verify that D_n is non-Abelian for $n \geq 3$. What about D_2?

54.16. Explain why the product $\rho\mu$ in the proof of Theorem 54.2 must be a proper rotation. (Notice that $\rho\mu$ must be a proper motion of finite order.)

54.17. Assume α and ρ are as in the proof of Theorem 54.2, with ρ denoting reflection through L. Use the following figure to give a geometrical proof that $\rho\alpha^k$ is the same as reflection through the line bisecting L and $\alpha^{-k}(L)$. [Compute $\rho\alpha^k(p)$.]

54.18. Use the notation (α and ρ) from the proof of Theorem 54.2 and compute the Cayley table for $D_6 = \{\iota, \alpha, \dots, \alpha^5, \rho, \rho\alpha, \dots, \rho\alpha^5\}$, the symmetry group of a regular hexagon.

54.19. With the vertices of a tetrahedron labeled as in Problem 52.12, write the permutation of $\{a, b, c, d\}$ corresponding to each of the 12 rotations in the tetrahedral group.

54.20. (Dodecahedral dice. This assumes knowledge of Section 52.) In how many distinguishable ways can the 12 faces of a dodecahedron be numbered 1 through 12 if two ways are considered indistinguishable if one can be obtained from the other by rotation of the dodecahedron? (Compare Example 52.3.)

54.21. (This assumes knowledge of Section 52.) A bead is placed at each of the 12 vertices of an icosahedron, and each bead is to be painted either red or blue. Under equivalence relative to the group of rotations of the icosahedron, how many distinguishable patterns are there?

54.22. Let μ be a proper motion of the plane, and let p be a point on a directed line L in the plane. (If you stay above the plane and look in the positive direction along L, and a point r is on your right, then $\mu(r)$ will also be on your right if you look in the positive direction along $\mu(L)$; $\mu(r)$ would be on your left if μ were improper.)
(a) Explain why μ is uniquely determined by $\mu(p)$ and $\mu(L)$.

(b) Explain why μ is a rotation if $p = \mu(p)$, and a translation if L and $\mu(L)$ have the same direction.

(c) Assume that $p \neq \mu(p)$, and that L and $\mu(L)$ do not have the same direction. Let q be a point equidistant from p and $\mu(p)$ such that angle $pq\mu(p)$ equals the angle between L and $\mu(L)$, as in the figure. Explain why μ is a rotation about q.

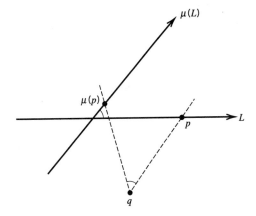

54.23. Let μ be an improper motion of the plane, and let p be a point on a directed line L in the plane. (Compare Problem 54.22.)

(a) Explain why μ is uniquely determined by $\mu(p)$ and $\mu(L)$.

(b) Let q be the midpoint of $p\mu(p)$, and let M be the line through q parallel to the bisector of the angle between the positive directions of L and $\mu(L)$, as in the following figure. Explain why μ is a glide-reflection with axis M.

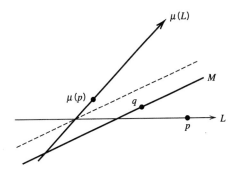

54.24. Prove Theorem 54.1, using Problems 54.22 and 54.23, and the remarks preceding the theorem.

SECTION 55   INFINITE TWO-DIMENSIONAL SYMMETRY GROUPS

If we imagine the designs on decorative borders or wallpaper to be repeated infinitely often, we obtain figures whose symmetry groups are discrete and infinite. These groups contain translations, and are conveniently divided into two classes. The groups in one class each leave a line invariant, and the groups in the other class do not. The groups in the first class are called *frieze groups*. There are seven frieze groups in all, and examples of the corresponding patterns can be seen in Figures 55.1–55.7. These are the symmetry groups for repeating patterns that appear on vases, bracelets, and decorative borders, and they will be described along with the examples of the patterns. Figure 55.8 gives one example of each pattern, all taken from Greek vases.

There are 17 groups that do not leave a line invariant. Figure 55.9 gives one pattern corresponding to each group. Patterns of this type will be familiar from designs on wallpaper, decorative ceilings, and tile and brick arrangements; the patterns in Figure 1 of the Introduction are of this type, for example. The 17 corresponding groups are discussed in Section 56.

Each type of pattern in this section can be found in the artwork of the past. The book [10] by Owen Jones contains an especially rich collection of such patterns. Another source is the drawings of M. C. Escher, which have been analyzed in terms of their symmetry in [13].

Descriptions of each of the seven frieze groups follow. The accompanying figures are assumed to extend infinitely far to the left and right.

Type I. The group in Figure 55.1 consists of translations only. If τ denotes translation through the smallest possible distance (to the right, say, in Figure 55.1), then the group is infinite cyclic with generator τ.

$$F_{\mathrm{I}} = \left\{ \tau^k : k \in \mathbb{Z} \right\}$$

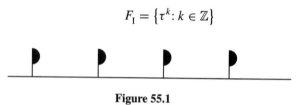

Figure 55.1

Type II. The group in Figure 55.2 is also infinite cyclic. It is generated by a glide-reflection; if this glide-reflection is denoted by γ, then the even powers of γ are translations.

$$F_{\mathrm{II}} = \left\{ \gamma^k : k \in \mathbb{Z} \right\}$$

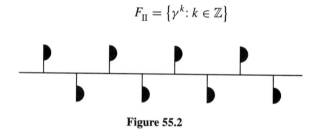

Figure 55.2

Type III. The group in Figure 55.3 is generated by a translation (say τ) and a reflection (say ρ) through a vertical line (such as the dotted line in the figure). This group is an *infinite dihedral group*. (Verify that $\tau\rho = \rho\tau^{-1}$.)

$$F_{\mathrm{III}} = \left\{ \tau^k \rho^m : k \in \mathbb{Z}, m = 0.1 \right\}$$

Figure 55.3

Type IV. The group in Figure 55.4 is generated by a translation (say τ) and a rotation (say α) through 180° around a point such as p in the figure. This is also an infinite dihedral group.

$$F_{IV} = \{\tau^k \alpha^m : k \in \mathbb{Z}, m = 0, 1\}$$

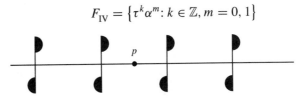

Figure 55.4

Type V. The group in Figure 55.5 is generated by a glide-reflection (say γ) and a rotation (say α) through 180° around a point such as p in the figure; F_V is another infinite dihedral group.

$$F_V = \{\gamma^k \alpha^m : k \in \mathbb{Z}, m = 0, 1\}$$

Figure 55.5

Type VI. The group in Figure 55.6 is generated by a translation (say τ) and a reflection (say β) through an axis of symmetry (which is horizontal in the figure). This group is Abelian. (Verify that $\tau\beta = \beta\tau$.)

$$F_{VI} = \{\gamma^k \beta^m : k \in \mathbb{Z}, m = 0, 1\}$$

Figure 55.6

Type VII. The group in Figure 55.7 is generated by a translation (say τ) and two reflections (say ρ, as in type III; and β, as in type VI). The elements τ and ρ multiply as in F_{III}; τ and β multiply as in F_{VI}; and $\rho\beta = \beta\rho$.

$$F_{VII} = \{\tau^k \beta^m \rho^n : k \in \mathbb{Z}, m = 0, 1, n = 0.1\}$$

Figure 55.7

Figure 55.8

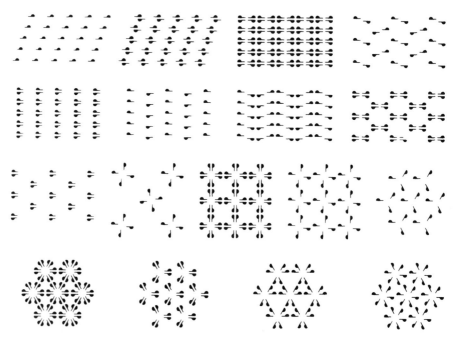

Figure 55.9

PROBLEMS

55.1. Match each part of Figure 55.8 with the appropriate group (F_I through F_{VII}.)

55.2. Each of the following figures has one of F_I through F_{VII} as symmetry group, and no two parts correspond to the same group. Match each part with its group.

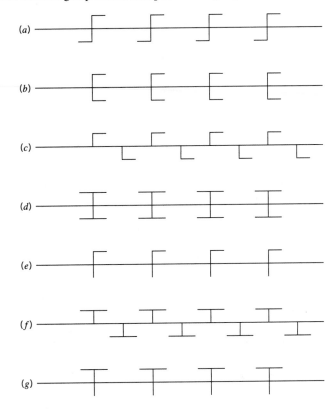

55.3. Each of the following figures has one of F_I through F_{VII} as its symmetry group. Find the appropriate group for each part.

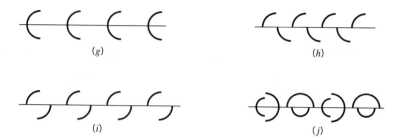

(g) *(h)*

(i) *(j)*

55.4. Draw seven figures, different from those in the book, illustrating the seven types of symmetry corresponding to the groups F_{I} through F_{VII}.

SECTION 56   ON CRYSTALLOGRAPHIC GROUPS

One of the most interesting applications of groups outside of mathematics is in crystallography. At the heart of this application are 32 finite (three-dimensional) symmetry groups, known as *crystallographic point groups*, and 230 infinite symmetry groups, known as *crystallographic space groups*, which can be constructed from translation groups and the 32 crystallographic point groups. A detailed description of these groups would take us too far afield, but we can get an idea of what they involve. We begin with the two-dimensional analogue.

Consider the following problem: Fill the plane with congruent polygons having no overlap except at the edges. (In this section "polygon" will mean "polygon and its interior.") Figure 56.1 suggests solutions with equilateral triangles, squares, and regular hexagons. It also shows why there is no solution with regular pentagons. In fact, 3, 4, and 6 are the only values of n for which there is a solution with regular n-gons (Problem 56.7). Figure 56.2 offers two solutions with congruent polygons that are not regular.

Now refine the problem by demanding that the congruent polygons needed to fill the plane be those obtained by applying the motions of some group to just a single polygon; this polygon is called a *fundamental region* for the corresponding group. In the case of squares [Figure 56.1b], translations in perpendicular directions would suffice—the plane can be filled by starting with one square and moving it repeatedly to the left and right and up and down. For regular hexagons [Figure 56.1c], translations in directions meeting at angles of 60° would suffice. For equilateral triangles [Figure 56.1a], we need the same translations as for hexagons, and we also need rotations or reflections—rotations through multiples of 60° about a vertex of a triangle will fill out a hexagon, whose translates will fill out the plane. In Figure 56.2 the parallelograms offer a solution, but the other example does not. (We return to this last point at the end of the section.)

The group associated with any solution of this problem will be the symmetry group of the figure formed by the edges of all the polygons making up the solution. The fundamental fact connecting group theory and two-dimensional crystallography is this: there are just 17 groups that arise in this way. These are the *two-dimensional crystallographic groups*, and it turns out that they are precisely the discrete symmetry groups (for plane figures) that leave neither a point nor a line invariant. Thus they are the groups of the patterns in Figure 55.9. (Each of the finite groups in Section 54 leaves a point invariant, and each of the frieze groups in Section 55 leaves a line invariant.)

Figure 56.1

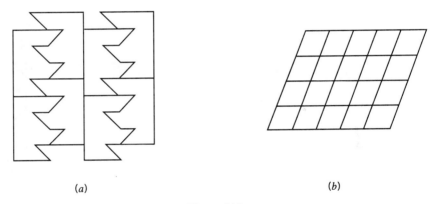

Figure 56.2

Figure 56.3 contains diagrams corresponding to each of the 17 groups. These diagrams are based on figures in [2] and give the following information.

1. The shaded area is a fundamental region.

2. The generators of the group are given by the following scheme:

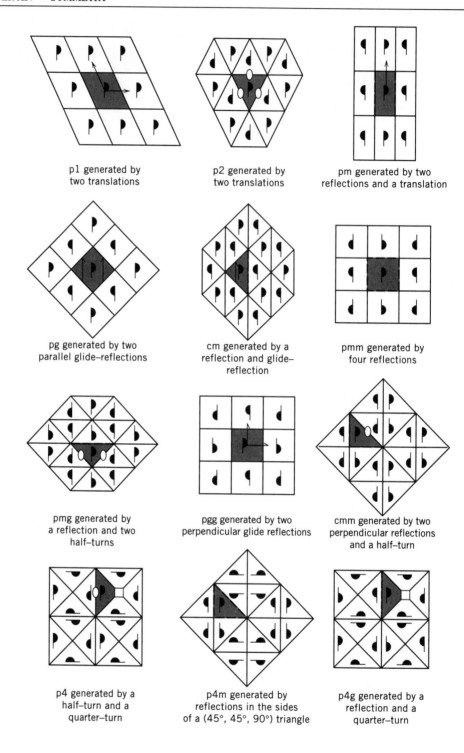

p1 generated by
two translations

p2 generated by
two translations

pm generated by two
reflections and a translation

pg generated by two
parallel glide–reflections

cm generated by a
reflection and glide–
reflection

pmm generated by
four reflections

pmg generated by
a reflection and two
half–turns

pgg generated by two
perpendicular glide reflections

cmm generated by two
perpendicular reflections
and a half–turn

p4 generated by a
half–turn and a
quarter–turn

p4m generated by
reflections in the sides
of a (45°, 45°, 90°) triangle

p4g generated by a
reflection and a
quarter–turn

Figure 56.3

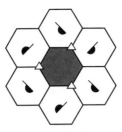

p3 generated by three
rotations through 120°

p31m generated by a reflection
and a rotation through 120°

p3m1 generated by three
reflections in the sides of an
equilateral triangle

p6 generated by a half–turn
and a rotation of 120°

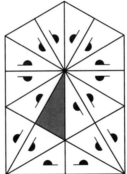

p6m generated by reflections in the
sides of a (30°, 60°, 90°) triangle

Figure 56.3 (continued)

By starting with a fundamental region and subjecting it repeatedly to the motions suggested by the generators, the entire plane will be filled. Each element of the group will correspond to a different congruent copy of the fundamental region. The notation **p1**, **p2**, ... for the groups is based on [8], and belongs to one of several different systems used to describe such groups. The notation is given here for convenience; it will not be explained.

Figure 56.4 gives an algorithm for identifying the two-dimensional crystallographic group of a plane figure. In Figure 56.4, *mirror line* refers to a reflection axis, slide line refers to the reflection axis of a glide-reflection, and C_n and D_n refer to cyclic and dihedral groups, respectively (Section 54). Figure 56.4 is taken from [17], which also gives algorithms for identifying frieze groups and symmetry groups of bounded plane figures.

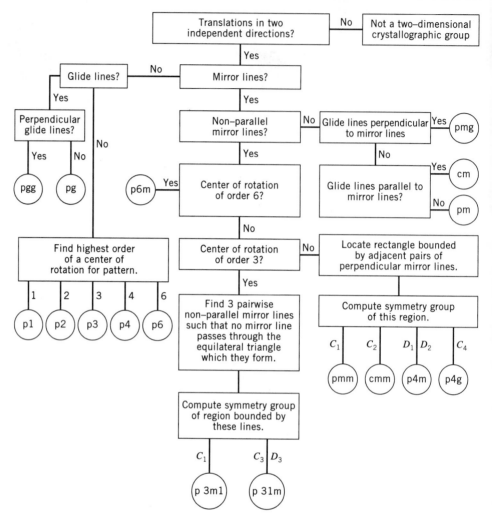

Figure 56.4   Algorithm for identifying two-dimensional crystallographic groups

The groups that are of interest in crystallography arise from the three-dimensional version of the problem just considered. These groups can be described as follows. First, a *lattice group* is a nontrivial discrete motion group whose elements are translations. If, for example, a rectangular coordinate system is chosen for three-dimensional space, then the three translations of unit length in the positive directions parallel to the coordinate axes generate a lattice group. If G is a lattice group, then any G-orbit is called a *lattice*.[†] (A G-orbit consists of all the points that can be reached by applying motions of G to a single point of the lattice. See Section 51.) For the example mentioned, the lattice containing the origin would consist of all the points with integral coordinates.

[†]This use of the word *lattice* is different from that in Chapter XVI, Both uses of the word are well-established, and confusion is unlikely.

The lattice and associated group are called one-, two-, or three-dimensional depending on whether the points of the lattice are collinear, coplanar but not collinear, or not coplanar, respectively. The points of intersection of the lines forming the parallelograms in Figure 56.2 form part of a two-dimensional lattice. The points of intersection of the lines in Figure 56.5 form part of a three-dimensional lattice.

A discrete group of motions is a *crystallographic point group* if it has an invariant point p and leaves invariant some three-dimensional lattice containing p (the individual lattice points other than p will in general not be invariant, of course). A discrete group of motions is a *crystallographic space group* if its translations form a three-dimensional lattice group.

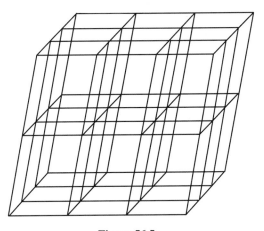

Figure 56.5

As stated at the outset, there are 32 crystallographic point groups, each of which is finite. Eleven of these groups contain only rotations and are among the groups in Theorem 54.3 (C_1, C_2, C_3, C_4, C_6, D_2, D_3, D_4, D_6, the tetrahedral group, and the octahedral group). Each of the other 21 point groups contains improper motions (reflections through a plane or a point) and contains one of the 11 listed groups of proper motions as a subgroup of index 2.

If G denotes a crystallographic space group, then the translations of G form a normal subgroup T that is a lattice group, and the quotient group G/T is one of the 32 crystallographic point groups. In the 1890s, E. S. Fedorov (Russian), A. Schönflies (German), and W. Barlow (British) showed that there are only 230 crystallographic space groups. Of these, 65 contain only proper motions and the remaining 165 contain improper motions. These groups are the basis for the classification of crystals by symmetry type and have been the subject of extensive study since their importance was first realized in the middle of the last century. The Introduction contains some brief remarks about this use of groups; for more details see [9] or [16].

The proof that there are only 230 crystallographic space groups is, as could be guessed, not particularly easy. One of the restricting factors is the following theorem, which will be proved in Section 57.

The Crystallographic Restriction. *Any nontrivial rotation in a two- or three-dimensional crystallographic point group is or order 2, 3, 4, or 6.*

At the International Congress of Mathematicians in 1900, the German mathematician David Hilbert (1862–1943) presented a collection of 23 problems that have attracted the interest of many mathematicians throughout this century. The eighteenth problem on Hilbert's list was divided into three parts; the second part was this: "Whether polyhedra also exist which do not appear as fundamental regions of groups of motions, by means of which nevertheless by a suitable juxtaposition of congruent copies a complete filling up of all [Euclidean] space is possible" A complicated three-dimensional example was given by K. Reinhardt in 1928. Figure 56.2a is a two-dimensional example, which was constructed by H. Heesch in 1935. Further references are given in the paper by John Milnor on Hilbert's Eighteenth Problem in [15], and also in [3].

PROBLEMS

All figures referred to in Problems 56.1–56.5 are assumed to be extended to fill the plane.

56.1. Determine the symmetry group of each of the designs in Figure 55.9. (Use Figure 56.4.)

56.2. (a-q) Determine the symmetry group of each design in Figure 56.6. Also find a fundamental region for each design. (Use Figures 56.3 and 56.4.)

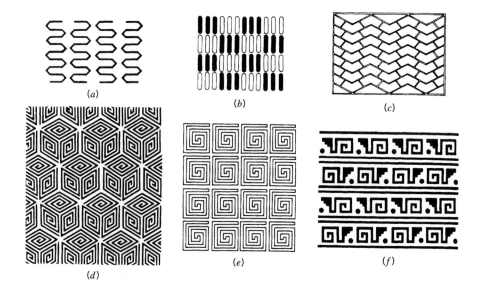

(a)

(b)

(c)

(d)

(e)

(f)

Figure 56.6

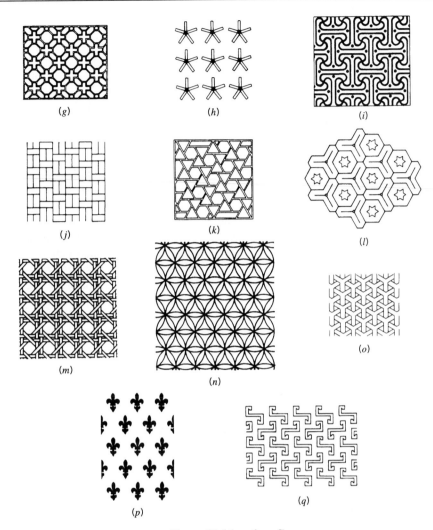

Figure 56.6 (continued)

56.3. Determine the symmetry group of each of the three designs in Figure 1 (of the Introduction).

56.4. Which of the 17 two-dimensional crystallographic groups contain rotations of order 4? 6? Which contain reflections? Which contain glide-reflections that are not reflections?

56.5. Draw 17 figures, different from those in the book, illustrating the 17 types of symmetry shown in Figure 56.3. (Be sure the figures do not have too much symmetry in each case.)

56.6. Verify that the rotations in the 32 crystallographic point groups are consistent with the crystallographic restriction.

56.7. Prove that the only values of n for which the plane can be filled with congruent regular n-gons having no overlap except at their edges are $n = 3, 4$, and 6. [Each of the interior angles of a regular n-gon has measure $(n - 2)\pi/n$. If r such polygons have a vertex in common, then $[r(n - 2)\pi/n] = 2\pi$. For which integral values of n and r is this possible?]

SECTION 57   THE EUCLIDEAN GROUP

The distance-preserving mappings of three-dimensional space form a group, called the *Euclidean group*, which is implicit in all of the mathematics and science based on Euclidean geometry. In agreeing not to distinguish between two congruent figures, we are in essence agreeing not to distinguish between the figures if there is an element of the Euclidean group that maps one of the figures onto the other. All of the symmetry groups of two- or three-dimensional figures are subgroups of the Euclidean group. In this section this group is studied in the context of coordinate geometry and matrices. It is shown that the Euclidean group is built up from several of its important subgroups. Also, and the crystallographic restriction is proved. The section uses ideas and notation from linear algebra that are reviewed in Appendix D.

Let $E(3)$ denote the Euclidean group, and let $\mathbb{R}^3$ denote the vector space of all 3-tuples of real numbers. We begin by studying those elements of $E(3)$ that are also linear transformations. It is convenient to identify the points of three-dimensional Euclidean space with the elements of $\mathbb{R}^3$ in the usual way. To do this, first choose a unit of length and three mutually perpendicular coordinate axes. Then associate the geometric unit vectors along the coordinate axes with the unit vectors $e_1 = (1, 0, 0)$, $e_2 = (0, 1, 0)$, and $e_3 = (0, 0, 1)$ in $\mathbb{R}^3$. Relative to the basis $\{e_1, e_2, e_3\}$, each linear transformation of $\mathbb{R}^3$ is represented by a unique matrix in $M(3, \mathbb{R})$. The elements of $E(3)$ are invertible, so if a linear transformation is in $E(3)$, then the corresponding matrix must be in $GL(3, \mathbb{R})$. Therefore, relative to a fixed coordinate system, the set of linear transformations in $E(3)$ is $E(3) \cap GL(3, \mathbb{R})$. This is an intersection of groups, so it is also a group. It is called the *(real) orthogonal group* and is denoted by $O(3)$. To say more about the elements of $O(3)$, we look at them in terms of coordinates and matrices.

The *inner product* of vectors $v = (x_1, x_2, x_3)$ and $w = (y_1, y_2, y_3)$ is defined by

$$\langle v, w \rangle = \sum_{i=1}^{3} x_i y_i.$$

In particular, then,

$$\langle e_i, e_j \rangle = \delta_{ij}, \tag{57.1}$$

where $\delta_{ii} = 1$ and $\delta_{ij} = 0$ for $i \neq j$.

The *length* of a vector $v = (x, y, z) \in \mathbb{R}^3$ is given by

$$\|v\| = \langle v, v \rangle^{\frac{1}{2}} = (x^2 + y^2 + z^2)^{\frac{1}{2}} \tag{57.2}$$

The *distance* between vectors v and w is given by

$$d(v, w) = \|v - w\|. \tag{57.3}$$

Therefore, a mapping $\alpha: \mathbb{R}^3 \to \mathbb{R}^3$ is in $E(3)$ iff

$$\|\alpha(v) - \alpha(w)\| = \|v - w\| \tag{57.4}$$

for all $v, w \in \mathbb{R}^3$. If the mapping α is linear, then $\alpha(v) - \alpha(w) = \alpha(v - w)$. It follows that a linear mapping is distance-preserving iff it is length-preserving (Problem 57.5). Therefore, $\alpha \in O(3)$ iff α is linear and

$$\|\alpha(v)\| = \|v\| \tag{57.5}$$

for all $v \in \mathbb{R}^3$. It can also be proved (Problem 57.7) that if $\alpha \in O(3)$, then α preserves inner products; that is,

$$\langle \alpha(v), \alpha(w) \rangle = \langle v, w \rangle \tag{57.6}$$

for all $v, w \in \mathbb{R}^3$. The matrix form of α is $A = [a_{ij}]$, where

$$\alpha(e_i) = \sum_{j=1}^{3} a_{ji} e_j \quad (1 \le i \le 3) \tag{57.7}$$

If we compute $\langle a(e_i), a(e_j) \rangle$, and make use of (57.1), (57.6), and (57.7), we obtain

$$\sum_{k=1}^{3} a_{ki} a_{kj} = \delta_{ij} \tag{57.8}$$

for $1 \le i \le 3$, $1 \le j \le 3$ (Problem 57.8). From these remarks and calculations it follows that

$$\alpha \in O(3) \quad \text{iff} \quad A'A = I_3, \tag{57.9}$$

where A is the matrix of α relative to $\{e_1, e_2, e_3\}$.

A matrix A such that $A'A = I_3$ is called *orthogonal*. Any such matrix is invertible, with $A^{-1} = A'$. If $\det A$ is used to denote the determinant of a matrix A, then $\det A = \pm 1$ if A is orthogonal, because if A is orthogonal then $\det(A'A) = (\det A')(\det A) = (\det A)^2$. We recall that if A_1 and A_2 are matrices representing the same linear transformation α relative to different bases, then $\det A_1 = \det A_2$ (Problem 57.10); therefore, $\det \alpha$ can be defined to be $\det A$, where A is the matrix representing α relative to any basis.

The mapping $\alpha \mapsto \det \alpha$ defines a homomorphism from $O(3)$ onto $\{1, -1\}$, because $\det(\alpha\beta) = (\det \alpha)(\det \beta)$. The kernel of this homomorphism is a normal subgroup of $O(3)$, called the *special orthogonal group* and denoted by $SO(3)$. Notice that $[O(3) : SO(3)] = 2$ because $O(3)/SO(3) \approx \{1, -1\}$. We shall now prove that $SO(3)$ consists of the rotations of Euclidean space about axes passing through the origin.

Lemma 57.1.   *If $\alpha \in SO(3)$, then there is a unit vector u_3 such that $\alpha(u_3) = u_3$ and α is a rotation with u_3 as the axis.*

PROOF.   Let A be the matrix of α relative to the basis $\{e_1, e_2, e_3\}$. Then

$$\begin{aligned}
\det(A - I_3) &= \det(A - I_3)' = \det(A' - I_3') = \det(A^{-1} - I_3) \\
&= (\det A^{-1})(\det(I_3 - A)) \\
&= (\det A^{-1})(\det(-I_3))(\det(A - I_3)) \\
&= -\det(A - I_3).
\end{aligned}$$

Therefore, $\det(A - I_3) = 0$ so that $\lambda = 1$ is a characteristic value of A. A characteristic vector u corresponding to $\lambda = 1$ will satisfy $a(u) = u$, as required.

Let u_3 be a unit vector in the direction of u, and choose u_1 and u_2 so that $\{u_1, u_2, u_3\}$ is an orthonormal basis for $\mathbb{R}^3$. Then

$$\langle a(u_i), a(u_j) \rangle = \langle u_i, u_j \rangle = \delta_{ij}. \tag{57.10}$$

The relations given by (57.10) can be used to show that the matrix of α relative to the basis $\{u_1, u_2, u_3\}$ is

$$\begin{bmatrix} \cos\theta & -\sin\theta & 0 \\ \sin\theta & \cos\theta & 0 \\ 0 & 0 & 1 \end{bmatrix} \tag{57.11}$$

for $0 \leq \theta < 2\pi$ (Problem 57.11). This shows that α is a clockwise rotation through angle θ about the axis u_3. ∎

Another important subgroup of $E(3)$ is $T(3)$, the group of *translations*. It can be proved that

$$T(3) \lhd E(3), \quad \frac{E(3)}{T(3)} \approx O(3), \quad \text{and} \quad |T(3) \cap O(3)| = 1.$$

In the language of Section 23, $E(3)$ is an extension of $T(3)$ by $O(3)$. (Proofs of these facts can be found in reference [14].) Instead of pursuing this, we return now to the following fact, which was stated without proof in Section 56.

The Crystallographic Restriction.   *Any nontrivial rotation in a two- or three-dimensional crystallographic point group is of order 2, 3, 4, or 6.*

PROOF.   The proof will be written for three-dimensional groups; the two-dimensional case follows from this (Problem 57.12).

Let α be a rotation in a three-dimensional crystallographic point group, and assume that α leaves the point p fixed and the lattice L invariant. Let $\{v_1, v_2, v_3\}$ be a basis for L; then $\{v_1, v_2, v_3\}$ will also be a basis for $\mathbb{R}^3$. The matrix for α relative to this basis is $A = [a_{ij}]$, defined by

$$\alpha(v_i) = \sum_{j=1}^{n} a_{ji} v_j; \tag{57.12}$$

and each a_{ji} must be an integer by the invariance of L. As shown in the proof of Lemma 57.1, there is a basis for $\mathbb{R}^3$ relative to which the matrix of α has the form (57.12). The trace of the matrix of a linear transformation is invariant under a change of basis; therefore, since the trace $a_{11} + a_{22} + a_{33}$ is an integer, $2\cos\theta + 1$ must be an integer, where θ is the angle of rotation for α (from Lemma 57.1). The only values of θ ($0 \leq \theta < 2\pi$) for which $2\cos\theta + 1$ is an integer are $\theta = 0, \pi/2, 3\pi/2, \pi/3, 2\pi/3, \pi, 4\pi/3$, and $5\pi/3$. The orders of the rotations through these angles are those in the crystallographic restriction. ∎

PROBLEMS

Compute $\langle v, w \rangle$, $\|v\|$, $\|w\|$, and $d(v, w)$.

57.1. $v = (1, 0, -2)$, $w = (-3, 4, 5)$.

57.2. $v = (1, 5, -1)$, $w = (2, 1, 3)$.

57.3. Let $A = \begin{bmatrix} 4/\sqrt{5} & -3/\sqrt{5} & 0 \\ 0 & 0 & -1 \\ 3/\sqrt{5} & 4/\sqrt{5} & 0 \end{bmatrix}$.

(a) Is A orthogonal?

(b) Is A in $SO(3)$?

57.4. Repeat Problem 57.3 for

$$A = \begin{bmatrix} 1/\sqrt{3} & 0 & -2/\sqrt{6} \\ 1/\sqrt{3} & 1/\sqrt{2} & 1/\sqrt{6} \\ -1/\sqrt{3} & 1/\sqrt{2} & -1/\sqrt{6} \end{bmatrix}.$$

57.5. Prove that a linear mapping is distance-preserving iff it is length-preserving. [See the discussion following Equation (57.4).]

57.6. Verify that the inner product used in this section has these properties (for all u, v, $w \in \mathbb{R}^3$ and all $a \in \mathbb{R}$).
   (a) $\langle u + v, w \rangle = \langle u, w \rangle + \langle v, w \rangle$ (b) $\langle u, v + w \rangle = \langle u, v \rangle + \langle u, w \rangle$
   (c) $\langle au, v \rangle = \langle u, av \rangle = a \langle u, v \rangle$ (d) $\langle u, v \rangle = \langle v, u \rangle$

57.7. Prove that if $\alpha \in O(3)$, then α preserves inner products. [See Equation (57.6). *Suggestion:* Use $\langle \alpha(v + w), \alpha (v + w) \rangle = \langle v + w, v + w \rangle$ and Problem 57.6.]

57.8. Verify Equation (57.8).

57.9. Use Equation (57.9) to prove that $O(3)$ is closed with respect to multiplication.

57.10. Prove that if A_1 and A_2 are matrices representing the same linear transformation relative to different bases, then $\det A_1 = \det A_2$. [Recall that $A_2 = B A_2 B^{-1}$ for an appropriate B, and $\det(XY) = \det(X) \det(Y)$.]

57.11. Prove the statement containing Equation (57.11).

57.12. Explain why the crystallographic restriction for three-dimensional groups implies the crystallographic restriction for two-dimensional groups. (*Suggestion:* Consider Figure 54.7.)

NOTES ON CHAPTER XIV

1. Coxeter, H. S. M., *Introduction to Geometry,* 2nd ed., Wiley, New York, 1989.
2. Coxeter, H. S. M., and W. O. J. Moser, *Generators and Relations for Discrete Groups,* 4th ed., Springer-Verlag, Berlin, 1984.
3. Farkas, D. R., Crystallographic groups and their mathematics, *Rocky Mountain Journal of Mathematics,* **11** (1981), 511–551.
4. Fejes Tóth, L., Regular Figures, Macmillan, New York, 1964.
5. Frey, Alexander H., Jr., and O. Singmaster, *Handbook of Cubik Math,* Enslow, Hillside, New Jersey, 1982.
6. Griesbach, C. B., *Historic Ornament: A Pictorial Archive,* Dover, New York, 1976.
7. Grove, L. C., and C. T. Benson, *Finite Reflection Groups,* 2nd ed., Springer-Verlag, Berlin, 1985.
8. Henry, N. F. M., and K. Lonsdale, eds., *International Tables for X-Ray Crystallography,* Kynock Press, Birmingham, England, 1965.
9. Hurlbut, C. S., Jr., and C. Klein, after J. D. Dana, *Manual of Mineralogy,* 21st ed., Wiley, New York, 1998.
10. Jones, O., *The Grammar of Ornament,* Dover, New York, 1989.
11. Kepler, J., *The Six-Cornered Snowflake,* English trans., Oxford University Press, London, 1966.
12. Lockwood, E. H., and R. H. Macmillan, *Geometric Symmetry,* Cambridge University Press, Cambridge, 1978.
13. MacGillavry, C. H., *Fantasy & Symmetry: The Periodic Drawings of M. C. Escher,* Harry N. Abrams, New York, 1976.
14. Miller, W., Jr., *Symmetry Groups and Their Applications,* Academic Press, New York, 1972.

15. Milnor, J., Hilbert's Problem 18: On crystallographic groups, fundamental domains, and on sphere packing, *Proceedings of Symposia in Pure Mathematics, Vol. XXVIII*, American Mathematical Society, Providence, R.I., 1976.

16. Phillips, F. C., *An Introduction to Crystallography*, 4th ed., Wiley, New York, 1971.

17. Rose, B. I., and R. D. Stafford, An elementary course in mathematical symmetry, *American Mathematical Monthly*, **88** (1981), 59–64.

18. Schnattschneider, D., The plane symmetric groups, *American Mathematical Monthly*, **85** (1978),439–450.

19. Schwarzenberger, R. L. E., The 17 plane symmetry groups, *Mathematical Gazette*, **58** (1974), 123–131.

20. _____, *N*-dimensional crystallography, Pitman, London, 1980.

21. Shubnikov, A. V., and V. A. Koptsik, *Symmetry in Art and Science*, Plenum Press, New York, 1974. (Translated from the 1972 Russian edition.)

22. Weyl, H., *Symmetry*, Princeton University Press, Princeton, N.J., 1989.

23. Yale, P. B., *Geometry and Symmetry*, Dover, New York, 1988.

CHAPTER XV

ALGEBRAIC CODING

Occasional errors are inevitable when information is transmitted from one place to another, as from a satellite to the earth, or when information is stored and then retrieved, as in the magnetic storage of a computer. These errors are likely to occur randomly, caused by such things as lightning or circuitry failure. The task for coding theory is to minimize the effect of these errors. This is done by introducing redundancy into the information before it is transmitted or stored, in such a way as to help detect when errors occur or even to correct errors that occur. Interestingly, the techniques used in doing this have been mostly algebraic, involving finite fields and ideas from linear algebra. The study of these algebraic techniques is the domain of algebraic coding theory. A thorough introduction to this theory would draw on much of the preceding material in this book. The treatment here is less thorough than that, to make it more easily accessible and of manageable length; the only prerequisites are parts of Chapters I to VI, material from linear algebra that is reviewed in Appendix D, and some elementary facts about probability that are used in the last section (and can be ignored by taking Theorem 61.1 on faith).

SECTION 58   INTRODUCTION

Example 58.1.   Suppose that a message consisting of a single binary digit, either 0 or 1, is to be sent from one place to another. There is some probability, however small, that 1 will be received if 0 is sent, and that 0 will be received if 1 is sent. One way to minimize the chance of error is to send the message three times: send 000 in place of 0, and 111 in place of 1. If 000 is received, it is reasonable to assume that 000 was sent. If either 100, 010, or 001 is received, it is still reasonable to assume that 000 was sent, because the probability of one error is smaller than the probability of two errors. (This is true if the probability of an error for any one symbol is small, an assumption we shall make throughout.) If either 111, 011, 101, or 110 is received, the assumption would be that 111 was sent. The correspondence (mapping) $0 \mapsto 000$, $1 \mapsto 111$ is called a *code*. This code will detect if two or less errors have occurred; this will be the case if any message other than 000 or 111 is received. The code cannot detect when three errors have occurred. Any message subjected to a single error will be interpreted correctly; this is not true for double errors.

In general, there will be several steps in transmitting information, as indicated schematically in Figure 58.1. We are not concerned with the engineering problems involved in this process, but rather with the idealized problem of the design of efficient codes. (Notice that this is not the same as constructing "secret" codes.) Our interest is in the *encoder* and the *decoder*. The encoder will take information from the source and produce a code word,

Figure 58.1

which will be a sequence of 0's and 1's, or more generally, elements from any specified finite field. The decoder will receive this code word after it has been subjected to possible alteration during transmittal or storage. The decoder must be designed to make a decision as to what message was sent based on what message is received, or perhaps to reject the message altogether if it differs substantially from anything that might have been sent.

The codes we shall consider are called *block codes*. Any such code has a sequence of *code words* (or *blocks*). The code words are themselves sequences of *digits*, which will be elements from some predetermined finite field F. If the digits are from $\mathbb{Z}_2$, the code is called a *binary code*. When working with $\mathbb{Z}_p$, we shall follow the practice, which we began in Chapter VIII, of omitting brackets from elements of $\mathbb{Z}_p$, writing $+$ instead of $\oplus$, using juxtaposition or $\cdot$ instead of $\odot$, and relying on the context to remind us that all calculations are to be done modulo n. In $\mathbb{Z}_5$, for example $3(4 + 3) = 3 \cdot 2 = 6 = 1$.

All code words will have the same length, n, called the *code length* (or *block length*) of the code. Code words of length n can be thought of as elements of F_n, the vector space of n-tuples over the field F. These words will be written as in Example 58.1; for example, 010 rather than (0, 1, 0).

Only code words will leave the encoder. (In Example 58.1, the code words are 000 and 111.) But any word (sequence of elements from F) of length n can be a *received word*, that is, a word received by the decoder. (In Example 58.1, there are eight possible received words.) The decision-making problem of the decoder is, given a received word, to either assign it a code word or declare it to be an error. The solution chosen for this problem can be conveniently presented by means of a *decoding table*, as in the following example. ■

Example 58.2. Table 58.1 is a decoding table for a code with code words 10001, 11100, 01011, and 00110. If any one of these four code words is received, it will be decoded as itself. If any other word is received, it will be decoded as the code word heading the column in which the received word appears. For instance, 01110 would be decoded as 00110. Each received word that is the result of an error in a single digit will be decoded currently by the table—in each column these are the five words above the dotted line. Some received words

Table 58.1

Code Word	10001	11100	01011	00110
	10000	11101	01010	00111
received words	10011	11110	01001	00100
with one error	10101	11000	01111	00010
	11001	10100	00011	01110
	00001	01100	11011	10110
....	....	....	....	....
received words	00000	11010	01101	10111
with two errors	00101	11111	01000	10010

that result from errors in two or more digits will also be decoded correctly—these are below the dotted line. However, some received words that are the result of errors in two digits will be decoded incorrectly; for example, if 11100 is sent and then received as 10110, it will be incorrectly decoded as 00110. If we arrange to have "error" signaled whenever the received word is not a code word, then all errors in one or two digits will be detected.   ■

The *distance* between two words is the number of positions in which they differ. The distance between 10001 and 01011 is 3, for example.

The smallest of the distances between pairs of code words of a code will be called the *minimum distance* of the code. In Example 58.2, the minimum distance is 3; that is why all errors in one or two digits can be detected—such errors cannot change one code word into a different code word.

We often use the following decoding rule: To decode an n-tuple, choose the code word closest to it in terms of the distance just defined; if this offers more than one choice, choose arbitrarily. This is called the *maximum-likelihood decoding rule*; under certain mild restrictions it maximizes the probability of correct decoding. (This is discussed in Section 61.) Table 58.1 is based on this rule.

Notice that a code will *detect* patterns of t or fewer errors if its minimum distance is $t + 1$, because in this case a code word will not be changed into a different code word through errors in t or fewer of its digits. Also, a code will correct patterns of t or fewer errors if its minimum distance is $2t + 1$, because in this case any word resulting from t or fewer errors will be decoded correctly, since no word is within distance t of more than one code word. This shows the desirability of having a code with code words as far apart as possible. One of the most effective methods for designing such codes makes use of finite fields and ideas from linear algebra, and is discussed in the next section.

PROBLEMS

58.1. Construct a decoding table for the code in Example 58.1. What is the minimum distance for the code?

58.2. Use the maximum-likelihood decoding rule to construct a decoding table for a binary code with code words 0000 and 1111. If 0000 is sent, which received words that result from errors in two digits will be decoded incorrectly?

58.3. Consider Example 58.2, and assume that the code word 10001 is sent. Which received words that result from errors in two digits will be decoded incorrectly?

58.4. Use the maximum-likelihood decoding rule to construct a decoding table for a binary code with code words 1001, 0101, 0010, 1110. What is the minimum distance for the code?

58.5. (a) Find three code words for a binary code of length 3 such that the minimum distance of the code is 2.
   (b) Is there a binary code with length 4 and three code words such that the minimum distance of the code is 3?

58.6. Denote the distance between two words x and y by $d(x, y)$. [Example: $d(10001, 01011) = 3$.]
   Verify that
   (a) $d(x, y) = 0$ iff $x = y$,   (b) $d(x, y) = d(y, x)$, and
   (c) $d(x, z) \leq d(x, y) + d(y, z)$ for all words x, y, z of a code.

58.7. Prove that a code with minimum distance d can correct all patterns of s or fewer errors and detect all patterns of t or fewer errors ($t \geq s$) if $t + s < d$. [*Suggestion*: For each word x, let $S_r(x)$ denote the *sphere* consisting of all words y such that $d(x, y) \leq r$. Draw a picture representing two spheres whose centers are code words a minimum distance apart. Problem 58.6 will support your geometrical intuition.]

SECTION 59   LINEAR CODES

The words in our codes have been arbitrary subsets of F^n, the vector space of n-tuples over a finite field F. One additional requirement—that the set of code words form a *subspace* of F^n—provides the bridge between the problems of coding and the techniques of matrix theory and linear algebra. The following definition singles out the resulting codes for fields of prime order.

Definition.   An (n, k) *linear code* over $\mathbb{Z}_p$ is a code whose code words form k-dimensional subspace of $\mathbb{Z}_p^n$.

If $\{w_1, w_2, \ldots, w_k\}$ is a basis for the subspace of code words, then this subspace is the set of all linear combinations

$$a_1 w_1 + a_2 w_2 + \cdots + a_k w_k,$$

with $a_1, a_2, \ldots, a_k \in \mathbb{Z}_p$. There are p choices for each a_i ($1 \leq i \leq k$), so there are p^k code words in an (n, k) linear code over $\mathbb{Z}_p$. Hereafter, we shall consider only linear codes, and our examples will even be restricted to linear *binary codes*, that is, linear codes over $\mathbb{Z}_2$. The code in Example 58.2 is not a linear code, because the code words do not form a subspace of $\mathbb{Z}_2^5$.

Example 59.1.   The set $\{10001, 11100, 01011\}$ is a basis for a $(5, 3)$ linear binary code. The set (subspace) of code words is

$$\{00000, 10001, 11100, 01011, 01101, 11010, 10111, 00110\}. \qquad \blacksquare$$

A $k \times n$ matrix whose rows form a basis for an (n, k) linear code is called a *generator matrix* for the code. The basis in Example 59.1 gives the generator matrix

$$G = \begin{bmatrix} 1 & 0 & 0 & 0 & 1 \\ 1 & 1 & 1 & 0 & 0 \\ 0 & 1 & 0 & 1 & 1 \end{bmatrix}. \qquad (59.1)$$

The code words come from forming all linear combinations of the rows of a generator matrix. These linear combinations can be obtained by multiplying the generator matrix on the left by the $p^k \times k$ matrix whose rows are the elements of $\mathbb{Z}_p^k$. For Example 59.1, an appropriate matrix is

$$A = \begin{bmatrix} 0 & 0 & 0 \\ 1 & 0 & 0 \\ 0 & 1 & 0 \\ 0 & 0 & 1 \\ 1 & 1 & 0 \\ 1 & 0 & 1 \\ 0 & 1 & 1 \\ 1 & 1 & 1 \end{bmatrix}.$$

Then

$$AG = \begin{bmatrix} 0 & 0 & 0 & 0 & 0 \\ 1 & 0 & 0 & 0 & 1 \\ 1 & 1 & 1 & 0 & 0 \\ 0 & 1 & 0 & 1 & 1 \\ 0 & 1 & 1 & 0 & 1 \\ 1 & 1 & 0 & 1 & 0 \\ 1 & 0 & 1 & 1 & 1 \\ 0 & 0 & 1 & 1 & 0 \end{bmatrix}.$$

The rows of AG are the code words in the same order as they are listed in Example 59.1. Notice that a generator matrix completely determines the code and, in general, will be much more compact than the list of all code words.

Another way to specify an (n, k) linear code is by giving an $(n - k) \times n$ matrix H whose rows form a basis for the null space of the generator matrix G. (The null space of a matrix is discussed in Appendix D.) This matrix H has the property that

$$GH' = 0, \tag{59.2}$$

where H' denotes the transpose of H. An element $w \in \mathbb{Z}_p^n$ is in the null space of G iff

$$Gw' = 0. \tag{59.3}$$

For the generator matrix in (59.1), $w = (a_1, a_2, \ldots, a_5)$ is in the null space iff

$$\begin{aligned} a_1 && && && + && a_5 &= 0 \\ a_1 &+ & a_2 &+ & a_3 && && &= 0 \\ && a_2 && && + & a_4 &+ & a_5 &= 0 \end{aligned} \tag{59.4}$$

In this case a_4 and a_5 can be assigned arbitrarily, and the equations in (59.4) will then determine a_1, a_2, a_3. The null space is {00000, 01110, 11001, 10111}, and we can use

$$H = \begin{bmatrix} 0 & 1 & 1 & 1 & 0 \\ 1 & 1 & 0 & 0 & 1 \end{bmatrix}.$$

The subspace of code words can be recovered from H, because an element $v \in \mathbb{Z}_p^n$ is a code word iff

$$vH' = 0, \tag{59.5}$$

or

$$Hv' = 0. \tag{59.6}$$

The conditions (59.3) and (59.6) exhibit the fact that the row space of H is the null space of G iff the row space of G is the null space of H. The matrix H is called a *parity-check matrix* for the code. The reason for this name can be seen from looking at (59.5) for H as in our example: if we write $v = (b_1, b_2, \ldots, b_5)$, then (59.5) is equivalent to the system of equations

$$\begin{aligned} b_2 + b_3 + b_4 \quad\quad\quad &= 0 \\ b_1 + b_2 \quad\quad\quad + b_5 &= 0 \end{aligned}.$$

The first equation says that an even number of the middle three components of $(b_1, b_2, \ldots, b_5)$ must be one; this is a parity check on the middle three components. The second equation is a parity check on the first, second, and fifth components.

The row space of a generator matrix G will not be changed if elementary row operations are performed on G. If columns of G are interchanged, the change in the code given by G will amount only to a permutation of the components of each code word. If one generator matrix can be obtained from another by a combination of elementary row operations and column permutations, then the codes corresponding to the two matrices are said to be *equivalent*.

Theorem 59.1.   *Every (n, k) linear code is equivalent to a code with a generator matrix of the form*

$$G = \begin{bmatrix} 1 & 0 & \cdots & 0 & p_{1,1} & \cdots & p_{1,n-k} \\ 0 & 1 & \cdots & 0 & p_{2,1} & \cdots & p_{2,n-k} \\ & & \cdots & & & & \\ 0 & 0 & \cdots & 1 & p_{k,1} & \cdots & p_{k,n-k} \end{bmatrix} = \begin{bmatrix} I_k \mid P \end{bmatrix}. \tag{59.7}$$

This is called the standard generator matrix *for the code.*

PROOF.   By elementary row operations a generator matrix for the given code can be put in row-reduced echelon form (Appendix D). The row space has dimension k so that each row of this row-reduced echelon matrix will have a nonzero entry. By appropriate permutation of the columns the matrix can be put in the form (59.7).   ■

One of the simplest ways to form an (n, k) linear code over $\mathbb{Z}_p$ is to begin with the elements of $\mathbb{Z}_p^k$ and transform them into elements of $\mathbb{Z}_p^n$ through right multiplication by a standard generator matrix. That is, with $v = (a_1, a_2, \ldots, a_k) \in \mathbb{Z}_p^k$, and G as in (59.7), we associate the code word

$$vG = (a_1, a_2, \ldots, a_k, b_1, b_2, \ldots, b_{n-k}),$$

where

$$b_j = \sum_{i=1}^{k} a_i p_{ij} \quad (1 \le j \le n - k).$$

A code of this type is called a *systematic code*. This first k components of each code word are called the *information symbols*, and the last $n - k$ components are called the *check symbols.* By Theorem 59.1 every linear code is equivalent to a systematic code. The next theorem shows how to get a parity-check matrix for a systematic code.

Theorem 59.2.   *The parity-check matrix for a code with generator matrix G as in (59.7) is*

$$H = \begin{bmatrix} -p_{1,1} & \cdots & -p_{k,1} & 1 & 0 & \cdots & 0 \\ -p_{1,2} & \cdots & -p_{k,2} & 0 & 1 & \cdots & 0 \\ & & \cdots & & & \cdots & \\ -p_{1,n-k} & \cdots & -p_{k,n-k} & 0 & 0 & \cdots & 1 \end{bmatrix} = \begin{bmatrix} -P' \mid I_{n-k} \end{bmatrix}.$$

PROOF.   It is easy to verify that $GH' = 0$. Therefore, the row space of H is contained in the null space of G. But the rows of H are linearly independent; thus dim(row space H) $= n - k$. Therefore, the row space of H must be all of the null space of G, because dim(null space G) $= n -$ dim(row space G) $= n - k$.   ∎

Example 59.2.   The standard generator matrix arising from the generator matrix G in (59.1) is

$$\begin{bmatrix} 1 & 0 & 0 & 0 & 1 \\ 0 & 1 & 0 & 1 & 1 \\ 0 & 0 & 1 & 1 & 0 \end{bmatrix}.$$

An element $(a_1, a_2, a_3) \in \mathbb{Z}_2^3$ gives rise to the code word

$$(a_1, a_2, a_3, a_2 + a_3, a_1 + a_2).$$

A parity-check matrix is

$$\begin{bmatrix} 0 & 1 & 1 & 1 & 0 \\ 1 & 1 & 0 & 0 & 1 \end{bmatrix}.$$ ∎

Other examples of linear codes are given by Hamming codes, which are described in Example 61.2. That example, and the corresponding Problems 61.5–61.7, can be covered now without a loss of continuity.

PROBLEMS

59.1. List a set of code words for a $(5, 2)$ linear binary code.

59.2. List a set of code words for a $(5, 2)$ linear code over $\mathbb{Z}_3$.

59.3. Write a matrix whose rows are the code words of the linear binary code with generator matrix
$$\begin{bmatrix} 1 & 0 & 0 & 0 \\ 0 & 1 & 1 & 0 \end{bmatrix}.$$

59.4. Determine a parity-check matrix for the code with the matrix in Problem 59.3 as a generator matrix. [Remember that if a code has a $k \times n$ generator matrix, then it has an $(n - k) \times n$ parity-check matrix.]

59.5. Write a matrix whose rows are the code words of the linear code over $\mathbb{Z}_3$ with generator matrix
$$\begin{bmatrix} 1 & 0 & 2 \\ 2 & 1 & 0 \end{bmatrix}.$$

59.6. Determine a generator matrix for the linear binary code with parity-check matrix $\begin{bmatrix} 0 & 1 & 0 \\ 1 & 1 & 0 \end{bmatrix}$.
(See the remark with Problem 59.4.)

59.7. (a) Write the standard generator matrix for the linear binary code with 00110 generator matrix
$$\begin{bmatrix} 0 & 0 & 1 & 1 & 0 \\ 0 & 1 & 0 & 1 & 0 \\ 1 & 1 & 1 & 0 & 0 \end{bmatrix}.$$
(b) Use Theorem 59.2 to write a parity-check matrix for the code in part (a).

59.8. Use the maximum-likelihood decoding rule to construct a decoding table for the linear binary code with standard generator matrix
$$\begin{bmatrix} 1 & 0 & 1 & 0 \\ 0 & 1 & 1 & 1 \end{bmatrix}.$$

59.9. Define the *weight*, $w(x)$, of a word $x \in F^n$ to be the number of nonzero components of x. [Example: $w(01101) = 3$.] Prove that for a linear binary code either all of the code words have even weight or half have even weight and half have odd weight. (*Suggestion:* First explain why the code words of even weight form a subgroup.)

SECTION 60   STANDARD DECODING

Given a choice, it is clearly desirable to choose a code and a procedure for decoding that will minimize the probability of error. This section and the next present some ways of satisfying this desire. We begin by describing a relevant decoding procedure. In Section 61 we'll see that for a large class of binary codes this procedure maximizes the average probability of correct decoding.

Let V be an (n, k) linear code over $\mathbb{Z}_p$. Then V is a subspace of $\mathbb{Z}_p$, and therefore, in particular, V is a subgroup of $\mathbb{Z}_p^n$ thought of as an Abelian group under vector addition. The cosets of V in $\mathbb{Z}_p^n$ form a partition of $\mathbb{Z}_p^n$, so they can be used as the rows of a decoding table. Notice that $|\mathbb{Z}_p^n| = p^n$, $|V| = p^k$, and by Lagrange's Theorem $[\mathbb{Z}_p^n : V] = p^{n-k}$; therefore, there will be p^{n-k} rows. A *standard array* is a table constructed with cosets for rows with the elements arranged according to the following scheme:

Assume $V = \{v_1, v_2, \ldots, v_{p^k}\}$, with $v_1 = 0$.

Let $\{u_1, u_2, \ldots, u_{p^{n-k}}\}$ be a complete set of coset representatives of V in $\mathbb{Z}_p^n$, with $u_1 = 0$.

The vector in the ith row and jth column of the standard array is $u_i + v_j$ (Table 60.1).

Table 60.1 Standard Array

v_1	v_2	v_3		v_p^k
u_2	$u_2 + v_2$	$u_2 + v_3$	$\cdots$	$u_2 + v_p^k$
u_3	$u_3 + v_2$	$u_3 + v_3$	$\cdots$	$u_3 + v_p^k$
$\vdots$	$\vdots$	$\vdots$		$\vdots$
u_p^{n-k}	$u_p^{n-k} + v_2$	$u_p^{n-k} + v_3$	$\cdots$	$u_p^{n-k} + v_p^k$

In coding theory the coset representatives $u_1, u_2, \ldots, u_p^{n-k}$ are called *coset leaders*. The *weight* of a vector $v \in \mathbb{Z}_p^n$ is defined as the number of nonzero components of v. We shall see later that there is an advantage in choosing each coset leader to be an element of minimum weight in its coset. This can be done by constructing the array one row at a time, from the top, choosing each successive coset leader to be a vector of smallest weight from among all the remaining vectors.

Example 60.1.   Table 60.2 gives a standard array for the (5, 3) code in Example 59.1.

Table 60.2

00000	10001	11100	01011	01101	11010	10111	00110
10000	00001	01100	11011	11101	01010	00111	10110
00010	10011	11110	01001	01111	11000	10101	00100
01000	11001	10100	00011	00101	10010	11111	01110

∎

If a code vector v is transmitted and a vector u is received, then $u - v$ is called the *error pattern*. For example, if the code vector 11010 is transmitted and 01110 is received, the error pattern is 10100. Notice that the weight of the error pattern $u - v$ is equal to the distance between u and v. Problem 60.6 asks you to show that for any linear code an error can be *detected* iff the error pattern is not a code vector. The following theorem tells which errors will be *corrected* if a standard array is used for decoding.

Theorem 60.1.   *If a standard array is used as the decoding table for a linear code, then a received vector u will be decoded correctly into the transmitted code vector v iff the error pattern $u - v$ is a coset leader.*

PROOF.   Adopt the notation in Table 60.1, and assume that v_j is transmitted, u is received, and u is decoded as v_t. We must prove that $v_j = v_t$ iff $u - v_j$ is a coset leader.

If $v_j = v_t$, then u is in the column of v_j, so that $u = u_i + v_j$ for some coset leader u_i. Therefore, the error pattern $u - v_j = u_i$ is a coset leader. Conversely, if $u - v_j = u_i$ is a coset leader, then $u = u_i + v_j$; therefore, u is in the column of v_j, and hence $v_j = v_t$.   ∎

Here is another way to look at this theorem. In using a standard array as a decoding table, we are in essence assuming that each received vector has its coset leader as an error pattern: if $u_i + v_j$ is received, it is decoded as v_j, which means that we are assuming an error pattern of $(u_i + v_j) - v_j = u_i$.

We now show that there is a more economical way of presenting the information given by a standard array. Let V be an (n, k) linear code over $\mathbb{Z}_p$, and let H denote the parity-check matrix for V. The *syndrome* of any received vector v is the $(n - k)$-dimensional vector vH'. By the very definition of parity-check matrix, a vector is a code word iff its syndrome is 0. This means that in a standard array the vectors of syndrome 0 are those in the first row. Theorem 60.2 shows that, more generally, two vectors have equal syndromes (not necessarily 0) iff they are in the same row of a standard array.

Theorem 60.2. *Two vectors are in the same coset of a linear code iff their syndromes are equal.*

PROOF. Two vectors v_1 and v_2 are in the same coset of a linear code V

$$\text{iff} \quad v_1 - v_2 = v \text{ for some } v \in V$$
$$\text{iff} \quad (v_1 - v_2)H' = 0$$
$$\text{iff} \quad v_1 H' - v_2 H' = 0$$
$$\text{iff} \quad v_1 H' = v_2 H' \qquad \blacksquare$$

To capitalize on Theorem 60.2 and the remark following Theorem 60.1, we first form a syndrome table, which has two columns, one for coset leaders and the other for their syndromes (Table 60.3). Assume that a vector u is received. Compute the syndrome of u; by Theorem 60.2 this equals the syndrome of the coset leader of u, so that we can get this coset leader from Table 60.3. By the remark following Theorem 60.1, this coset leader (call it u_i) is the assumed error pattern in u. Therefore, u would be decoded by a standard array as $v_j = u - u_i$. To summarize, decoding by a standard array can be done by computing a syndrome and using a syndrome table.

Table 60.3

Coset Leaders	Syndromes
u_1	$u_1 H'$
u_2	$u_2 H'$
u_3	$u_3 H'$
$\vdots$	$\vdots$
u_p^{n-k}	$u_p^{n-k} H'$

Example 60.2. Table 60.4 is the syndrome table for the standard array in Table 60.2. The parity-check matrix was computed for this code in Example 59.2, and is

$$H = \begin{bmatrix} 0 & 1 & 1 & 1 & 0 \\ 1 & 1 & 0 & 0 & 1 \end{bmatrix}.$$

Therefore, the syndrome of 01000, for example, is

$$[0 \quad 1 \quad 0 \quad 0 \quad 0] \begin{bmatrix} 0 & 1 \\ 1 & 1 \\ 1 & 0 \\ 1 & 0 \\ 0 & 1 \end{bmatrix} = [1 \quad 1].$$

To decode 11011, we first compute

$$[1 \quad 1 \quad 0 \quad 1 \quad 1] \begin{bmatrix} 0 & 1 \\ 1 & 1 \\ 1 & 0 \\ 1 & 0 \\ 0 & 1 \end{bmatrix} = [0 \quad 1].$$

This is the syndrome of 10000, which is also the error pattern. Therefore, 11011 would be decoded as $11011 - 10000 = 01011$. This agrees with Table 60.2.

Table 60.4

Coset Leaders	Syndromes
00000	00
10000	01
00010	10
01000	11

■

PROBLEMS

60.1. Construct a standard array for a (4, 2) linear binary code with generator matrix
$$\begin{bmatrix} 1 & 1 & 0 & 0 \\ 0 & 0 & 1 & 1 \end{bmatrix}.$$

60.2. Construct the syndrome table for the standard array in Problem 60.1.

60.3. Construct a standard array for a (3, 2) linear code over $\mathbb{Z}_3$ with generator matrix
$$\begin{bmatrix} 1 & 2 & 0 \\ 0 & 1 & 2 \end{bmatrix}.$$

60.4. Construct the syndrome table for the standard array in Problem 60.3.

60.5. Suppose that the vectors in the second row of a standard array for a linear binary code are 1000, 1100, 1111, and 1011 (in order). Complete the array. (There is more than one solution.)

60.6. Prove that for a linear code an error can be detected iff the error pattern is not a code vector.

60.7. How many error patterns can be detected by an (n, k) linear code over $\mathbb{Z}_p$? (See Problem 60.6.)

60.8. If a standard array is used for decoding, how many error patterns can be corrected by an (n, k) linear code over $\mathbb{Z}_p$?

60.9. If a standard array is used for a linear binary code, and if each vector of weight w or less appears as a coset leader, then each error pattern of weight w or less will be corrected. Why?

60.10. Construct a standard array, generator matrix, parity-check matrix, and syndrome table for a (3, 1) linear binary code that will correct all error patterns of weight 1. (See Problem 60.9.)

60.11. Construct a generator matrix and list of coset leaders for a (9, 2) linear binary code that will correct all error patterns of weight 2 or less.

SECTION 61 ERROR PROBABILITY

We now restrict our attention to linear binary codes. We assume, moreover, that there is a constant probability p that any symbol (binary digit) will be altered between transmission and reception; it is said in this case that the code is being transmitted over a *binary symmetric channel*. Theorem 61.1 adds the assumption that all code words are equally likely to be transmitted, and then gives a "best" decoding rule. Theorem 61.1 applies when a code (any code) has already been chosen; Example 61.2 goes further, and describes how to choose a "best" code as well as decoding procedure. To make these statements precise, we must use some elementary facts from probability theory. If you prefer to omit the proof of Theorem 61.1, simply move ahead now to the statement of the theorem, read that, and then pass on to Example 61.2.

If the probability is p that a digit will be altered, then the probability is $q = 1 - p$ that it will be transmitted without alteration. We assume that $p < \frac{1}{2}$. The probabilities of alteration are assumed to be independent for the different components of a code word; therefore, the probability that a word of length n will be altered in m specified positions is $p^m q^{n-m}$. In particular, the probability of a specified error pattern of weight w is $p^w q^{n-w}$. For instance, the probability is $p^2 q^3$ that if 01110 is sent, then 10110 will be received, because the error pattern 11000 has weight 2.

Question: If we assume that the code word v_j is transmitted, and that decoding is done with the standard array in Table 60.1, what is the probability of correct decoding? *Answer*: There will be correct decoding iff the received word is in the column headed by v_j. Therefore, the probability is $p_1 + p_2 + \cdots + p_{2^{n-k}}$, where p_i denotes the probability of receiving $u_i + v_j$ ($1 \leq i \leq 2^{n-k}$). But u_i is the error pattern, so that $p_i = p^{w_i} q^{n-w_i}$, where w_i denotes the weight of u_i. Thus the probability of correct decoding is

$$\sum_{i=1}^{2^{n-k}} p^{w_i} q^{n-w_i}.$$

Example 61.1. In Example 60.1, the code word 11100 will be decoded correctly if it is received as either 11100, 01100, 11110, or 10100. The error patterns here are 00000, 10000, 00010, and 01000, respectively; and these have weights 0, 1, 1, 1, respectively. Therefore, the probability of correct decoding is

$$q^5 + 3pq^4.$$ ∎

The probability of correct decoding will be the same for each of the other code words in Example 60.1, because the probability depends only on the weights of the error patterns. And these are the weights of the coset leaders (since Table 60.2 is a standard array).

If one error pattern has weight w_1, and another has weight w_2, and $w_1 < w_2$, then the first error pattern is more likely to occur, because

$$p^{w_1} q^{n-w_1} > p^{w_2} q^{n-w_2} \tag{61.1}$$

if $p < \frac{1}{2}$ and $w_1 < w_2$ (Problem 61.3). In particular, the error pattern $000 \cdots 0$ is most likely of all; that is, if $p < \frac{1}{2}$, the most probable received word is the transmitted word. (This does not mean that the probability of no altered digit is always larger than the probability of an alteration. See Problem 61.4.) We are now ready for the following theorem.

Theorem 61.1. *If an (n, k) binary code is transmitted over a binary symmetric channel, and if all code words are equally likely to be transmitted, then the average probability of correct decoding will be maximum if the decoding table is chosen to be a standard array with each coset leader of minimum weight in its coset.*

PROOF.   Consider a decoding table with ij-entry v_{ij} ($1 \le i \le 2^{n-k}$, $1 \le j \le 2^k$), and with code words v_{1j} ($1 \le j \le 2^k$). Let w_{ij} denote the weight of the assumed error pattern of v_{ij}, that is, the weight of $v_{ij} - v_{1j}$. Then the probability of correct decoding, if v_{1j} is transmitted, is

$$\sum_{i=1}^{2^{n-k}} p^{w_{ij}} q^{n-w_{ij}}. \tag{61.2}$$

Because the 2^k code words are equally likely to be transmitted, the average probability of correct decoding is

$$\left(\frac{1}{2}\right)^k \sum_{i,j} p^{w_{ij}} q^{n-w_{ij}}. \tag{61.3}$$

By (61.1), the individual terms $p^{w_{ij}} q^{n-w_{ij}}$ in the sum (61.3) will be maximum if the weights w_{ij} are minimum. Therefore, the average probability given in (61.3) will certainly be maximum if the table can be arranged so that the weights w_{ij} are simultaneously minimum. The theorem asserts that this is achieved by a standard array (Table 60.1) with each coset leader of minimum weight in its coset. To verify this, it suffices to show that if $u_i + v_j$ is a received vector, and v' is the code vector closest to it, then

$$w[(u_i + v_j) - v'] \ge w[(u_i + v_j) - v_j] = w[u_i].$$

But this is true because

$$(u_i + v_j) - v' = u_i + (v_j - v') \in u_i + V,$$

so that $(u_i + v_j) - v'$ is in the same coset as u_i, and u_i has minimum weight in its coset. ∎

Example 61.2.   For each positive integer m, there is a special $(2^m - 1, 2^m - m - 1)$ binary code, known as the *Hamming code*, which yields the maximum probability of correct decoding among all $(2^m - 1, 2^m - m - 1)$ binary codes. This code can be defined in terms of its parity-check matrix H: the columns of H are all of the nonzero m-dimensional column vectors over $\mathbb{Z}_2$. A parity-check matrix for the Hamming code with $m = 2$ is

$$H = \begin{bmatrix} 0 & 1 & 1 \\ 1 & 0 & 1 \end{bmatrix}. \tag{61.4}$$

A parity-check matrix for the Hamming code with $m = 3$ is

$$H = \begin{bmatrix} 0 & 0 & 0 & 1 & 1 & 1 & 1 \\ 0 & 1 & 1 & 0 & 0 & 1 & 1 \\ 1 & 0 & 1 & 0 & 1 & 0 & 1 \end{bmatrix}. \tag{61.5}$$

Let us look more closely at the case $m = 3$. The resulting code is a $(2^3 - 1, 2^3 - 3 - 1) = (7, 4)$ code, so there will be $2^4 = 16$ code words. By (59.5), a vector $v = (a_1, a_2, \ldots, a_7)$ is a code word iff $vH' = 0$. This will be true iff

$$\begin{aligned} a_4 + a_5 + a_6 + a_7 &= 0 \\ a_2 + a_3 \quad\quad + a_6 + a_7 &= 0 \\ a_1 \quad + a_3 \quad + a_5 \quad\quad + a_7 &= 0. \end{aligned} \tag{61.6}$$

The symbols a_3, a_5, a_6, a_7 can be assigned arbitrarily, and then a_1, a_2, a_4 will be determined by the system (61.6). We can think of the 16 different 4-tuples $a_3 a_5 a_6 a_7$ as the messages to be handled by this code, with $a_3 a_5 a_6 a_7$ encoded as $a_1 a_2 \cdots a_7$. For example, 1010 would be encoded as $a_1 a_2 1 a_4 010$, with a_1, a_2, a_4 determined by

$$
\begin{aligned}
a_4 + 1 &= 0 \\
a_2 + 0 &= 0 \\
a_1 + 1 &= 0.
\end{aligned}
$$

Therefore $a_1 = 1, a_2 = 0, a_4 = 1$; and 1010 would be encoded as 1011010.

Each Hamming code will correct all single errors. These codes were introduced by R. W. Hamming in 1950. The comparable double-error- correcting codes—known as BCH codes—were not developed until 10 years later. ∎

PROBLEMS

61.1. Assume that the code word 10101 is transmitted over a binary symmetric channel, with probability $\frac{1}{4}$ that a digit will be altered.
(a) What is the probability that 11011 will be received?
(b) What is the probability that the error pattern will have weight 2?

61.2. Assume that the code in Table 58.1 is being transmitted over a binary symmetric channel, with probability p that a digit will be altered. Compute the probability of correct decoding for each of the four different code words.

61.3. Verify inequality (61.1) for $p < \frac{1}{2}, q = 1 - p$, and $w_1 < w_2$.

61.4. Consider Example 61.1. Prove that if $p > \frac{1}{4}$, then for each code word the probability of an altered digit exceeds the probability of no alteration.

61.5. For the Hamming code considered at the end of Example 61.2, how would 0111 be encoded?

61.6. Is 0100110 a code word for the Hamming code with parity-check matrix given by (61.5)?

61.7. Determine the generator matrix for the code with parity-check matrix given by (61.4).

61.8. Construct a standard array for the code with parity-check matrix given by (61.4). (See Problem 61.7.) Also construct the corresponding syndrome table.

61.9. Determine the standard generator matrix (61.7) for the code with parity-check matrix given by (61.5).

61.10. Determine a standard array for the code with parity-check matrix given by (61.5). (See Problem 61.9.) Also construct the corresponding syndrome.

NOTES ON CHAPTER XV

1.   Berlekamp, E. R., *Algebraic Coding Theory*, rev. ed., Aegean Park Press, Laguna Hills, CA., 1984

2.   Blake, I. F., and R. C. Mullin, *An Introduction to Algebraic and Combinatorial Coding Theory*, Academic Press, New York, 1976.

3.   Levinson, N., Coding theory: a counterexample to G. H. Hardy's conception of applied mathematics, *American Mathematical Monthly* **77** (1970), 249–258.

4. Peterson, W. W., and E. J. Weldon, Jr., *Error Correcting Codes*, 2nd ed., MIT Press, Cambridge, Mass., 1972.
5. Sloane, N. J. A., *The Theory of Error-Correcting Codes*, Elsevier Science, Ltd., 1978.

CHAPTER XVI

LATTICES AND BOOLEAN ALGEBRAS

Just as the axioms for groups and rings reflect properties of addition and multiplication in the familiar number systems, the axioms for lattices and Boolean algebras reflect properties of inclusion, union, and intersection in the theory of sets. But we shall see with lattices and Boolean algebras, as with groups and rings, there are other important examples beyond those furnishing our original motivation. In particular, Boolean algebras provide appropriate algebraic settings for formal logic and the theory of switching networks. Section 65 contains the most interesting theorem in this chapter, but it is not a prerequisite for the application in Section 66.

SECTION 62   PARTIALLY ORDERED SETS

The first of several fundamental ideas for this chapter is an abstraction from $\leq$ for numbers and $\subseteq$ for sets.

Definition.   A *partially ordered set* is a set S together with a relation $\leq$ on S such that each of the following axioms is satisfied:

reflexive

$$\text{if } a \in S, \text{ then } a \leq a,$$

antisymmetric

$$\text{if } a, b \in S, a \leq b, \text{ and } b \leq a, \text{ then } a = b,$$

transitive

$$\text{if } a, b, c \in S, a \leq b, \text{ and } b \leq c, \text{ then } a \leq c.$$

Example 62.1.   The integers form a partially ordered set with respect to $\leq$. Here $\leq$ has its usual meaning. In other examples $\leq$ is replaced by whatever is appropriate for the relation involved.   ∎

Example 62.2.   For each set S, let $\mathcal{P}(S)$ denote the set of all subsets of S. Then $\mathcal{P}(S)$ is a partially ordered set with $A \leq B$ defined to mean $A \subseteq B$. Notice that $\subseteq$ is a relation on $\mathcal{P}(S)$, not on S. [The set $\mathcal{P}(S)$ is called the *power set* of S; if S is finite, then $|\mathcal{P}(S)| = 2^{|S|}$, "two to the *power* $|S|$."]   ∎

Example 62.3.   The set of all subgroups of any group is a partially ordered set with respect to ⊆. ∎

Example 62.4.   The set $\mathbb{N}$ of all natural numbers (positive integers) is a partially ordered set with $a \leq b$ defined to mean $a \mid b$. The set of all positive divisors of a fixed positive integer n is also a partially ordered set with this relation. ∎

The notation $a < b$ means that $a \leq b$ and $a \neq b$; $b \geq a$ means that $a \leq b$; and $b > a$ means that $a < b$. An element b in a partially ordered set S is said to *cover* an element a in S if $a < b$ and there is no x in S such that $a < x < b$.

It is often helpful to represent a finite partially ordered set by a diagram, in the following way. Each element of the set is represented by a small circle (or other appropriate symbol), and if b covers a, then the circle for b is placed above the circle for a and a line is drawn connecting the two circles. In this way $c < d$ iff d is above c and there is a sequence of segments connecting c to d.

Example 62.5.   Figure 62.1 shows the diagram for the set of subsets of $\{x, y, z\}$ with the relation ⊆. ∎

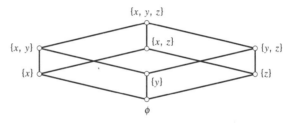

Figure 62.1

Example 62.6.   Figure 62.2 shows the diagram for the set of positive divisors of 12, with the relation $a \leq b$ defined to mean that $a \mid b$. ∎

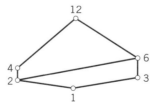

Figure 62.2

An element u in a partially ordered set S is said to be an *upper bound* for a subset A of S if $x \leq u$ for each $x \in A$. The element u is a *least upper bound (l.u.b.)* for A if it is an upper bound and $u \leq v$ for each upper bound v of A. *Lower bound* and *greatest lower bound (g.l.b.)* are defined by replacing $\leq$ by $\geq$.

If a subset has an l.u.b., then it is unique, for if u_1 and u_2 are both l.u.b's for A, then $u_1 \leq u_2$ and $u_2 \leq u_1$, and therefore $u_1 = u_2$. Similarly, if there is a g.l.b. for A, then it is unique; simply replace l.u.b. by g.l.b. and $\leq$ by $\geq$ throughout the preceding sentence.

Example 62.7. Let S be a set, and consider the partially ordered set $\mathcal{P}(S)$ with $\subseteq$. If $A, B \in \mathcal{P}(S)$ (that is, if $A \subseteq S$ and $B \subseteq S$), then $A \cup B$ is an l.u.b. for $\{A, B\}$, and any set containing $A \cup B$ is an upper bound for $\{A, B\}$. Also, $A \cap B$ is a g.l.b. for $\{A, B\}$. More generally, if C is any subset of $\mathcal{P}(S)$, then the union of all the subsets in C is in $\mathcal{P}(S)$ and is the l.u.b. for C; and the intersection of all the subsets in C is in $\mathcal{P}(S)$ and is the g.l.b. for C. ∎

Example 62.8. Consider Example 62.4. If A is any finite subset of $\mathbb{N}$, then the least common multiple of the integers in A is an l.u.b. for A, and the greatest common divisor of the integers in A is a g.l.b. for A. ∎

Example 62.9. The diagram in Figure 62.3 represents a partially ordered set in which $\{a, b\}$ has no upper bound (and therefore no l.u.b.), and in which $\{c, d\}$ has no lower bound (and therefore no g.l.b.). ∎

Figure 62.3

The question of the existence of least upper bounds and greatest lower bounds will be critical in Section 63. Here is another important example in which they always exist.

Example 62.10. Let G be a group, and consider the partially ordered set of subgroups of G with $\subseteq$ (Example 62.3). The subgroup generated by any collection of subgroups of G is an l.u.b. for that collection of subgroups. (Apply Theorem 15.2, with S being the set union of the subgroups in the given collection.) The point here is that the *set union* of subgroups will not in general be a subgroup, but there is an l.u.b. nonetheless. The intersection of any collection of subgroups is a subgroup (Theorem 15.1), and this intersection is a g.l.b. for the collection. Figure 62.4 shows the diagram for the subgroups of $\mathbb{Z}_6$; Figure 62.5 shows the diagram for the subgroups of S_3.

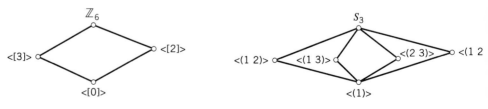

Figure 62.4   **Figure 62.5**

If either $a \leq b$ or $b \leq a$ for every pair of elements $\{a, b\}$ in a partially ordered set S, then S is said to be *linearly ordered* or to be a *chain*. With the usual meaning of $\leq$, each of $\mathbb{Z}$, $\mathbb{Q}$, and $\mathbb{R}$ is a chain. None of Figures 62.1 through 62.5 represents a chain.   ■

Example 62.11.   The subgroups of $\mathbb{Z}_8$ form a chain, as shown in Figure 62.6. In fact, the subgroups of any cyclic group of prime power order form a chain (Problem 62.13).   ■

$\mathbb{Z}_8$

$\{[0], [2], [4], [6]\}$

$\{[0], [4]\}$

$\{[0]\}$

Figure 62.6

PROBLEMS

62.1. The set of all nonzero integers is *not* partially ordered with $a \leq b$ defined to mean $a \mid b$. Why? (Compare Example 62.4.)

62.2. Draw the diagram for the set of subsets of $\{x, y\}$ with the relation $\subseteq$.

62.3. Draw the diagram for the set of subsets of $\{w, x, y, z\}$ with the relation $\subseteq$.

62.4. Draw the diagram for the set of positive divisors of 20, with $a \leq b$ defined to mean $a \mid b$.

62.5. Assume that p and q are distinct primes, and let $n = p^2 q$. Draw the diagram for the set of positive divisors of n, with $a \leq b$ defined to mean $a \mid b$. (Compare Figure 62.2 and Problem 62.4.)

62.6. For which $n \in \mathbb{N}$ does the set of all positive divisors of n form a chain, with $a \leq b$ defined to mean $a \mid b$?

62.7. For $\mathbb{N}$ partially ordered as in Example 62.4, determine the l.u.b. and the g.l.b. of the subset $\{12, 30, 126\}$.

62.8. Consider the partially ordered set $\mathbb{N}$ with $a \leq b$ defined to mean $a \mid b$ (Example 62.4).
(a) Which integers are covered by 6?
(b) Which integers cover 6? (There are a lot.)
(c) For $m \in \mathbb{N}$, which integers are covered by m?
(d) For $m \in \mathbb{N}$, which integers cover m?

62.9. Consider the partially ordered set of all subgroups of the group of integers (Examples 62.3 and 62.10 with $G = \mathbb{Z}$).
(a) For which $m, n \in \mathbb{Z}$ is $\langle m \rangle \subseteq \langle n \rangle$?
(b) For $m, n \in \mathbb{Z}$, determine k so that $\langle k \rangle$ is the greatest lower bound of $\langle m \rangle$ and $\langle n \rangle$.
(c) For $m, n \in \mathbb{Z}$, determine k so that $\langle k \rangle$ is the least upper bound of $\langle m \rangle$ and $\langle n \rangle$.

62.10. Assume that A and B are subgroups of a group G. Prove that $A \cup B$ (set union) is a subgroup iff either $A \subseteq B$ or $B \subseteq A$. (This explains why $\langle A, B \rangle$ rather than $A \cup B$ is used in Example 62.10.)

62.11. A mapping $\theta: L_1 \to L_2$ of one partially ordered set onto another is called *order preserving* if $a \le b$ implies $\theta(a) \le \theta(b)$ for all $a, b \in L_1$. An invertible mapping $\theta: L_1 \to L_2$ is an *isomorphism* if both θ and θ^{-1} are order preserving. (Finite partially ordered sets are isomorphic iff their diagrams are the same except possibly for labeling.)

    (a) Verify that isomorphism is an equivalence relation for partially ordered sets.

    (b) There are two isomorphism classes for partially ordered sets with two elements. Draw a diagram corresponding to each class. (Remember that two elements need not be "comparable.")

    (c) There are five isomorphism classes for partially ordered sets with three elements. Draw a diagram corresponding to each class.

    (d) There are 16 isomorphism classes for partially ordered sets with four elements. Draw a diagram corresponding to each class.

    (e) Formulate a necessary and sufficient condition on positive integers m and n for the partially ordered sets of positive divisors of m and n to be isomorphic, with $a \le b$ defined to mean $a \mid b$. (*Suggestion*: See Problem 62.5.) Give several examples to support your condition, but do not write a formal proof.

62.12. Prove that $|\mathcal{P}(S)| = 2^{|S|}$. (See Example 62.2. Assume $|S| = n$ and $x \in S$. The subsets of S can be put in two classes, those that do contain x and those that do not. The induction hypothesis will tell you how many subsets there are in each class.)

62.13. Prove that the subgroups of every cyclic group of prime power order form a chain, as claimed in Example 62.11.

SECTION 63   LATTICES

Definition.   A *lattice* is a partially ordered set in which each pair of elements has a least upper bound and a greatest lower bound.

The l.u.b. of elements a and b in a lattice will be denoted $a \vee b$, and the g.l.b. will be denoted $a \wedge b$. The operations $\vee$ and $\wedge$ are called *join* and *meet*, respectively.

Example 63.1.   The partially ordered sets in Examples 62.1 through 62.4 are all lattices. The proof for Example 62.1 is obvious, and proofs for the other cases follow from remarks in Section 62. We can now speak, for instance, of "the lattice of subgroups" of a group. ■

The definition of lattice demands that each *pair* of elements have an l.u.b. and a g.l.b. It follows from this that each *finite subset* has an l.u.b. and a g.l.b. For example, the l.u.b. of $\{a, b, c\}$ is $(a \vee b) \vee c$, which can be seen as follows. Let $u = (a \vee b) \vee c$. Then $a \vee b \le u$, and therefore $a \le u$ and $b \le u$; also $c \le u$. On the other hand, if $a \le v$, $b \le v$, and $c \le v$, then $a \vee b \le v$ and $c \le v$; therefore $u = (a \vee b) \vee c \le v$. Thus u is an l.u.b. for $\{a, b, c\}$, as claimed. A similar argument shows that $(a \wedge b) \wedge c$ is a g.l.b. for $\{a, b, c\}$ (replace l.u.b. by g.l.b. and $\le$ by $\ge$ throughout). The l.u.b. and g.l.b. of a finite subset $\{a_1, a_2, \ldots, a_n\}$ will be denoted

$$a_1 \vee a_2 \vee \cdots \vee a_n \quad \text{and} \quad a_1 \wedge a_2 \wedge \cdots \wedge a_n,$$

respectively. The inductive proofs of the existence of these elements are left to Problem 63.16.

In the paragraph preceding Example 62.7 we proved that if a subset of a partially ordered set has an l.u.b., then that l.u.b. is unique; we then stated that the uniqueness of the g.l.b's could be proved similarly: replace l.u.b. by g.l.b. and $\leq$ by $\geq$. In the same way, the proof that $(a \vee b) \vee c$ is an l.u.b. for $\{a, b, c\}$ in a lattice, given previously, can be transformed into a proof that $(a \wedge b) \wedge c$ is a g.l.b. for $\{a, b, c\}$. These are applications of a very useful principle that, for lattices, has the following form.

Principle of Duality.   *Any statement that is true for every lattice remains true if $\leq$ and $\geq$ are interchanged throughout the statement and $\vee$ and $\wedge$ are interchanged throughout the statement.*

The Principle of Duality is valid because of three factors. First, $\geq$, as well as $\leq$, is reflexive, antisymmetric, and transitive. Second, g.l.b. is defined by replacing $\leq$ by $\geq$ in the definition of l.u.b. Third, a statement is true for every lattice only if it can be proved from the reflexive, antisymmetric, and transitive properties and the existence of l.u.b's and g.l.b's of finite sets.

If $\leq$ and $\geq$ are interchanged and l.u.b. and g.l.b. are interchanged in a statement, then the new statement obtained is called the *dual* of the original statement. For example, the dual of $a \vee a = a$ is $a \wedge a = a$. The Principle of Duality simply says that if a statement is true for every lattice, then so is its dual.

Although each finite subset of a lattice has both an l.u.b. and a g.l.b., an infinite subset of a lattice need not have an l.u.b. or a g.l.b. For example, in $\mathbb{Z}$, with the relation $\leq$, the set $\mathbb{Z}$ itself has neither an l.u.b. nor a g.l.b. In $\mathbb{N}$, with $a \leq b$ defined to mean $a \mid b$, no infinite subset has an l.u.b. A lattice is said to be *complete* if each subset (finite or infinite) has both an l.u.b. and a g.l.b. An element 1 in a lattice L is called a *unity* (or *identity*) for L if $a \leq 1$ for each $a \in L$. And an element 0 in L is called a *zero* for L if $0 \leq a$ for each $a \in L$. Notice that if 1 and 0 exist in L, then

$$a \vee 0 = 0 \vee a = a \quad \text{and} \quad a \wedge 1 = 1 \wedge a = a \tag{63.1}$$

for each $a \in L$. Also, any finite lattice (one with a finite number of elements) has both a unity and a zero. We shall always require $0 \neq 1$.

Example 63.2.
   (a)   In $\mathbb{Z}$, with the relation $\leq$, there is neither a lattice unity nor a lattice zero.
   (b)   In $\mathcal{P}(S)$ (Example 62.2), S is a unity and $\varnothing$ (the empty set) is a zero.
   (c)   In the lattice of subgroups of a group G, G is a lattice unity and $\{e\}$ is a lattice zero.
   (d)   In $\mathbb{N}$, with $a \leq b$ defined to mean $a \mid b$, there is no lattice unity, but 1 is a lattice zero.   ■

If L is a lattice, then both $\vee$ and $\wedge$ are operations on L, and they have properties much like those for $+$ and $\cdot$ in a ring. The next theorem lists the properties that are essential and shows that they give an alternative way to define a lattice.

Theorem 63.1.   *The operations $\vee$ and $\wedge$ on a lattice L satisfy each of the following laws (for all $a, b, c \in L$):*

commutative laws

$$a \vee b = b \vee a, \quad a \wedge b = b \wedge a,$$

associative laws

$$a \vee (b \vee c) = (a \vee b) \vee c, \quad a \wedge (b \wedge c) = (a \wedge b) \wedge c,$$

idempotent laws

$$a \vee a = a, \quad a \wedge a = a,$$

absorption laws

$$(a \vee b) \wedge a = a, \quad (a \wedge b) \vee a = a.$$

Conversely, if operations $\vee$ and $\wedge$ on a set L satisfy each of these laws, and if a relation $\leq$ is defined on L by

$$a \leq b \quad iff \quad either \quad a \vee b = b \quad or \quad a \wedge b = a, \tag{63.2}$$

then L is a lattice.

PROOF.   The proof of the first half of the theorem is left to Problem 63.11. To prove the converse, we first show that in the presence of the assumptions being made on $\vee$ and $\wedge$, the conditions $a \vee b = b$ and $a \wedge b = a$ in (63.2) are equivalent (that is, either one implies the other). Assume that $a \vee b = b$. Then

$$
\begin{aligned}
a \wedge b &= a \wedge (a \vee b) & (a \vee b = b) \\
&= (a \vee b) \wedge a & \text{(commutative law)} \\
&= a & \text{(absorption law)}.
\end{aligned}
$$

The proof of the other implication ($a \wedge b = a$ implies that $a \vee b = b$) is similar (Problem 63.12).

Now we shall prove that L is partially ordered relative to $\leq$. If $a \in L$, then $a \wedge a = a$ by one of the idempotent laws, and therefore $a \leq a$ by (63.2); thus $\leq$ is reflexive. If $a \leq b$ and $b \leq a$, then $a \wedge b = a$ and $b \wedge a = b$, so that $a = b$ by one of the commutative laws; thus $\leq$ is antisymmetric. If $a \leq b$ and $b \leq c$, then $a \wedge b = a$ and $b \wedge c = b$; therefore

$$a \wedge c = (a \wedge b) \wedge c = a \wedge (b \wedge c) = a \wedge b = a, \tag{63.3}$$

so that $a \leq c$. Thus $\leq$ is transitive.

To complete the proof that L is a lattice, we must verify that $a \vee b$ is an l.u.b. for a and b, and that $a \wedge b$ is a g.l.b. for a and b. Consider $a \vee b$. First, $a \leq a \vee b$ because $a \wedge (a \vee b) = (a \vee b) \wedge a = a$ by commutativity and absorption. Similarly, $b \leq a \vee b$. Now assume that $a \leq c$ and $b \leq c$. Then $a \vee c = c$ and $b \vee c = c$, so that $(a \vee b) \vee c = a \vee (b \vee c) = a \vee c = c$, which means $a \vee b \leq c$. This proves that $a \vee b$ is an l.u.b. for a and b. The proof that $a \wedge b$ is a g.l.b. for a and b now follows by using duality.   ∎

Although $\vee$ and $\wedge$ satisfy the commutative and associative laws, they need not satisfy the *distributive law*

$$a \wedge (b \vee c) = (a \wedge b) \vee (a \wedge c) \tag{63.4}$$

for all $a, b, c \in L$. A lattice for which (63.4) is satisfied is called a *distributive lattice*. The lattice of subsets of a set is distributive. Each of the lattices in Figure 63.1 is nondistributive. In the one on the left, for instance, $a \wedge (b \vee c) = a \wedge 1 = a$ while $(a \wedge b) \vee (a \wedge c) = 0 \vee c = c$.

A subset M of a lattice L is called a *sublattice* of L if M is closed relative to the operations $\vee$ and $\wedge$ of L. It can be proved that any nondistributive lattice contains a sublattice whose diagram is like one of the two in Figure 63.1.

Figure 63.1

Another important property satisfied by the lattice of subsets of a set is that each of its elements has a complement: In a lattice with 0 and 1, an element a' is a *complement* of an element a if

$$a \wedge a' = 0 \quad \text{and} \quad a \vee a' = 1. \tag{63.5}$$

A lattice with 0 and 1 in which each element has a complement is called a *complemented lattice*. The divisors of 12 (Example 62.6) form a lattice that is not complemented (Problem 63.4). The diagrams in Figure 63.1 represent complemented lattices; they also show that the complement of an element need not be unique (in the example on the right, for instance, each of a, b, c is a complement of the other two).

All of these properties are used together in the next section.

PROBLEMS

63.1. Show that the lattice on the right in Figure 63.1 is nondistributive.

63.2. Construct the diagram for the lattice of subgroups of $\langle (1 \quad 2), (3 \quad 4) \rangle$, and verify that it is nondistributive.

63.3. Prove: If a partially ordered set is a chain, then it is a distributive lattice.

63.4. Show that the lattice of divisors of 12 (Example 62.6) is not complemented. For which $n \in \mathbb{N}$ is the lattice of divisors of n complemented?

63.5. Is the lattice of subgroups of $\mathbb{Z}_6$ distributive? Is it complemented? (See Figure 62.4.)

63.6. Is the lattice of subgroups of S_3 distributive? Is it complemented? (See Figure 62.5.)

63.7. Construct the diagram for the lattice of subgroups of the group of rotations of a cube (Example 52.3).

63.8. Prove that if $n \in \mathbb{N}$, then the set of all positive divisors of n is a sublattice of the lattice in Example 63.2(d). This sublattice has a lattice unity. What is it?

63.9. If S is a set, what is the complement of an element in the lattice $\mathcal{P}(S)$?

63.10. Construct the diagram for the lattice of subgroups of the group of symmetries of a square (Example 8.1).

63.11. Prove the first half of Theorem 63.1.

63.12. Supply the proof that $a \wedge b = a$ implies $a \vee b = b$, which was omitted from the proof of Theorem 63.1.

63.13. The lattice $\mathbb{N}$, with $a \leq b$ defined to mean $a \mid b$, is distributive. Determine and then prove the property of integers that makes this true. (*Suggestion*: See Problem 13.6.).

63.14. Explain why each distributive lattice satisfies the law $a \vee (b \wedge c) = (a \vee b) \wedge (a \vee c)$ for all a, b, c, as well as the law in Equation (63.4).

63.15. Lattices are *isomorphic* iff they are isomorphic as partially ordered sets (see Problem 62.11). Draw a diagram corresponding to each isomorphism class of lattices that have four or fewer elements. [See Problem 62.11 (b), (c), (d).]

63.16. Prove that each finite subset of a lattice has a g.l.b. and an l.u.b.

SECTION 64   BOOLEAN ALGEBRAS

In 1854 the British mathematician George Boole published a book entitled *An Investigation of the Laws of Thought, on Which Are Founded the Mathematical Theories of Logic and Probabilities*. This book amplified ideas Boole had introduced in a shorter work published in 1847, and brought the study of logic clearly into the domain of mathematics. Boolean algebra, which originated with this work, can now be seen as the proper tool for the study not only of algebraic logic, but also such things as the theory of telephone switching circuits and computer design.

Definition I.   A *Boolean algebra* is a lattice with zero (0) and unity (1) that is distributive and complemented.

Lattices were presented in two forms in Section 63: first, in the definition, in terms of a partial ordering $\leq$; then, in Theorem 63.1, in terms of two operations $\vee$ and $\wedge$. Boolean algebras are most often discussed in the second of these forms. Because of this we next give an alternative to Definition I. Theorem 64.1 establishes the equivalence of the two definitions. Hereafter, you may work only from Definition II, if you like. All that is required from Theorem 64.1 is the definition of $\leq$ given in (64.1), and the fact that this gives a partial ordering in a Boolean algebra as defined in Definition II.

Definition II.   A *Boolean algebra* is a set B together with two operations $\vee$ and $\wedge$ on B such that each of the following axioms is satisfied (for all $a, b, c \in B$):

commutative laws

$$a \vee b = b \vee a, \quad a \wedge b = b \wedge a,$$

associative laws

$$a \vee (b \vee c) = (a \vee b) \vee c, \quad a \wedge (b \wedge c) = (a \wedge b) \wedge c,$$

distributive laws

$$a \wedge (b \vee c) = (a \wedge b) \vee (a \wedge c), \quad a \vee (b \wedge c) = (a \vee b) \wedge (a \vee c),$$

existence of zero and unity
  there are elements 0 and 1 in B such that

$$a \vee 0 = a, \quad a \wedge 1 = a,$$

existence of complements
  for each a in B there is an element a' in B such that

$$a \vee a' = 1 \quad \text{and} \quad a \wedge a' = 0.$$

Theorem 64.1. *(Definitions I and II are equivalent) In any Boolean algebra as defined in Definition I, the lattice operations $\vee$ and $\wedge$ satisfy the axioms of Definition II. Conversely, any Boolean algebra as defined in Definition II is a Boolean algebra as defined in Definition I, with $\leq$ given by*

$$a \leq b \quad \textit{iff} \quad a \vee b = b. \tag{64.1}$$

PROOF.   Assume that B is a Boolean algebra according to Definition I. Then B is a lattice, so its operations $\vee$ and $\wedge$ satisfy the commutative and associative laws by Theorem 63.1. The other axioms in Definition II are part of Definition I. (Also see Problem 63.14.) Therefore B satisfies all of the axioms of Definition II.

Assume that B is a Boolean algebra according to Definition II. To show that B is a Boolean algebra according to Definition I, it suffices, by Theorem 63.1, to show that the operations $\vee$ and $\wedge$ satisfy the idempotent and absorption laws. For the idempotent law $a \vee a = a$, write

$$a = a \vee 0 = a \vee (a' \wedge a) = (a \vee a') \wedge (a \vee a) = 1 \wedge (a \vee a) = a \vee a.$$

For the absorption law $(a \vee b) \wedge a = a$, write

$$(a \vee b) \wedge a = (a \vee b) \wedge (a \vee 0) = a \vee (b \wedge 0) = a \vee 0 = a.$$

The laws $a \wedge a = a$ and $(a \wedge b) \vee a = a$ can be proved by interchanging $\vee$ and $\wedge$, and 0 and 1 (see the Principle of Duality below).   ∎

The *dual* of a statement in a Boolean algebra is the statement that results by interchanging $\vee$ and $\wedge$, and 0 and 1. Because the dual of each axiom in Definition II is also an axiom, we have the following Principle of Duality for Boolean algebras.

Principle of Duality.   *If a statement is true for every Boolean algebra, then so is its dual statement.*

For example, $(a \wedge b) \vee a = a$ is true in every Boolean algebra because $(a \vee b) \wedge a = a$ is true in every Boolean algebra, a fact that was used at the end of the proof of Theorem 64.1.

Example 64.1.   If S is any nonempty set, then $\mathcal{P}(S)$, the power set of S, is a Boolean algebra relative to $\cup$ and $\cap$. The unity is S, the zero is $\varnothing$, and the complement of a subset is the ordinary set complement (Appendix A). We shall prove later that any finite Boolean algebra is essentially of this type (Theorem 65.1).   ■

Example 64.2.   A careful treatment of the connection between Boolean algebras and logic would require a separate chapter, but roughly the idea is as follows. By a *proposition* is meant a statement that is either true or false. For example, "Triangle ABC is equilateral" is a proposition. Propositions p and q are *logically equivalent* if either both are true or both are false. Example: In Euclidean geometry, "ABC is equilateral" is logically equivalent to "ABC is equiangular." Logical equivalence is an equivalence relation on the set of propositions. For convenience we can treat equivalent propositions as being equal. Then by "the set of all propositions" is meant the set of equivalence classes of all propositions, and by a "proposition" is meant the equivalence class of that proposition. The logical connectives "or" and "and" are operations on the set of all propositions, and it can be verified that, used for $\vee$ and $\wedge$, respectively, they yield a Boolean algebra. If p and q denote propositions, then

$$p \vee q \text{ is true iff } p \text{ is true or } q \text{ is true or both are true}$$

and

$$p \wedge q \text{ is true iff } p \text{ is true and } q \text{ is true.}$$

This algebra is called the *algebra of propositions.*

The commutative, associative, and distributive laws are consequences of the usual meaning of "or" and "and." The negation of a proposition, "not p," is denoted by p'. Then $p \vee p'$ is true for every p, and $p \wedge p'$ is false for every p. (Example: "ABC is equilateral *or ABC* is not equilateral" is true. "ABC is equilateral *and ABC* is not equilateral" is false.) For a zero, we require a proposition 0 such that $q \vee 0 = q$ for every statement q. That is, $q \vee 0$ must be true iff q is true. This will be the case iff 0 is a false statement. Thus for 0 we can use (the equivalence class of) $p \wedge p'$ for any statement p ($p \wedge p'$ is called a *contradiction*).

For a unity, we require a proposition 1 such that $q \wedge 1 = q$ for every statement q. That is, $q \wedge 1$ must be true iff q is true. This will be the case iff 1 is a true statement. Thus for 1 we can use (the equivalence class of) $p \vee p'$ for any statement p ($p \vee p'$ is called a *tautology*). With these choices for 0 and 1, we also have p' ("not p") for the (Boolean algebra) complement of each statement p.

It is interesting to interpret $\leq$, as given by (64.1), in the algebra of propositions. For propositions p and q, $p \leq q$ iff $p \vee q = q$. Reflection shows, then, that $p \leq q$ iff p is true and q is true, or p is false (no matter what q). (Remember the meaning of $\vee$ and $=$.) Therefore $p \leq q$ is logically equivalent to "if p then q" or "p implies q" The property "$p \leq 1$ for every p" means that every statement implies a true statement. The property "$0 \leq p$ for every p" means that every statement is implied by a false statement.   ■

The following theorem brings together some properties of Boolean algebras that are not listed in Definition II. The laws $(a \vee b)' = a' \wedge b'$ and $(a \wedge b)' = a' \vee b'$ in this theorem are called *DeMorgan's Laws*, after the British mathematician Augustus DeMorgan (1806–1871). DeMorgan helped lay the foundations for mathematical logic, and was one of the first to stress the purely symbolic nature of algebra.

Theorem 64.2.   *If B is a Boolean algebra and $a, b, c \in B$, then*

$$a \vee a = a, \qquad\qquad a \wedge a = a,$$
$$(a \vee b) \wedge a = a, \qquad\qquad (a \wedge b) \vee a = a,$$
$$a \vee 1 = 1, \qquad\qquad a \wedge 0 = 0,$$

a has a unique complement, a',   *and* $(a')' = a,$

$$(a \vee b)' = a' \wedge b', \qquad\qquad (a \wedge b)' = a' \vee b'$$
$$0' = 1, \qquad\qquad 1' = 0,$$

and

$$a \leq b \quad iff \quad a \vee b = b \quad iff \quad a \wedge b = a.$$

PROOF.   The first two pairs of properties (the idempotent and absorption laws) hold by Theorem 63.1 because B is a lattice. Certainly $a \vee 1 = 1$, because $a \leq 1$ for every a, and $a \wedge 0 = 0$ because $0 \leq a$ for every a.

To prove that a has a unique complement, assume that $a \vee x = 1$ and $a \wedge x = 0$, and that $a \vee y = 1$ and $a \wedge y = 0$; we shall show that $x = y$:

$$x = x \wedge 1 = x \wedge (a \vee y) = (x \wedge a) \vee (x \wedge y)$$
$$= 0 \vee (x \wedge y) = 0 \vee (y \wedge x) = (y \wedge a) \vee (y \wedge x)$$
$$= y \wedge (a \vee x) = y \wedge 1 = y.$$

The law $(a')' = a$ follows from the definition and uniqueness of complements.

To prove $(a \vee b)' = a' \wedge b'$ it suffices by uniqueness of complements to show that $(a \vee b) \wedge (a' \wedge b') = 0$ and $(a \vee b) \vee (a' \wedge b') = 1$. By the distributive laws,

$$(a \vee b) \wedge (a' \wedge b') = (a \wedge a' \wedge b') \vee (b \wedge a' \wedge b')$$
$$= 0 \vee 0$$
$$= 0$$

and

$$(a \vee b) \vee (a' \wedge b') = (a \vee b \vee a') \wedge (a \vee b \vee b')$$
$$= 1 \wedge 1$$
$$= 1.$$

The law $(a \wedge b)' = a' \vee b'$ now follows by the Principle of Duality.

The laws $0' = 1$ and $1' = 0$ are both consequences of $0 \wedge 1 = 0$ and $0 \vee 1 = 1$. The last part of the theorem is a consequence of the definition in Equation (64.1), and the first part of the proof of Theorem 63.1.   ∎

PROBLEMS

64.1.   (a)   An axiom in Definition II requires that $a \vee 0 = a$ for each $a \in B$. Why is $a \wedge 0 = 0$ also true? [*Suggestion:* To show that $a \wedge 0 = 0$, start with $a \wedge 0 = (a \wedge 0) \vee (a \wedge a')$.]

(b)   An axiom in Definition II requires that $a \wedge 1 = a$ for each $a \in B$. Why is $1 \wedge a = a$ also true? Why is $a \vee 1 = 1$ also true?

64.2.   Justify each step in the proof of the idempotent law $a \vee a = a$ in the proof of Theorem 64.1. Also write a proof of the dual statement $a \wedge a = a$.

64.3. Justify each step in the proof of the absorption law $(a \vee b) \wedge a = a$ in the proof of Theorem 64.1. Also write a proof of the dual statement $(a \wedge b) \vee a = a$.

64.4. Justify each step in the proof of the uniqueness of a complement for a in the proof of Theorem 64.2.

64.5. Assume that B is a Boolean algebra and $a \in B$. Prove that $a = 0$ iff $b = (a \wedge b') \vee (a' \wedge b)$ for each $b \in B$.

64.6. Assume that B is a Boolean algebra and that $a, b \in B$.
(a) Prove that $a \leq b$ iff $b' \leq a'$.   (b) Prove that $a \leq b'$ iff $a \wedge b = 0$.
(c) Prove that $a \leq b$ iff $a' \vee b = 1$.

64.7. Construct Cayley tables for the operations $\vee$ and $\wedge$ on the lattice in Figure 64.1. Verify that this lattice is not a Boolean algebra by finding an element without a unique complement. (This will show that B is not a Boolean algebra because a condition in Theorem 64.2 is violated.) Is this lattice distributive?

Figure 64.1

64.8. (a) Give an example to show that $a \vee c = b \vee c$ does not imply $a = b$ in Boolean algebras.
(b) Prove that if B is a Boolean algebra and $a, b, c \in B$, then $a \vee c = b \vee c$ and $a \vee c' = b \vee c'$ imply $a = b$.

64.9. Restate and do Problem 64.8 with $\vee$ replaced by $\wedge$ throughout.

64.10. Prove that if B is a Boolean algebra and $a, b, c \in B$, then $a \vee c = b \vee c$ and $a \wedge c = b \wedge c$ imply $a = b$. (Compare Problems 64.8 and 64.9.)

64.11. By considering the possible diagrams for partially ordered sets, show that there is no Boolean algebra containing exactly three elements. [See Problem 62.11(c).]

64.12. For which $n \in \mathbb{N}$ will the lattice of all positive divisors of n be a Boolean algebra, with $a \leq b$ defined to mean $a \mid b$? (Problems 64.4, 64.8, and 64.13 will help. Begin by looking at specific examples.)

SECTION 65   FINITE BOOLEAN ALGEBRAS

The goal of this section is to prove Theorem 65.1, which characterizes all finite Boolean algebras. Boolean algebras, like groups and other algebraic structures, are classified according to isomorphism.

Definition.   If A and B are Boolean algebras, an *isomorphism* of A onto B is a mapping $\theta: A \rightarrow B$ that is one-to-one and onto and satisfies

$$\theta(a \vee b) = \theta(a) \vee \theta(b)$$

and

$$\theta(a \wedge b) = \theta(a) \wedge \theta(b)$$

for all $a, b \in A$. If there is an isomorphism of A onto B, then A and B are said to be *isomorphic*, and we write $A \approx B$.

Theorem 65.1.   *Every finite Boolean algebra is isomorphic to the Boolean algebra of all subsets of some finite set.*

Example 65.1.   The divisors of 30 form a Boolean algebra with $a \leq b$ defined to mean $a \mid b$. Its diagram is shown in Figure 65.1. Here $a \vee b$ is the least common multiple of a and b, and $a \wedge b$ is the greatest common divisor of a and b.

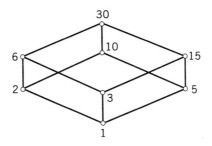

Figure 65.1

A comparison of Figure 65.1 with Figure 62.1, the diagram for the Boolean algebra of subsets of $\{x, y, z\}$, suggests an isomorphism determined by $\theta(2) = \{x\}, \theta(3) = \{y\}$, and $\theta(5) = \{z\}$. The condition $\theta(a \vee b) = \theta(a) \vee \theta(b)$ forces $\theta(6) = \{x, y\}, \theta(10) = \{x, z\}, \theta(15) = \{y, z\}$, and $\theta(30) = \{x, y, z\}$. Also, the condition $\theta(a \wedge b) = \theta(a) \wedge \theta(b)$ forces $\theta(1) = \varnothing$. This mapping θ *is* an isomorphism. The idea here is to match the elements covering 1 (the prime divisors of 30) with the elements covering $\varnothing$ (the single-element subsets of $\{x, y, z\}$). This simple and important idea is the key to Theorem 65.1.

(Although the divisors of 30 form a Boolean algebra relative to $a \mid b$, the divisors of 12 (Figure 62.2) do not, because 6 has no complement among the divisors of 12. See Problem 65.7 for a more general statement.)   ■

We lead up to the proof of Theorem 65.1 with a definition and several lemmas. This amounts to showing that every finite Boolean algebra has elements that play the same role as the single-element subsets in the Boolean algebra of subsets of a finite set. *In the remainder of this section B denotes a finite Boolean algebra.*

An element $a \in B$ is an *atom* of B if a covers 0 (the zero of B), that is, if $0 < a$ and there is no $x \in B$ such that $0 < x < a$ (Section 62). Equivalently, a is an atom iff

$$a \neq 0, \quad \text{and} \quad x \wedge a = a \quad \text{or} \quad x \wedge a = 0 \tag{65.1}$$

for each $x \in B$. The atoms in the Boolean algebra of all subsets of a set are the single-element subsets of that set. The atoms in the Boolean algebra of divisors of 30 are 2, 3, and 5, the prime divisors of 30 (Example 65.1).

Lemma 65.1.   *If $b \in B$ and $b \neq 0$, then there is an atom $a \in B$ such that $a < b$.*

PROOF.   If b is an atom, take $a = b$. Otherwise, choose an element $a_1 \in B$ such that $0 < a_1 < b$; there is such an element a_1 if b is not an atom. If a_1 is an atom, take $a = a_1$. Otherwise, choose an element $a_2 \in B$ such that $0 < a_2 < a_1 < b$. If a_2 is an atom, take $a = a_2$. Continue, if necessary, to get

$$0 < \cdots < a_3 < a_2 < a_1 < b.$$

This cannot continue indefinitely because B is finite. Therefore, a_k must be an atom for some k; take $a = a_k$. ■

Lemma 65.2.   *If a_1 and a_2 are atoms in B and $a_1 \wedge a_2 \neq 0$, then $a_1 = a_2$.*

PROOF.   Using (65.1), first with $a = a_1$ and $x = a_2$, and then with $a = a_2$ and $x = a_1$, we conclude that $a_2 \wedge a_1 = a_1$ and $a_1 \wedge a_2 = a_2$. Therefore $a_1 = a_2$ ■

Lemma 65.3.   *For $b, c \in B$, the following conditions are equivalent.*

(a) $b \leq c$

(b) $b \wedge c' = 0$

(c) $b' \vee c = 1.$

PROOF.   To prove the equivalence, it suffices to prove that (*a*) implies (*b*), (*b*) implies (*c*), and (*c*) implies (*a*).

(*a*) implies (*b*): If $b \leq c$, then $b \vee c = c$, so that (using substitution and one of De-Morgan's Laws)

$$\begin{aligned}
b \wedge c' &= b \wedge (b \vee c)' \\
&= b \wedge (b' \wedge c') \\
&= (b \wedge b') \wedge c' \\
&= 0 \wedge c' \\
&= 0.
\end{aligned}$$

(*b*) implies (*c*): If $b \wedge c' = 0$, then $(b \wedge c')' = 0'$. Therefore, by one of DeMorgan's Laws, $b' \vee c = 1$.

(*c*) implies (*a*): If $b' \vee c = 1$, then $b \wedge (b' \vee c) = b$, $(b \wedge b') \vee (b \wedge c) = b$, $0 \vee (b \wedge c) = b$, $b \wedge c = b$, and $b \leq c$. ■

Lemma 65.4.   *If b, $c \in B$ and $b \nleq c$, then there is an atom $a \in B$ such that $a \leq b$ and $a \nleq c$.*

PROOF.   If $b \nleq c$, then $b \wedge c' \neq 0$ by Lemma 65.3. Therefore, by Lemma 65.1, there is an atom $a \in B$ such that $a \leq b \wedge c'$. For this a, $a \leq b$ and $a \nleq c$.   ■

Lemma 65.5.   *If $b \in B$, and $a_1, a_2, \ldots, a_m$ are all the atoms $\leq b$, then $b = a_1 \vee a_2 \vee \cdots \vee a_m$.*

PROOF.   Let $c = a_1 \vee a_2 \vee \cdots \vee a_m$. Then $c \leq b$ since $a_i \leq b$ for $1 \leq i \leq m$. Therefore, it suffices to show that $c \geq b$. Assume, to the contrary, that $c \ngeq b$. Then by Lemma 65.4 there is an atom a such that $a \leq b$ and $a \nleq c$. But $a \leq b$ implies $a = a_i$ for some i, by the definition of the a_i $(1 \leq i \leq m)$. The conditions $a = a_i$ and $a \wedge c = 0$ are contradictory, so that we must have $c \geq b$.   ■

The next lemma shows that if any of the atoms a_i $(1 \leq i \leq m)$ in Lemma 65.5 are omitted, then b is not the l.u.b. of the remaining atoms.

Lemma 65.6.   *If $b \in B$, and $a, a_1, a_2, \ldots, a_m$ are atoms of B, with $a \leq b$ and $b = a_1 \vee a_2 \vee \cdots \vee a_m$, then $a = a_i$ for some i.*

PROOF.   We have $a \wedge b = a$ because $a \leq b$. Therefore

$$a \wedge (a_1 \vee a_2 \vee \cdots \vee a_m) = (a \wedge a_1) \vee (a \wedge a_2) \vee \cdots \vee (a \wedge a_m) = a,$$

so $a \wedge a_i \neq 0$ for some i. Lemma 65.2 gives $a = a_i$.   ■

PROOF OF THE THEOREM 65.1.   Let B be a finite Boolean algebra, and let S be the set of all atoms in B. We shall prove that $B \approx \mathcal{P}(S)$.

If $b \in B$, then by Lemmas 65.5 and 65.6, b can be written uniquely in the form $b = a_1 \vee a_2 \vee \cdots \vee a_m$ with $a_1, a_2, \ldots, a_m \in S$; moreover, $\{a_1, a_2, \ldots, a_m\} = \{a \in S : a \leq b\}$. Therefore, we can define a mapping $\theta : B \to \mathcal{P}(S)$ by

$$\theta(a_1 \vee a_2 \vee \cdots \vee a_m) = \{a_1, a_2, \ldots, a_m\}.$$

The mapping θ is clearly onto.

To show that θ is one-to-one, it suffices to show that if $b, c \in B$ and $b \neq c$, then either there is $a \in S$ such that $a \leq b$ and $a \nleq c$, or else there is $a \in S$ such that $a \nleq b$ and $a \leq c$. But this follows from Lemma 65.4, since $b \neq c$ only if $b \nleq c$ or $b \ngeq c$.

Assume that $b, c \in B$. Then $\theta(b \vee c) = \theta(b) \cup \theta(c)$ because if a is an atom then

$$a \leq b \vee c \quad \text{iff} \quad a \leq b \quad \text{or} \quad a \leq c \tag{65.2}$$

(Problem 65.5). And $\theta(b \wedge c) = \theta(b) \cap \theta(c)$ because if a is an atom then

$$a \leq b \wedge c \quad \text{iff} \quad a \leq b \quad \text{and} \quad a \leq c \tag{65.3}$$

(Problem 65.6).   ■

Corollary.   *If B is a finite Boolean algebra, then $|B| = 2^n$ for some positive integer n.*

Theorem 65.1 is not true if we remove the requirement that the Boolean algebra be finite. However, in 1936 the American mathematician M. H. Stone proved that if B is a Boolean algebra, then there is a set S such that B is isomorphic to a Boolean algebra formed by some collection $\mathcal{C}$ of subsets of S. The operations on $\mathcal{C}$ are union and intersection; the collection $\mathcal{C}$ is not necessarily all of $\mathcal{P}(S)$, but $\mathcal{C}$ is closed with respect to union, intersection, and complementation relative to S. (See [5].)

PROBLEMS

65.1. Prove that for Boolean algebras the definition of isomorphism in Problem 62.11 is equivalent to the definition in this section.

65.2. Prove: If $\theta: A \to B$ is a Boolean algebra isomorphism, then $\theta(1_A) = 1_B$ and $\theta(0_A) = 0_B$.

65.3. Prove: If $\theta: A \to B$ is a Boolean algebra isomorphism, then $\theta(a') = \theta(a)'$ for each $a \in A$.

65.4. The positive divisors of 1155 form a Boolean algebra when $a \leq b$ is taken to mean $a \mid b$. What are the atoms?

65.5. Assume that B is a Boolean algebra, $a, b, c \in B$, and a is an atom. Prove that $a \leq b \vee c$ iff $a \leq b$ or $a \leq c$. Also show that if a is not an atom then this is not necessarily true.

65.6. Assume that B is a Boolean algebra and $a, b, c \in B$. Prove that $a \leq b \wedge c$ iff $a \leq b$ and $a \leq c$. (Notice that in contrast to Problem 65.5, a need not be an atom here.)

65.7. Assume $n \in \mathbb{N}$. Using only Problem 13.12 and the corollary of Theorem 65.1, prove that if the lattice of divisors of n is a Boolean algebra (with $a \leq b$ iff $a \mid b$), then n must be a product of distinct primes. (Compare Problem 64.12.)

65.8. (This problem puts Boolean algebras in the context of rings, as discussed earlier in the book. This problem is longer than most of the problems in the book.)

(a) A ring R is called a *Boolean ring* if $x^2 = x$ for each $x \in R$. Every Boolean ring is commutative and satisfies $2x = 0$ for each $x \in R$ (Problem 27.21). Let R be a Boolean ring with unity 1, and define operations $\vee$ and $\wedge$ on R by

$$a \vee b = a + b - ab$$

and

$$a \wedge b = ab.$$

Prove that with these operations R is a Boolean algebra, with unity 1 and zero 0, and with $1 - a$ for the complement a' of a for each $a \in R$.

(b) Let B be a Boolean algebra and define two operations on B by

$$a + b = (a \wedge b') \vee (a' \wedge b)$$

and

$$ab = a \wedge b.$$

(The first of these operations is called the *symmetric difference* of a and b.) Prove that with these operations B is a Boolean ring with unity.

SECTION 66   SWITCHING

Boolean algebras have drawn increasing interest in recent years because of their usefulness in such applications as computer design and the simplification of telephone switching circuits. This section introduces some of the ideas underlying these applications. Although the terminology is that of switches and ordinary electrical circuits, the ideas also apply to circuits with various kinds of magnetic or electronic two-state devices in place of conventional two-way switches.

Think of a *switch* as a device that can be located at a point of an electrical circuit and can be either *closed* or *open*. Current can pass through the point if the switch is closed, but not if the switch is open. If two switches in a circuit are always closed or open simultaneously, we denote them by the same letter. A switch that is always open when a is closed and always closed when a is open will be denoted by a'; a' is *opposite* to a.

Two switches such as a and b in Figure 66.1 are said to be in *series*—we denote this combination by $a \wedge b$. Current can pass from p to q (the *terminals*) in Figure 66.1 iff both a and b are closed. Two switches such as a and b in Figure 66.2 are said to be in *parallel*—we denote this combination by $a \vee b$. Current can pass from p to q in Figure 66.2 iff either a is closed or b is closed. *Series-parallel circuits* are circuits that can be built up by beginning with a circuit with a single switch and repeatedly replacing switches by combinations of the type in either Figure 66.1 or Figure 66.2. Series-parallel circuits will always have two terminals, and the circuit will be called *closed* or *open* depending on whether current can or cannot pass between these terminals. All of the figures in this section represent series-parallel circuits. Two circuits are said to be *equivalent* if the switch positions that close one are the same as the switch positions that close the other.

| Figure 66.1 | Figure 66.2 |

Example 66.1.   The two circuits in Figure 66.3 are equivalent. If a is open in either circuit, then the circuit will be open. If a is closed, then both circuits will be closed iff either b or c is closed. ∎

Figure 66.3

If equivalent circuits are treated as being equal, then the series-parallel circuits form a Boolean algebra with $\vee$ and $\wedge$ as operations. The equivalence (equality) of the two circuits in Figure 66.3 amounts to one of the distributive laws: $a \wedge (b \vee c) = (a \wedge b) \vee (a \wedge c)$. The left-hand circuit in Figure 66.4 is $a \wedge a'$, the zero of the Boolean algebra: $a \wedge a'$

is always open; therefore $b \vee (a \wedge a') = b$ for every b. The right-hand circuit in Figure 66.4 is $a \vee a' = 1$, the unity of the Boolean algebra: $a \vee a'$ is always closed; therefore $b \wedge (a \vee a') = b$ for every b. Problem 66.1 asks you to draw diagrams for circuits representing most of the Boolean algebra laws from Definition II in Section 64.

$$a \wedge a' = 0 \qquad\qquad a \vee a' = 1$$

Figure 66.4

Boolean algebras can help in the simplification of circuits and also in the design of circuits with specified properties. We shall have to be content with an illustration of the first of these two applications. References for full discussions of both applications are found in the notes at the end of this chapter.

If we begin with a series-parallel circuit, then write the algebraic expression corresponding to that circuit, and then change this algebraic expression by using the laws of a Boolean algebra, the result will be an expression that corresponds to a circuit equivalent to the one with which we began. A circuit can often be simplified in this way.

Example 66.2.   The circuit in Figure 66.5 is represented by

$$(a \vee b \vee c) \wedge (a' \vee b \vee c) \wedge (a' \vee b \vee c').$$

The laws of a Boolean algebra can be used to simplify this expression as follows:

$$
\begin{aligned}
&(a \vee b \vee c) \wedge (a' \vee b \vee c) \wedge (a' \vee b \vee c') &&\text{associative law}\\
&= [a \vee (b \vee c)] \wedge [a' \vee (b \vee c)] \wedge (a' \vee b \vee c') &&\text{distributive law}\\
&= [(a \wedge a') \vee (b \vee c)] \wedge (a' \vee b \vee c') &&\text{complement}\\
&= [0 \vee (b \vee c)] \wedge (a' \vee b \vee c') &&\text{zero and commutative law}\\
&= (b \vee c) \wedge [b \vee (a' \vee c')] &&\text{distributive law}\\
&= b \vee [c \wedge (a' \vee c')] &&\text{distributive law}\\
&= b \vee [(c \wedge a') \vee (c \wedge c')] &&\text{zero and complement}\\
&= b \vee (c \wedge a').
\end{aligned}
$$

Therefore, the circuit in Figure 66.5 is equivalent to the circuit in Figure 66.6.

Figure 66.5

Figure 66.6

PROBLEMS

66.1. Draw diagrams for pairs of circuits representing the commutative laws, the associative laws, and the second distributive law from Definition II in Section 64.

66.2. Draw circuits equivalent to but simpler than each of the following circuits, by using Boolean algebra laws to simplify the algebraic expression corresponding to the circuit.

(a)   (b)

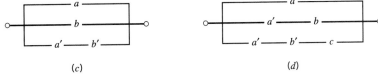

(c)   (d)

66.3. Draw a circuit equivalent to but simpler than the following diagram.

Compare Example 66.2, and use duality to state a general principle suggested by the two diagrams.

66.4. (a) Draw a diagram for a circuit that will be closed iff exactly one of a and b is closed.

   (b) Draw a diagram for a circuit that will be closed iff at least one of a and b is closed.

   (c) Draw a diagram for a circuit that will be closed iff at most one of a and b is closed.

66.5. (a) Draw a diagram for a circuit that will be closed iff exactly one of a, b, and c is closed.

   (b) Draw a diagram of a circuit that will be closed iff at least one of a, b, and c is closed.

   (c) Draw a diagram for a circuit that will be closed iff at most one of a, b, and c is closed.

NOTES ON CHAPTER XVI

1.   Donnellan, T., *Lattice Theory*, Pergamon Press, Oxford, 1968.

2.   Halmos, P. R., and S. R. Givant, Mathematical Association of America, 1998.

3.   Hohn, F. R., *Applied Boolean Algebra*, 2nd ed., Macmillan, New York, 1966.

4.   Roth, C. H., Jr., *Fundamentals of Logic Design*, 2nd ed., West, St. Paul, Minn., 1979.

5.   Stone, M. H., The theory of representations of Boolean algebras, *Transactions of the American Mathematical Society* **40** (1936), 37–111.

6.   Whitesitt, J. E., *Boolean Algebra and Its Applications*, Addison-Wesley, Reading, Mass., 1961.

APPENDIX A

SETS

This appendix contains a summary of basic facts and notation about sets.

A set is a collection of objects, called its *elements* or *members*. To indicate that x is an element of a set A, we write

$$x \in A.$$

To indicate that x is *not* an element of A, we write

$$x \notin A.$$

There are three commonly used methods of defining a set. First, simply by describing its elements:

The set of all positive integers.

Second, by listing its elements in braces:

$$\{a, b, c\} \qquad \text{or} \qquad \{1, 2, 3, \ldots\}.$$

(Not all elements in the second set can be listed, or course, but there should be no doubt that the set of all positive integers is intended.) Third, a set can be defined by *set-builder notation:*

$$\{x : x \text{ has property } P\} \text{ means "the set of all } x$$
$$\text{such that } x \text{ has property } P."$$

Thus $\{x : x \text{ is a positive integer}\}$ also denotes the set of all positive integers.

If A and B are sets, and each element of A is an element of B, then A is a *subset* of B; this is denoted by

$$A \subseteq B \qquad \text{or} \qquad B \supseteq A.$$

Notice that $A \subseteq B$ does not preclude the possibility that $A = B$. In fact,

$$A = B \qquad \text{iff}^\dagger \qquad A \subseteq B \qquad \text{and} \qquad A \supseteq B.$$

The following three statements are equivalent:

(*i*) $A \subseteq B$.
(*ii*) If $x \in A$, then $x \in B$.
(*iii*) If $x \notin B$, then $x \notin A$.

---

†We use "iff" to denote "if and only if."

Statement (iii) is the contrapositive of (ii) (see Appendix B); it is sometimes easier to prove that $A \subseteq B$ by using (iii) rather than (ii).

The *empty (null, vacuous) set* contains no elements, and is denoted by $\varnothing$. Thus,

$$x \in \varnothing \text{ is always false.}$$

If A is not a subset of B, we write $A \nsubseteq B$. This is true, of course, iff A contains at least one element that is not in B. Set inclusion ($\subseteq$) has the following properties:

$\varnothing \subseteq$ for every set A.

$A \subseteq A$ for every set A.

If $A \subseteq B$ and $B \subseteq C$, then $A \subseteq C$.

If A and B are sets, then $A \cap B$, the *intersection* of A and B, is defined by

$$A \cap B = \{x : x \in A \quad \text{and} \quad x \in B\}.$$

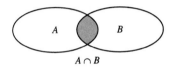

$$A \cap B$$

If A and B are sets, then $A \cup B$, the *union* of A and B, is defined by

$$A \cup B = \{x : x \in A \quad \text{or} \quad x \in B\}.$$

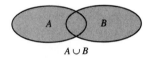

$$A \cup B$$

The diagrams above are called *Venn diagrams*.

Example A.1.   Let $S = \{a, b, c\}$, $T = \{c, d, e\}$, $U = \{d, e\}$. Then

$$
\begin{aligned}
& a \in S & & a \notin T \\
& S \cap T = \{c\} & & S \cap U = \varnothing \\
& S \nsubseteq T & & T \supseteq U \\
& S \cup T = S \cup U = \{a, b, c, d, e\}
\end{aligned}
$$

If A, B, and C denote any sets, then

$$
\begin{aligned}
& A \cap B = B \cap A & & A \cup B = B \cup A \\
& A \cap B \subseteq A & & A \subseteq A \cup B \\
& A \subseteq B \text{ implies } A \cap B = A \\
& A \subseteq B \text{ implies } A \cup B = B \\
& A \cap A = A & & A \cup A = A \\
& A \cap (B \cap C) = (A \cap B) \cap C & & A \cup (B \cup C) = (A \cup B) \cup C \\
& A \cap (B \cup C) = (A \cap B) \cup (A \cap C) \\
& A \cup (B \cap C) = (A \cup B) \cap (A \cup C). \quad\quad\quad\quad\quad\quad\quad\quad\quad\quad \blacksquare
\end{aligned}
$$

The *complement* of a set A in a set S that contains A is $A^c = \{x : x \in S \text{ and } x \notin A\}$.

The *Cartesian product* of sets A and B is denoted by $A \times B$, and is defined by

$$A \times B = \{(x, y) : x \in A \quad \text{and} \quad y \in B\}.$$

Here each (x, y) is an *ordered pair*. The term *ordered* is used because (x, y) is to be distinguished from (y, x) if $x \neq y$. Ordered pairs (x, y) and (x', y') are *equal* iff $x = x'$ and $y = y'$.

Example A.2.   If $A = \{1, 2\}$ and $B = \{u, v, w\}$, then

$$A \times B = \{(1, u), (1, v), (1, w), (2, u), (2, v), (2, w)\}.$$

Notice that in this case $A \times B \neq B \times A$. In general, $A \times B = B \times A$ iff $A = B$.   ■

Plane coordinate (analytic) geometry begins with a one-to-one correspondence between the points of the plane and the set of pairs of real numbers, that is, elements of the Cartesian product $\mathbb{R} \times \mathbb{R}$. This example explains the choice of the name "Cartesian product"— Descartes was one of the developers of analytic geometry.

The notion of Cartesian product can be used to give a definition of mapping that is preferred by some to that given in Section 1: A *mapping* from a set S to a set T is a subset of $S \times T$ such that each $x \in S$ is a first member of precisely one pair in the subset. The connection between this definition and that given in Section 1 is that if $\alpha : S \to T$ (in the sense of Section 1), then each $x \in S$ contributes the pair $(x, \alpha(x))$ to the subset of $S \times T$ in the definition of mapping as a subset of $S \times T$. For example, the mapping α in Example 1.1 corresponds to the subset $\{(x, 2), (y, 1), (z, 3)\}$ of $\{x, y, x\} \times \{1, 2, 3\}$.

PROBLEMS

A.1. Assume $A = \{t, u, y\}$, $B = \{w, x\}$, and $C = \{x, y, z\}$. Find each of the following sets.
  (a) $A \cap B$ (b) $A \cap C$ (c) $B \cup C$
  (d) $A \backslash B$ (e) $B \backslash C$ (f) $A \times B$
  (g) $C \times B$ (h) $A \cup (B \cap C)$ (i) $(A \cup B) \cap C$

A.2. Find each of the following if A, B, and C are as in Problem A.1.
  (a) $|A \times C|$ (b) $|\emptyset \backslash B|$ (c) $|B \times \emptyset|$

A.3. Here is a Venn diagram for $A \cap (B \cup C)$.

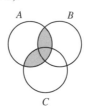

Draw a similar diagram for $(A \cap B) \cup (A \cap C)$. How do the two diagrams compare?

A.4. Draw Venn diagrams for $A \cup (B \cap C)$ and $(A \cup B) \cap (A \cup C)$. Are the results consistent with the property $A \cup (B \cap C) = (A \cup B) \cap (A \cup C)$?

A.5. Assume that $S = \{1, 2, 3, \ldots, 10\}$, $A = \{1, 2, 3, 4, 5\}$, and $B = \{2, 4, 6, 8, 10\}$. Find A^c, B^c, $(A \cup B)^c$, and $(A \cap B)^c$, where each complement is taken relative to the set S.

APPENDIX B

PROOFS

A distinguished feature of mathematics is the use of deductive reasoning. In the deductive method we begin with a collection of statements (*axioms*) that are assumed to be true, and then use logical reasoning (*rules of inference*) to prove other statements (*theorems*).

An important example is Euclidean geometry. In Book I of the *Elements* (written c. 300 B.C.), Euclid began with ten exioms (divided into five postulates and five common notions) and then deduced the fundamental theorems of Euclidean plane geometry, including the Pythagorean Theorem and its converse.

In many parts of mathematics the underlying formal structure (axioms, theorems, and the connections between them) is not always made clear. In a typical calculus course, for example, some theorems are given without proof, some proofs are only sketched, and even the proofs that are complete rely on ideas from geometry and algebra for which many students may not have seen formal proofs. For some purposes this way of presenting mathematics is sufficient, since the goal may be only a working knowledge of the basic ideas and their applications. However, the underlying formal structure is an integral part of this book. For example, nearly all of the central ideas—such as group, ring, and field—are defined by axioms.

This appendix is meant to help with some of the ideas and techniques that are used in deductive reasoning.

In understanding or constructing proofs we are invariably concerned with *conditional* statements, that is, with statements of the form

$$\text{if } p \text{ then } q.$$

Any such statement is logically equivalent to each of a number of other statements, where to say that two statements are *logically equivalent* means that they are either both true or both false. In particular, the conditional statement above is logically equivalent to each of the following statements:

p implies q.

p only if q.

q if p.

p is sufficient for q.

q is necessary for p.

Example B.1. Let A and B denote sets. The statement

$$A \text{ is a subset of } B$$

is logically equivalent to the conditional statement

$$\text{if } x \in A \text{ then } x \in B.$$

This conditional statement, in turn, is logically equivalent to each of the following statements:

$$x \in A \text{ implies } x \in B.$$
$$x \in A \text{ only if } x \in B.$$
$$x \in B \text{ if } x \in A.$$
$$x \in A \text{ is sufficient for } x \in B.$$
$$x \in B \text{ is necessary for } x \in A. \qquad \blacksquare$$

A *direct proof* of a conditional statement "If p then q" will consist of a finite sequence of statements $p = p_1, p_2, \ldots, p_n = q$, leading from the *hypothesis* p to the *conclusion* q. Each of the statements $p_1, p_2, \ldots, p_n$ will be an axiom, a hypothesis, a statement that has already been proved, or a statement derivable from such statements by using rules of inference (which are covered in books that present introductory logic). Following is an example (half of Problem 1.27). The ideas used are in Section 1 and Appendix A.

Example B.2.   Prove that if $\alpha : S \to T$ and A and B are subsets of S, then $\alpha(A \cup B) \subseteq \alpha(A) \cup \alpha(B)$.

PROOF.   To begin, we must have a clear idea of what must be done. In this case, we must demonstrate (given $\alpha : S \to T$ and A and B are subsets of S) that every element of $\alpha(A \cup B)$ is also an element of $\alpha(A) \cup \alpha(B)$, that is, an element of $\alpha(A)$ or an element of $\alpha(B)$.

Thus assume that $y \in \alpha(A \cup B)$. Then $y = \alpha(x)$ for some $x \in A \cup B$, by the definition of image of a subset (Section 1). Since $x \in A \cup B$ we know $x \in A$ or $x \in B$, by the definition of union of sets. If $x \in A$, then $y = \alpha(x) \in \alpha(A)$, and if $x \in B$ then $y = \alpha(x) \in \alpha(B)$. Thus in either case $y \in \alpha(A) \cup \alpha(B)$, which is what we were to prove. $\qquad \blacksquare$

Another statement logically equivalent to the conditional statement

$$\text{if } p \text{ then } q$$

is its *contrapositive*,

$$\text{if not } q \text{ then not } p.$$

This is the basis for the method of *indirect proof*, in which a conditional statement is proved by proving its contrapositive.

Example B.3.   The statement

$$\text{if } x \in A \text{ then } x \in B$$

is logically equivalent to

$$\text{if } x \notin B \text{ then } x \notin A.$$

Therefore, to prove "$A \subseteq B$" is the same as to prove "if $x \notin B$ then $x \notin A$" $\qquad \blacksquare$

Another method, similar to indirect proof, is *proof by contradiction* (or *reductio ad absurdum*). With this method, a conditional statement "if p then q" is proved by showing that if p were true and q were not true, then some contradiction (absurdity) would result. For an example of a proof by contradiction see Theorem 31.1.

The statement

$$\text{if } q \text{ then } p$$

is the *converse* of

$$\text{if } p \text{ then } q.$$

A statement and its converse are *not* logically equivalent. For example, if we were to say that "if $x \in A$ then $x \in B$" is logically equivalent to "if $x \in B$ then $x \in A$," we would be saying that $A \subseteq B$ is the same as $B \subseteq A$, which is manifestly false.

Putting a conditional statement and its converse together, we get a *biconditional* statement:

$$p \text{ if and only if } q,$$

which we are writing

$$p \text{ iff } q.$$

It is an immediate consequence of earlier remarks that this biconditional statement is logically equivalent to

$$p \text{ is necessary and sufficient for } q.$$

Sometimes, "if p then q" or one of its equivalents in written when in fact "p iff q" is true. This happens especially with definitions. For example, the "if" in

$$\text{An integer } p \text{ is a } prime \text{ if } p > 1 \text{ and } p$$
$$\text{is divisible by no positive integer other than}$$
$$1 \text{ and } p \text{ itself,}$$

really means "iff."

The connectives "or," "and," and "not" arise often in forming compound statements. Unless it is explicitly stated otherwise, the word "or" should always be taken in the *inclusive* sense:

$$p \quad \text{or} \quad q$$

means

$$p \quad \text{or} \quad q \quad \text{or both.}$$

The words "and" and "not" have their usual meanings.

The following rules are used frequently:

1. not (p and q) is logically equivalent to (not p) or (not q).
2. not (p or q) is logically equivalent to (not p) and (not q).

Example B.4.   Let A and B denote sets, and let

$$p \quad \text{denote} \quad x \in A$$

and

$$q \quad \text{denote} \quad x \in B.$$

Then

$$\text{"}p \text{ and } q\text{" means } x \in A \cap B,$$

and

$$\text{"}p \text{ or } q\text{" means } x \in A \cup B.$$

This leads to the following two lists of logically equivalent statements:

$$\left. \begin{array}{l} \text{not } (p \text{ and } q) \\ x \notin A \cap B \\ x \notin A \text{ or } x \notin B \\ (\text{not } p) \text{ or } (\text{not } q) \end{array} \right\} \quad \begin{array}{l} \text{logically} \\ \text{equivalent} \end{array}$$

$$\left. \begin{array}{l} \text{not } (p \text{ or } q) \\ x \notin A \cup B \\ x \notin A \text{ and } x \notin B \\ (\text{not } p) \text{ and } (\text{not } q) \end{array} \right\} \quad \begin{array}{l} \text{logically} \\ \text{equivalent.} \end{array} \qquad \blacksquare$$

In statements such as

$$x^2 - 4 = 0,$$

and

$$x \text{ is nonnegative,}$$

the letter "x" is called a *variable*. Variables are assumed to belong to some previously agreed-upon universal set. In the two examples, this might be the set of all real numbers, for instance. A statement with a variable can be combined with either a *universal quantifier* or an *existential quantifier*:

Universal quantifier: "for each x."
Existential quantifier: "there is an x such that."

Once a quantifier has been introduced it becomes meaningful to ask whether a statement with a variable is true or false.

Example B.5. Let the variable x represent a real number.

For each x, $x^2 - 4 = 0$.	False
There is an x such that $x^2 - 4 = 0$.	True
For each x, x^2 is nonnegative.	True

$\blacksquare$

The terms "for every" and "for all" are often used in place of "for each." Also, "there exists an x such that" and "for some x" are often used in place of "there is an x such that." Moreover, statements that can be made using quantifiers are often made without their explicit use. For instance, "for each (real number) x, x^2 is nonnegative," is logically equivalent to "the square of every real number is nonnegative."

In a statement with more than one variable, each variable requires quantification. Example: For all real numbers a, b, and c, $a(b + c) = ab + ac$.

The negation of a statement with a universal quantifier is a statement with an existential quantifier, and the negation of a statement with an existential quantifier is a statement with a universal quantifier.

Example B.6.

Statement I: For each x, x^2 is nonnegative.

Negation of I: There is an x such that x^2 is negative.

Statement II: There is an x such that $x^2 - 4 = 0$.

Negation of II: For each x, $x^2 - 4 \neq 0$.

An example showing that a statement with a universal quantifier is false is called a *counterexample*. Thus $x = 2$ is a counterexample to the statement "for each prime number x, x is odd." ∎

The preceding remarks were concerned with the technical aspects of proofs. A slightly different problem is that of how to *discover* proofs; the second and third books below are recommended for their advice on this problem. The first book is recommended for advice on writing mathematics.

1. Gillman, L., *Writing Mathematics Well: A Manual for Authors*, The Mathematical Association of America, 1987.
2. Pólya, G., *How to Solve It*, Princenton University Press, Princeton, N.J., 1988.
3. Solow, D., *How to Read and Do Proofs: An Introduction to Mathematical Thought Process*, 2nd ed., Wiley, New York, 1990.

APPENDIX C

MATHEMATICAL INDUCTION

PRINCIPLE OF MATHEMATICAL INDUCTION

For each positive integer n, let $P(n)$ represent a statement depending on n. If

(a) $P(1)$ is true, and

(b) $P(k)$ implies $P(k+1)$ for each positive integer k,

then $P(n)$ is true for every positive integer n.

Intuitively, the principle is valid because $P(1)$ is true by (a), and then $P(2)$ is true by (b) with $k = 1$, and then $P(3)$ is true by (b) with $k = 2$, and so on. Logically, the principle is equivalent to the Least Integer Principle, stated in Section 10. (Also see Section 26, especially Problem 26.7.)

Example C.1. If $S(n)$ denotes the sum

$$a + ar + ar^2 + \cdots + ar^{n-1}$$

of the first n terms of a geometric progression with first term a and common ratio $r \neq 1$, then

$$S(n) = \frac{a - ar^n}{1 - r} \tag{C.1}$$

To prove this, for each positive integer n let $P(n)$ be the statement that the formula (C.1) is correct.

(a) For $n = 1$, (C.1) gives

$$S(1) = \frac{a - ar}{1 - r} = a,$$

which is clearly correct.

(b) Assume that $P(k)$ is true. Then

$$S(k) = \frac{a - ar^k}{1 - r}$$

is correct. Therefore

$$S(k+1) = S(k) + ar^k$$
$$= \frac{a - ar^k}{1 - r} + ar^k$$
$$= \frac{a - ar^k + (1 - r)ar^k}{1 - r}$$
$$= \frac{a - ar^{k+1}}{1 - r}$$

and the statement $P(k + 1)$ is true. Therefore $P(k)$ does imply $P(k + 1)$, as required.   ■

Example C.2.   Let a denote a real number. Define positive integral powers of a as follows: $a^1 = a$, $a^2 = aa$, $a^3 = a^2a$, and, in general, $a^{n+1} = a^n a$. We shall prove that

$$a^m a^n = a^{m+n} \tag{C.2}$$

for all positive integers m and n. In doing this we shall assume the law

$$a(bc) = (ab)c \tag{C.3}$$

for all real numbers a, b and c. [The law in Equation (C.2) will be familiar for real numbers from elementary algebra. The proof here will show that it is valid, more generally, in any system satisfying Equation (C.3). See Section 14.]

Here we let $P(n)$ be the statement that Equation (C.2) is correct for n and every positive integer m.

(a)   With $n = 1$, Equation (C.2) is $a^m a = a^{m+1}$, which is correct because $a^m a$ is the definition of a^{m+1}. Therefore $P(1)$ is true.

(b)   Assume that $P(k)$ is true. Then $a^m a^k = a^{m+k}$ is correct. Therefore

$$
\begin{aligned}
a^m a^{k+1} &= a^m (a^k a^1) & \text{[by the definition of } a^{k+1}] \\
&= (a^m a^k)a & \text{[by (C.3)]} \\
&= a^{m+k} a & \text{[by } P(k)] \\
&= a^{m+k+1} & \text{[by the definition of } a^{m+k+1}],
\end{aligned}
$$

and the statement $P(k + 1)$ is true.   ■

Example C.3 Binomial Theorem.   If a and b denote any real numbers and n denotes a positive integer, then

$$(a + b)^n = C(n, n)a^n + C(n, n - 1)a^{n-1}b + C(n, n - 2)a^{n-2}b^2 + \cdots$$
$$+ C(n, r)a^r b^{n-r} + \cdots + C(n, 0)b^n \tag{C.4}$$

$C(n, r)$ is the *binomial coefficient*, defined by

$$C(n, r) = \frac{n!}{r!(n - r)!}$$

for $0 \leq r \leq n$ [with $r! = r(r - 1) \cdots 1$ for $r > 0$, and $0! = 1$].

To prove this, let $P(n)$ be the statement that Equation (C.4) is correct.

(a) For $n = 1$, the right-hand side of Equation (C.4) gives

$$C(1, 1)a + C(1, 0)b = a + b,$$

which is equal to $(a + b)^1$, as required.

(b) Assume that $P(k)$ is true. If both sides of Equation (C.4) (with n replaced by k) are multiplied by $(a + b)$, then equality will result, and, moreover, the coefficient of $a^r b^{k+1-r}$ on the right will be

$$C(k, r - 1) + C(k, r).$$

(It may take pencil and paper to verify that.) The coefficient of $a^r b^{k+1-r}$ on the right when n is replaced by $k + 1$ in (C.4) will be $C(k + 1, r)$. The truth of $P(k + 1)$ is a consequence of the identity

$$C(k + 1, r) = C(k, r - 1) + C(k, r),$$

which can be verified by simple addition. ∎

The statement $P(k)$, assumed in the second part of proofs by mathematical induction, is sometimes called the *induction hypothesis*. In the following alternative form of the Principle of Mathematical Induction an apparently stronger induction hypothesis is used: In place of assuming only $P(k)$, we assume $P(i)$ for all $i \le k$. The proof that the two forms of the principle are logically equivalent will be left as an exercise; the appropriate context for this exercise is Section 26.

SECOND PRINCIPLE OF MATHEMATICAL INDUCTION

For each positive integer n, let $P(n)$ represent a statement depending on n. If

(a) $P(1)$ is true and

(b) the truth of $P(i)$ for all $i \le k$ implies the truth of $P(k + 1)$, for each positive integer k.

then $P(n)$ is true for every positive integer n.

APPENDIX D

LINEAR ALGEBRA

This appendix contains a concise review of the facts about vector spaces, linear transformations, and matrices that are needed elsewhere in this book. It also presents important examples of groups and rings that arise from linear transformations and matrices. Ideas from Chapters I to VI are used freely, and proofs are omitted.

Definition. A *vector space* over a field F is a set V together with an operation $+$ on V and a mapping $F \times V \to V$ $[(a, v) \mapsto av]$ such that each of the following axioms is satisfied:

1. V is an Abelian group with respect to $+$,
2. $(ab)v = a(bv)$,
3. $(a + b)v = av + bv$,
4. $a(v + w) = av + aw$,
5. $ev = v$,

for all $a, b \in F$ and all $v, w \in V$, with e the unity of F.

The elements of V and F are called *vectors* and *scalars*, respectively. And av is called the *scalar multiple* of $v \in V$ by $a \in F$. Throughout this section V will denote a vector space over a field F.

Example D.1. Let F^n denote the set of n-tuples $(a_1, a_2, \ldots, a_n)$ with each $a_i \in F$. With the operations

$$(a_1, a_2, \ldots, a_n) + (b_1, b_2, \ldots, b_n) = (a_1 + b_1, a_2 + b_2, \ldots, a_n + b_n)$$

and

$$a(a_1, a_2, \ldots, a_n) = (aa_1, aa_2, \ldots, aa_n),$$

F^n is a vector space over F. The elements of $\mathbb{R}^2$ and $\mathbb{R}^3$ will sometimes be identified with geometric vectors in the usual way. If p is a prime, then $\mathbb{Z}_p^n$ is a vector space with p^n elements. ∎

Example D.2. The complex field $\mathbb{C}$ can be thought of as a vector space over the real field $\mathbb{R}$. The elements of $\mathbb{C}$ are the vectors; the addition of vectors is the addition in $\mathbb{C}$; and $a(b + ci) = ab + aci$ $(a \in \mathbb{R}, b + ci \in \mathbb{C})$. That is, the scalar multiples are just multiples in $\mathbb{C}$ except that the first factors are restricted to $\mathbb{R}$. The same ideas can be extended to show that if E is any field and F is any subfield of E, then E is a vector space over F.

A subset of W of V is a *subspace* of V if W is itself a vector space with respect to the addition of V and the scalar multiplication $aw(a \in F, w \in W)$. It is useful to know that a nonempty subset W of V is a subspace of V iff

1. $v + w \in W$ for all $v, w \in W$, and
2. $av \in W$ for all $a \in F, v \in W$.

If $a_1, a_2, \ldots, a_n \in F$ and $v_1, v_2, \ldots, v_n \in V$, then

$$a_1 v_1 + a_2 v_2 + \cdots + a_n v_n$$

is called a *linear combination* of $v_1, v_2, \ldots, v_n$. If $\{v_1, v_2, \ldots, v_n\}$ is a subset of V, let

$$\langle v_1, v_2, \ldots, v_n \rangle = \{a_1 v_1 + a_2 v_2 + \cdots + a_n v_n : a_1, a_2, \ldots, a_n \in F\},$$

the set of all linear combinations of $v_1, v_2, \ldots, v_n$. Then $\langle v_1, v_2, \ldots, v_n \rangle$ is a subspace of V, called the subspace *generated* (or *spanned*) by $\{v_1, v_2, \ldots, v_n\}$. The vectors $v_1, v_2, \ldots, v_n$ are said to be *linearly independent* if

$$a_1 v_1 + a_2 v_2 + \cdots + a_n v_n = 0$$

implies that

$$a_2 = a_2 = \cdots = a_n = 0$$

for $a_1, a_2, \ldots, a_n \in F$; otherwise $v_1, v_2, \ldots, v_n$ are *linearly dependent*.

If $v_1, v_2, \ldots, v_n$ are linearly independent and $\langle v_1, v_2, \ldots, v_n \rangle = V$, then $\{v_1, v_2, \ldots, v_n\}$ is said to be a *basis* for V. We shall be concerned only with vector spaces having finite bases. Any two bases for a vector space V have the same number of elements; this number is called the *dimension* of V and is denoted dim V. ∎

Example D.3. If

$$e_1 = (1, 0, 0, \ldots, 0)$$
$$e_2 = (0, 1, 0, \ldots, 0)$$
$$\vdots$$
$$e_n = (0, 0, 0, \ldots, 1),$$

then $\{e_1, e_2, \ldots, e_n\}$ is a basis for $\mathbb{R}^n$. Thus dim $\mathbb{R}^n = n$. The vectors $e_1, e_2, \ldots, e_n$ will be called *standard unit vectors*. ∎

Any set that generates V contains a basis for V. Any linearly independent subset of V is contained in a basis for V. The dimension of a subspace of V cannot exceed the dimension of V.

Example D.4. The set $\{1, i\}$ is a basis for $\mathbb{C}$ as a vector space over $\mathbb{R}$ (Example D.2). It generates $\mathbb{C}$ because each complex number can be written as $a \cdot 1 + b \cdot i = a + bi$ with $a, b \in \mathbb{R}$. It is linearly independent because if $a + bi = 0$ with $a, b \in \mathbb{R}$, then $a = 0$ and $b = 0$. Therefore, as a vector space over $\mathbb{R}$, dim $\mathbb{C} = 2$. ∎

Let V denote a vector space over a field F. A mapping $\alpha : V \to V$ is a *linear transformation* if

$$\alpha(av + bw) = a\alpha(v) + b\alpha(w) \tag{D.1}$$

for all $a, b \in F$, $v, w \in V$. A linear transformation is *nonsingular* (or *invertible*) iff it is invertible as a mapping, that is, iff it is one-to-one and onto. The inverse of a nonsingular linear transformation is necessarily linear. Let

$$L(V) \text{ denote the set of all linear transformation from } V \text{ to } V$$

and

$$GL(V) \text{ denote the set of all nonsingular elements of } L(V).$$

If $\alpha, \beta \in L(V)$, then $\alpha + \beta$ and $\beta\alpha$ are defined as follows:

$$(\alpha + \beta)(v) = \alpha(v) + \beta(v)$$
$$(\beta\alpha)(v) = \beta(\alpha(v)) \tag{D.2}$$

for all $v \in V$. It can be verified that both $\alpha + \beta$ and $\beta\alpha$ are linear, so we have two operations on $L(V)$.

Theorem D.1.
  (a) $L(V)$ *is a ring with respect to the operations defined in* (D.2).
  (b) $GL(V)$ *is a group with respect to the product (composition) defined in* (D.2).

If A and B are both $m \times n$ matrices over F (that is, with entries in F), then their sum, $A + B$, is defined to be the $m \times n$ matrix with ij-entry $a_{ij} + b_{ij}$, where a_{ij} and b_{ij} are the ij-entries of A and B, respectively. If A is an $m \times n$ matrix over F, and B is an $n \times p$ matrix over F, then their product, AB, is defined to be the $m \times p$ matrix with ij-entry

$$a_{i1}b_{1j} + a_{i2}b_{2j} + \cdots + a_{in}b_{nj} = \sum_{k=1}^{n} a_{ik}b_{kj}.$$

Let I_n denote the $n \times n$ *identity matrix*, which is defined by $a_{ii} = 1$ for $1 \le i \le n$ and $a_{ij} = 0$ for $i \ne j$. (Here 1 denotes the unity of F.) An $n \times n$ matrix A is *nonsingular* or (*invertible*) iff there is an $n \times n$ matrix A^{-1} such that $A^{-1}A = AA^{-1} = I_n$. Let

$$M(n,F) \text{ denote the set of all } n \times n \text{ matrices over } F$$

and

$$GL(n, F) \text{ denote the set of all nonsingular elements of } M(n, F).$$

Theorem D.2.
  (a) $M(n, F)$ *is a ring with respect to matrix addition and multiplication.*
  (b) $GL(n, F)$ *is a group with respect to matrix multiplication.*

Let $\{v_1, v_2, \ldots, v_n\}$ be a basis for V and let $\alpha \in L(V)$. Then each $\alpha(v_i)$ is in $V(1 \le i \le n)$, and can be written as a linear combination of $v_1, v_2, \ldots, v_n$:

$$\alpha(v_i) = \sum_{j=1}^{n} a_{ji} v_j = a_{1i} v_1 + a_{2i} v_2 + \cdots + a_{ni} v_n, \tag{D.3}$$

with all $a_{ij} \in F$. The matrix with ij-entry a_{ij} is the *matrix of α relative to the basis* $\{v_1, v_2, \ldots, v_n\}$. This matrix depends on the order of the vectors in the basis $\{v_1, v_2, \ldots, v_n\}$. Conversely, given an ordered basis for V, and an $n \times n$ matrix $A = [a_{ij}]$ over F, there is a unique $\alpha \in L(V)$ having the matrix A; this is defined by (D.3).

Example D.5.   Assume that V is a vector space over $\mathbb{R}$, $\{v_1, v_2, v_3\}$ is a basis for V, $\alpha \in L(V)$, and

$$\begin{aligned}
\alpha(v_1) &= 5v_1 + v_2 - 2v_3 \\
\alpha(v_2) &= \quad\;\; - 2v_2 + v_3 \\
\alpha(v_3) &= v_1 + 4v_2.
\end{aligned}$$

Then the matrix for α relative to $\{v_1, v_2, v_3\}$ is

$$\begin{bmatrix} 5 & 0 & 1 \\ 1 & -2 & 4 \\ -2 & 1 & 0 \end{bmatrix}.$$

If $v = a_1 v_1 + a_2 v_2 + a_3 v_3 \in V$, then the representation of $\alpha(v)$ as a linear combination of v_1, v_2, v_3 can be obtained by matrix multiplication, as follows:

$$\begin{bmatrix} 5 & 0 & 1 \\ 1 & -2 & 4 \\ -2 & 1 & 0 \end{bmatrix} \begin{bmatrix} a_1 \\ a_2 \\ a_3 \end{bmatrix} = \begin{bmatrix} 5a_1 + a_3 \\ a_1 - 2a_2 + 4a_3 \\ -2a_1 + a_2 \end{bmatrix}$$

and

$$\alpha(v) = (5a_1 + a_3)v_1 + (a_1 - 2a_2 + 4a_3)v_2 + (-2a_1 + a_2)v_3. \qquad \blacksquare$$

The correspondence $\alpha \leftrightarrow [a_{ij}]$ determined by (D.3) can be used to prove the following theorem.

Theorem D.3.
(a) If $\dim V = n$, then $L(V) \approx M(n, F)$, as rings.
(b) If $\dim V = n$, then $GL(V) \approx GL(n, F)$, as groups. Both $GL(V)$ and $GL(n, F)$ are called general linear groups.

The following ideas are used in Chapter XV, on algebraic coding. The *row space* of an $m \times n$ matrix A over F is the subspace of F^n generated by the rows of A (thought of as elements of F^n). The *column space* of A is the subspace of F^m generated by the columns of A (thought of as elements of F^m). The dimensions of the row space and column space of A are called the *row rank* and *column rank* of A, respectively. It can be proved that these two ranks are equal for each matrix A; their common value is called the *rank* of A.

The *transpose* of an $m \times n$ matrix A is the $n \times m$ matrix, denoted A', obtained by interchanging the rows and columns of A. Thus each ij-entry of A' is the ji-entry of A.

The transpose of a $1 \times n$ (*row*) matrix is an $n \times 1$ (*column*) matrix. We shall identify each $1 \times n$ matrix with the corresponding vector in F^n.

Let A be an $m \times n$ matrix over F. The *null space* of A is the set of all $v \in F^n$ such that

$$Av' = 0.$$

The null space of A is a subspace of F^n, and its dimension is called the *nullity* of A. It can be proved that

$$\text{rank } A + \text{ nullity } A = n.$$

The *elementary row operations* on a matrix are of three types:

I. Interchange two rows.

II. Multiply a row by a nonzero scalar.

III. Add a multiple of one row to another row.

If a matrix B can be obtained from a matrix A by a finite sequence of elementary row operations, then A and B are said to be *row equivalent*. Row equivalence is an equivalence relation on the set of $m \times n$ matrices over F. Each matrix is row equivalent to a unique matrix in *row-reduced echelon form*, that is, a matrix such that

1. the first nonzero entry (the *leading entry*) of each row is 1,

2. the other entries in any column containing such a leading entry are all 0,

3. the leading entry in each row is to the right of the leading entry in each preceding row, and

4. rows containing only 0's are below rows with nonzero entries.

APPENDIX E

SOLUTIONS TO SELECTED PROBLEMS

This appendix contains answers or solutions to nearly all the odd-numbered problems preceding the double line in each section.

1.1. (a) α is neither one-to-one nor onto; β is both one-to-one and onto.

(b) $\alpha(A) = \{1, 2\}$; $\beta(B) = \{1, 2, 3\}$; $\alpha(A \cap B) = \{1\}$; $\beta(A \cup B) = T$.

1.3. 6   **1.5.** 8

1.7. One answer: $\ldots y \in T$ there is no $x \in S$ such that $\alpha(x) = y$. Another answer: $\ldots y \in T$, $\alpha(x) \neq y$ for each $x \in S$. (See Example B.6 in Appendix B.)

1.9. $\ldots \alpha(x) \neq \beta(x)$ for some $x \in S$.

1.11. Both one-to-one and onto. And $f(P) = P$.

1.13. Both one-to-one and onto. And $f(P) = P$.

1.15. One-to-one but not onto. And $f(P) = \{x \in \mathbb{R} : \quad x > 1\}$.

1.17. $\alpha(A) = \{2n : n \in \mathbb{N} \quad \text{and} \quad n \text{ is odd}\}$

$\alpha(B) = C$

$\alpha(C) = \{8n : \quad n \in \mathbb{N}\}$

1.19. 1 and -1.

2.1. $(\alpha \circ \alpha)(n) = 4n$. Image = all integral multiples of 4.

2.3. $(\alpha \circ \beta)(n) = 2n + 2$. Image = all even integers.

2.5. $(\beta \circ \gamma)(n) = n^2 + 1$. Image = $\{n^2 + 1 : \quad n \in \mathbb{Z}\}$.

2.7. If $x \in S$, then $(\alpha \circ \iota_S)(x) = \alpha(\iota_S(x)) = \alpha(x)$, so $\alpha \circ \iota_S = \alpha$. And if $x \in S$, then $(\iota_T \circ \alpha)(x) = \iota_T(\alpha(x)) = \alpha(x)$, so $\iota_T \circ \alpha = \alpha$.

2.9. (a) $g(x) = x/5$.   (b) $g(x) = x + 4$.   (c) $g(x) = \log_{10} x \, (x > 0)$.

2.11. (a) False.   (b) True.   (c) True.   (d) False.

2.13. No. Check $x = \pi/2$, for example. [In fact, $(f \circ g)(x) = \sin 2x = 2 \sin x \cdot \cos x$, but $(g \circ f)(x) = 2 \sin x$.]

2.15. $\ldots \alpha$ is not onto. (See Example B.4 in Appendix E.)

2.17. (a) No. For one example, use $S = \{v\}$, $T = \{w, x\}$, $U = \{y, z\}$, $\alpha(v) = x$, $\beta(w) = y$, and $\beta(x) = z$.

(b) If $\beta \circ \alpha$ is one-to-one and onto, then α is one-to-one and β is onto. True. (Give reasons.)

(c) If $\beta \circ \alpha$ is not one-to-one and onto, then α is not one-to-one or β is not onto. (Or: If $\beta \circ \alpha$ is not one-to-one or $\beta \circ \alpha$ is not onto, then α is not one-to-one or β is not onto.) This contrapositive statement is false. (Give reasons.) See Example B.4 in Appendix B.

SECTION 3

3.1. Operation. Commutative but not associative. No identity.

3.3. Operation. Associative but not commutative. No identity.

3.5. Operation. Commutative but not associative. No identity.

3.7. Operation. Commutative and associative. No identity.

3.9. No. (give a reason.)

3.11. For k to be an identity, we would need $m * k = m^k = m$ for every positive integer m, so $k = 1$. But $k * m = m$ is also required, and $1 * m = 1^m = 1$ for every positive integer m, so 1 is not an identity.

3.13. (a) and (c) are true.

3.15. ... for some $a, b \in S$. (Compare Example 1.2.)

3.17. ... $e * a \neq a$ or $a * e \neq a$.

3.19. The set of all positive even integers.

SECTION 4

4.1. (a)

$\circ$	π	ρ	σ	τ
π	π	π	π	π
ρ	π	ρ	σ	τ
σ	τ	σ	ρ	π
τ	τ	τ	τ	τ

(b) ρ.   (c) No.   (d) ρ and σ.   (e) Yes.

4.3.

$\circ$	ρ_1	ρ_2	ρ_3	ρ_4
ρ_1	ρ_1	ρ_2	ρ_3	ρ_4
ρ_2	ρ_2	ρ_3	ρ_4	ρ_1
ρ_3	ρ_3	ρ_4	ρ_1	ρ_2
ρ_4	ρ_4	ρ_1	ρ_2	ρ_3

ρ_1 is an identity element and each element does have an inverse.

SECTION 5

5.1. Group.   **5.3.** Group.   **5.5.** Group.

5.7. Not a group. (Only 1 and -1 have inverses.)

5.9. Not a group. (Division is not associative. Also, there is no identity element.)

5.11. Let S denote the given set.
Operation: $2^m \cdot 2^n = 2^{m+n} \in S$ if $2^m 2^n \in S$ because $m + n \in \mathbb{Z}$ if $m, n, \in \mathbb{Z}$.
Associativity: True because multiplication is associative on $\mathbb{R}$.
Identity element: $1 = 2^0 \in S$ because $0 \in \mathbb{Z}$.
Inverses: The inverse of 2^m is 2^{-m}, and $2^{-m} \in S$ if $2^m \in S$ because $-m \in \mathbb{Z}$ if $m \in \mathbb{Z}$.

5.13. Associativity: Assume $f, g, h \in F$. If $x \in \mathbb{R}$, then

$$(f + (g + h))(x) = f(x) + (g + h)(x) = f(x) + (g(x) + h(x))$$
$$= (f(x) + g(x)) + h(x) = (f + g)(x) + h(x)$$
$$= ((f + g) + h)(x),$$

so $f + (g + h) = (f + g) + h$.
Identity: Define $0_F \in F$ by $0_F(x) = 0$ for each $x \in \mathbb{R}$. (Verify that 0_F is an identity element.)
Inverse: For $f \in F$, define $-f \in F$ by $(-f)(x) = -(f(x))$. (Verify that $-f$ is an inverse of f.)

SECTION 6

6.1. (a) $\begin{pmatrix} 1 & 2 & 3 & 4 \\ 3 & 2 & 4 & 1 \end{pmatrix}$ (b) $\begin{pmatrix} 1 & 2 & 3 & 4 \\ 3 & 1 & 2 & 4 \end{pmatrix}$

(c) $\begin{pmatrix} 1 & 2 & 3 & 4 \\ 1 & 4 & 3 & 2 \end{pmatrix}$ (d) $\begin{pmatrix} 1 & 2 & 3 & 4 \\ 2 & 4 & 1 & 3 \end{pmatrix}$

(e) $\begin{pmatrix} 1 & 2 & 3 & 4 \\ 2 & 3 & 1 & 4 \end{pmatrix}$ (f) $\begin{pmatrix} 1 & 2 & 3 & 4 \\ 4 & 2 & 1 & 3 \end{pmatrix}$

(g) $\begin{pmatrix} 1 & 2 & 3 & 4 \\ 4 & 2 & 1 & 3 \end{pmatrix}$ (h) $\begin{pmatrix} 1 & 2 & 3 & 4 \\ 2 & 3 & 1 & 4 \end{pmatrix}$

6.3. (a) $(1 \quad 3 \quad 6)(2 \quad 5)$ (b) $(1 \quad 4 \quad 3 \quad 2)$
(c) $(1 \quad 3 \quad 4 \quad 2 \quad 5)$ (d) $(2 \quad 3)(4 \quad 5)$

SECTION 7

7.1. (a) and (c) are subgroups.
(b) is not a subgroup. For example $(1 \quad 2 \quad 3)(2 \quad 3 \quad 4)$ is not the set, so closure is not satisfied.
(d) is not a subgroup. For example $(1 \quad 2 \quad 3 \quad 4)(1 \quad 2 \quad 3 \quad 4)$ is not in the set, so closure us not satisfied.

7.3. (a) $G_T = \{(1), (2 \quad 3)\}$. $G_{(T)} = \{(1), (2 \quad 3)\}$.
(b) $G_T = \{(1)\}$. $G_{(T)} = \{(1), (2 \quad 3)\}$.

7.5. Consider the three parts of Theorem 7.1, and let H denote the given set.
(a) $H \neq \varnothing$ since $\alpha_{1,0} \in H$.
(b) If $\alpha_{a,0} \in H$ and $\alpha_{b,0} \in H$, then $\alpha_{a,0} \circ \alpha_{b,0} = \alpha_{ab,0} \in H$.
(c) If $\alpha_{a,0} \in H$, then $\alpha_{a,0}^{-1} = \alpha_{a^{-1},0} \in H$.
The subgroup consists of the identity, all magnifications, and all contractions.

SECTION 8

8.1.

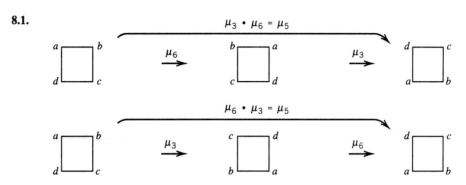

8.3. The group has order six, containing the identity, two (other) rotations, and three reflections. (Describe the elements more explicitly.)

SECTION 9

9.1. (a), (b), and (c).

9.3. (a) and (b).

9.5. (a) Reflexive: $(x_1, y_1) \sim (x_1, y_1)$ because $y_1 = y_1$ by the reflexive property of $=$ on $\mathbb{R}$.
Symmetric: $(x_1, y_1) \sim (x_2, y_2)$ iff $y_1 = y_2$ iff $y_2 = y_1$ iff $(x_2, y_2) \sim (x_1, y_1)$, where we have used the symmetric property of $=$ on $\mathbb{R}$.
Transitive: Use the transitive property of $=$ on $\mathbb{R}$.
(b) The x-axis and the lines parallel to the x-axis.
(c) The set of points on any line not parallel to the x-axis would do.

9.7. (a) The proof requires the reflexive, symmetric, and transitive properties of $=$ on $\mathbb{R}$.
(b) The set of all nonnegative real numbers would do.

9.9. In the first case, symmetric and transitive. In the second case, reflexive and symmetric.

9.11. The interval $[-\pi/2, \pi/2]$ would do. By restricting the domain of sine to this interval, we make the function one-to-one, and therefore invertible; the inverse is then the inverse sine function.

9.13. (a) There are five.   (b) Five.

9.15. The equivalence class to which a point (number) belongs consists of all the points an integral distance away on the line. The points on the interval $[0, 1)$ would be a complete set of equivalence class representatives.

SECTION 10

10.1. (a) There are six.   (b) There are six.   (c) There are two.

10.3. (a) 5   (b) 2   (c) 3

10.5. The set of all multiples of 5.

10.7. $x = 2$ and $x = 5$.

10.9. $a \sim b$ iff $a \equiv b \pmod{10}$.

10.11. (a) $q = 3, r = 4$ (b) $q = -2, r = 3$ (c) $q = 0, r = 11$

10.13. If $a|b$ and $b|c$, then there are integers u and v such that $b = au$ and $c = bv$. Hence $c = auv$, so $a|c$, since uv is an integer.

10.15. There are integers u and v such that $a - b = un$ and $c - d = vn$. Hence $(a + b) - (b + d) = (u + v)n$, and so $a + c \equiv b + d \pmod{n}$, since $u + v$ is an integer.

10.17. $a = 1$, $b = -1$, and $n = 3$ would do.

SECTION 11

11.1. $-7, -2, 3, 8,$ and 13 would do.

11.3. (a) [2]   (b) [0]   (c) [3]   (d) [2]
      (e) [1]   (f) [2]   (g) [3]   (h) [3]

11.5.

$\oplus$	0	1	2
0	0	1	2
1	1	2	0
2	2	0	1

k is written for [k].

11.7. $[a] \oplus [b] = [a + b] = [b + a] = [b] \oplus [a]$ for all $[a]$, $[b] \in \mathbb{Z}_n$. (Give a reason for each step.)

SECTION 12

12.1. $(6, 14) = 2$ and $[6, 14] = 42$

12.3. $(36, -60, 90) = 6$ and $[36, -60, 90] = 180$

12.5. $(1001, 33) = 11 = 1001(1) + 33(-30)$

12.7. $(-90, 1386) = 18 = -90(-31) + 1386(-2)$

12.9. There are four.

SECTION 13

13.1. $105 = 3^1 \cdot 5^1 \cdot 7^1$ **13.3.** $1375 = 5^3 \cdot 11$

13.5. First, if p is a prime and $a, b \in \mathbb{Z}$, then the highest power of p dividing ab is the sum of the highest powers of p dividing a and b. Therefore, if $m|n$, so that $n = mu$ with $u \in \mathbb{Z}$, then for each $p_i (1 \le i \le k)$ the highest power of p_i dividing n is the sum of the highest powers of p_i dividing m and u. Thus $s_i \le t_i$ for $1 \le i \le k$.

Now, suppose $s_i \le t_i$ for $1 \le i \le k$, and let $u_i = t_i - s_i$ for $1 \le i \le k$. Let $w = p_1^{u_1} p_2^{u_2} \ldots p_k^{u_k}$. Then $n = mw$ so $m|n$.

13.7. $(10, 105) = 5$ and $[10, 105] = 210$. (Show why.)

13.9. $(56, 126) = 14$ and $[56, 126] = 504$. (Show why.)

13.11. (a) There are five.   (b) There are four.
      (c) There are twelve.   (d) There are $(r + 1)(s + 1)$.

SECTION 14

14.1. $x = (2 \quad 3)$

14.3. (Partial answer.) The order of (1) is one. The order of (1 2), (1 3), and (2 3) is two. The order of (1 2 3) and (1 3 2) is three.

14.5. Two.

14.7. (a) There are four. (b) There are five. (c) n (Why?)

14.9. Four. (Why?)

SECTION 15

15.1. $\langle -20, 8 \rangle \subseteq \langle 4 \rangle$ because $-20 = -(5 \cdot 4)$ and $8 = 2 \cdot 4$. $\langle -20, 8 \rangle \supseteq \langle 4 \rangle$ because $4 = -(-20) - (2)8$.

15.3. Use $(1 \quad 4 \quad 6 \quad 2 \quad 3 \quad 5)^3 = (1 \quad 2)(3 \quad 4)(5 \quad 6)$,
$(1 \quad 4 \quad 6 \quad 2 \quad 3 \quad 5)^4 = (1 \quad 3 \quad 6)(2 \quad 4 \quad 5)$, and
$(1 \quad 2)(3 \quad 4)(5 \quad 6)(1 \quad 3 \quad 6)(2 \quad 4 \quad 5) = (1 \quad 4 \quad 6 \quad 2 \quad 3 \quad 5)$
[or $((1 \quad 2)(3 \quad 4)(5 \quad 6))^{-1}(1 \quad 3 \quad 6)(2 \quad 4 \quad 5) = (1 \quad 4 \quad 6 \quad 2 \quad 3 \quad 5)$].

15.5. The subgroup has order eight.

15.7. 28 **15.9.** $([1], (1 \quad 4))$

15.11. Let $e = ([0], (1))$, $v = ([0], (1 \quad 2))$, $w = ([1], (1))$, $x = ([1], (1 \quad 2))$, $y = ([2], (1))$, and $z = ([2], (1 \quad 2))$. Then either x or z generates the group. The Cayley table follows.

	e	v	w	x	y	z
e	e	v	w	x	y	z
v	v	e	x	w	z	y
w	w	x	y	z	e	v
x	x	w	z	y	v	e
y	y	z	e	v	w	x
z	z	y	v	e	x	w

SECTION 16

16.1. $\langle [4] \rangle = \{[0], [4]\}$
$\langle [4] \rangle \oplus [1] = \{[1], [5]\}$
$\langle [4] \rangle \oplus [2] = \{[2], [6]\}$
$\langle [4] \rangle \oplus [3] = \{[3], [7]\}$

16.3. $\langle \mu_5 \rangle = \{\mu_1, \mu_5\}$
$\langle \mu_5 \rangle \mu_2 = \{\mu_2, \mu_8\}$
$\langle \mu_5 \rangle \mu_3 = \{\mu_3, \mu_6\}$
$\langle \mu_5 \rangle \mu_4 = \{\mu_4, \mu_7\}$

16.5. $\langle (1 \quad 2 \quad 3) \rangle = \{(1), (1 \quad 2 \quad 3), (1 \quad 3 \quad 2)\}$
$\langle (1 \quad 2 \quad 3) \rangle (1 \quad 2) = \{(1 \quad 2), (1 \quad 3), (2 \quad 3)\}$

16.7. (Partial answer.) In the Cartesian plane, H corresponds to the line whose equation is $y = x$. The right cosets correspond to the lines with slope 1. (Each line corresponds to a coset.) Why?

SECTION 17

17.1. 3 (Why?)   **17.3.** 2 (Why?)   **17.5.** 4 (Why?)

17.7. $|G| = 20$ or 40. (Why?)

17.9. $|G| = 30, 36, 42$ or 48. (Why?)

17.11. There are only four subgroups. (Notice $\langle[1]\rangle = \langle[5]\rangle$ and $\langle[2]\rangle = \langle[4]\rangle$.)

17.13. There are five.

SECTION 18

18.1. Verify that if θ is defined by $\theta(m) = 2^m$, then θ is an isomorphism.

18.4.

$*$	a	b	c	d
a	a	b	c	d
b	b	c	d	a
c	c	d	a	b
d	d	a	b	c

18.5. (a) y (b) v (c) w (d) u

SECTION 19

19.1. Theorem 19.1(a).   **19.3.** Theorem 19.1(b) or 19.1(c).

19.5. Theorem 19.1(e). (Why?)

19.7. Theorem 19.1(c). [Give a proof. If you know about infinite cardial numbers, you could also use Theorem 19.1(a).]

19.9. Theorem 19.1(d).

19.11. No.   **19.13.** $\mathbb{Z}_7 \times \mathbb{Z}_7$.

SECTION 20

20.1. $\theta([k]) = \begin{pmatrix} [0] & [1] & [2] & [3] & [4] \\ [0+k] & [1+k] & [2+k] & [3+k] & [4+k] \end{pmatrix}$

SECTION 21

21.1. Definition of the operation on $A \times B$, definition of π_1, definition of π_1.

21.3. Use $x = \begin{bmatrix} a & b \\ c & d \end{bmatrix}$, $y = \begin{bmatrix} e & f \\ g & h \end{bmatrix}$, $\begin{bmatrix} a & b \\ c & d \end{bmatrix} \begin{bmatrix} 1 & 0 \\ 0 & 0 \end{bmatrix} = \begin{bmatrix} a & 0 \\ c & 0 \end{bmatrix}$, and so on.

21.5. (a) $[a]_6 = [b]_6$ implies $6|(b - a)$ implies $3|(b - a)$ implies $[a]_3 = [b]_3$.
(b) Straightforward.
(c) $\langle [3]_6 \rangle$

21.7. Similar to Theorem 18.1.

21.9. Use Theorem 7.1. Here is part (b): If $x, y \in \theta(A)$, then $x = \theta(a)$ and $y = \theta(b)$ for some $a, b \in A$. Since $ab \in A$ (why?), $xy = \theta(a)\theta(b) = \theta(ab) \in \theta(A)$.

SECTION 22

22.1. (a) 4   (b) 4

22.3. If we write k in place of $[[k]]$, and both outside row and outside column in the order 0, 1, 2, 3, the rows (in order) are 0, 1, 2, 3; 1, 2, 3, 0; 2, 3, 0, 1; 3, 0, 1, 2.

22.5. Use $(Na)(Nb) = N(ab) = N(ba) = N(b)N(a)$.

SECTION 23

22.1. $\mathbb{Z}_1, \mathbb{Z}_2, \mathbb{Z}_3, \mathbb{Z}_6$

22.3. $\mathbb{Z}_1, \mathbb{Z}_5$

23.5. $\mathbb{Z}_4$ and $\mathbb{Z}_2 \times \mathbb{Z}_2$

23.7. For $a \in G$, $(\phi \circ \eta)(a) = \phi(\eta(a)) = \phi(Na) = \theta(a)$.

SECTION 24

24.1. $[3] \odot ([4] \oplus [5]) = [3] \odot [3] = [3]$ and $([3] \odot [4]) \oplus ([3] \odot [5]) = [0] \oplus [3] = [3]$.

24.3. $\begin{bmatrix} 2 & 7 \\ 11 & 0 \end{bmatrix}$

24.5. Associativity of addition and multiplication, commutativity of addition, and the distributive laws are all satisfied in $\mathbb{Z}[\sqrt{2}]$ because they are satisfied in $\mathbb{R}$. The zero element is $0 + 0\sqrt{2} \in \mathbb{Z}[\sqrt{2}]$. The negative of $a + b\sqrt{2}$ is $(-a) + (-b)\sqrt{2}$, which is in $\mathbb{Z}[\sqrt{2}]$ if $a + b\sqrt{2}$ in $\mathbb{Z}[\sqrt{2}]$, because $-a, -b \in \mathbb{Z}$ if $a, b \in \mathbb{Z}$.

24.7. (a) and (b) are true in every ring. In a commutative ring, (c) is true also.

SECTION 25

25.1. Only [2].

25.3. $([2], [0])([0], [1]) = ([0], [0])$

25.5. $\{3k : k \in \mathbb{Z}\}$

25.7. Subring.

SECTION 26

Only one example is given in each case.

26.1. $\mathbb{Z}$

26.3. $\mathbb{Z}_4$

26.5. $\mathbb{Z}_2$

26.7. There is none.

26.9. Even integers.

SECTION 27

27.1. If $b \in S$, then $\theta(a) = b$ for some $a \in R$, and $b = \theta(a) = \theta(ae) = \theta(a)\theta(e) = b\theta(e)$. Similarly, $b = \theta(e)b$.

27.3. By Problems 27.1 and 27.2, it suffices to prove that S has not zero divisors. If $a, b \in S$, then $\theta(x) = a$ and $\theta(y) = b$ for some $x, y \in R$, and $ab = \theta(x)\theta(y) = \theta(xy)$; therefore $ab = 0$ iff $xy = 0$. Because θ also maps nonzero elements to nonzero elements, it follows that S has no zero divisors if R does not.

27.5. For $m > 0$, $m(ab) = ab + ab + \cdots + ab$ (m terms)
$$= (a + a + \cdots + a)b \text{(distributivity)}$$
$$= (ma)b.$$
For $m = 0$, $m(ab) = 0(ab) = 0 = 0b = (0a)b = (ma)b$.
For $m < 0$, $m(ab) = -[(-m)(ab)]$ (Section 24)
$$= -[(-m)a]b \quad \text{(case } m > 0)$$
$$= [-(-m)a]b \quad \text{(Theorem 24.2)}$$
$$= (ma)b \quad \text{(Section 24)}$$

27.7. For $m > 0$ and $n > 0$, the result follows from the generalized associative law and the definition of mn. For $m \le 0$ or $n \le 0$, use the case $m > 0$ and $n > 0$ and follow the ideas in the solution of Problem 27.5.

SECTION 28

28.1. By the corollary of Lemma 28.1, $e > 0$. Therefore $-e < 0$, and if x^2 is a solution of $x^2 + e = 0$, then $x^2 < 0$, contradicting Lemma 28.1. (Clearly $x = 0$ is not a solution.)

28.3. Assume otherwise. If $a = b$, then $ac = bc$; if $a < b$ and $c > 0$, then $ac < bc$. Either possibility contradicts $ac > bc$.

28.5. $a > b$ iff $a - b > 0$ iff $b - a < 0$ iff $-a < -b$.

SECTION 29

29.1. An example of a set of positive rationals with no least element is $\{1, \ 1/2, \ 1/3, \ 1/4, \ldots\}$.

29.3. Use $a - (a - e) = e > 0$.

SECTION 30

30.1. Transitivity fails. [Consider $(2, \ 1)$, $(0, \ 2)$, and $(2, \ 2)$.]

30.3. Verify that $[a + b\sqrt{2},\ c + d\sqrt{2}] \mapsto (ac - 2bd)/(c^2 - 2d^2) + (bc - ad)\sqrt{2}/(c^2 - 2d^2)$ defines an isomorphism. Or verify that $\mathbb{Q}[\sqrt{2}]$ has no proper subfield containing $\mathbb{Z}[\sqrt{2}]$, and use Theorem 30.1.

SECTION 31

31.1. Multiply each part of $b > a > 0$ by $(ab)^{-1}$ to get $b(ab)^{-1} > a(ab)^{-1} > 0$, that is, $a^{-1} > b^{-1} > 0$.

31.3. If $a = r/s$ and $a + b = u/v$ were both rational, then $b = u/v - r/s = (us - vr)/vs$ would be rational.

31.5. Use the fact that $x \leq u$ iff $2x \leq 2u$.

31.7. Sum the geometric series $0.9 + 0.09 + 0.009 + \cdots$.

SECTION 32

32.1. (a) $3 + i$ (b) $-i$ (c) $1/5 - 2i/5$

32.3. $x^2 + 1 = 0$ has its coefficients in $\mathbb{Q}$, but no solution in $\mathbb{Q}$.

32.5. $x^2 + x + 1 = 0$ has no solution in $\mathbb{Z}_2$.

32.7. Multiply the first equation by i and the second by -3, and add. This will lead to $w = 2i$. Substitute in the first equation to get $z = i$.

SECTION 33

33.1. $\sqrt{3} + 1$

33.5. $32i$

33.9. $-1/4$

33.13. $r = \sqrt{2},\ \theta = \pi/4$

33.17. $r = 2\sqrt{2},\ \theta = 7\pi/4$

33.3. $(5\sqrt{2}/2) + (5\sqrt{2}/2)i$

33.7. $8i$

33.11. $\cos(k\pi/4) + i\,\sin(k\pi/4)$, $k = 0, 1, \ldots, 7$

33.15. $r = 5,\ \theta = \pi$

SECTION 34

34.1. $x^2, x^2 + x, x^2 + 1, x^2 + x + 1$

34.4. (a) $3 + x + 2x^2$

    (c) $2 + 3x + 4x^3$

    (e) $8x^3 + 12x^4 + 6x^5 + x^6$

    (b) $3x + x^3 + 2x^4$

    (d) $2x^2 + x^4 + 4x^5 + 2x^7$

SECTION 35

35.1. $q(x) = x^2 + x + 2$, $r(x) = 3$ **35.3.** $q(x) = 0$, $r(x) = x^3$

35.5. $q(x) = 4x^2 + 2x + 2$, $r(x) = 2x + 4$ **35.7.** The remainder is 109.

35.9. 3 in $\mathbb{Z}_7$

SECTION 36

36.1. $x - 2$ **36.3.** 1   **36.5.** $x^3 - (\frac{1}{2})x + \frac{1}{2}$ **36.7.** $x^5 + 2x^2 + 4$

36.9. Check that $x^3 - 3 \neq 0$ in $\mathbb{Z}_7$ for $x = 0, 1, \ldots, 6$.

36.11. $(x + 3)(x^2 + 3)$

SECTION 37

37.1. Multiply by each of the units in $\mathbb{Z}_5[x]$ (that is, by 1, 2, 3, 4) to get four associates.

37.3. $q = 2 - i$, $r = -1 + i$

SECTION 38

38.1. No.

38.3. Yes. Ker $\theta = \{[0], [3]\}$.

38.5. No.

SECTION 39

39.1. (a), (b), (c) Similar to Problem 21.5.

39.3. $(I + a)(I + b) = I + ab = I + ba = (I + b)(I + a)$

SECTION 40

40.7. $x - 3$ **40.9.** $x^2 + 1$

SECTION 41

41.1. Follow the proof of Theorem 40.3, using $d(a)$ in place of deg $f(x)$.

41.3. Similar to Problem 12.23

41.5. Similar to Lemma 13.1.

SECTION 42

42.1. Use Problem 26.20.

42.3. Prove that the set of elements described is a field. Since it is clearly contained in $F(S)$, it must equal $F(S)$.

SECTION 43

43.3. Use the suggestion: $1 + x^2 \neq 0$ for $1, 2 \in \mathbb{Z}_3$.

SECTION 44

44.1. $(x - 2)(x - 3)^2 = x^3 - 8x^2 + 21x - 18$

44.3. x^3 (three), $x^3 + 1$ (one), $x^3 + x$ (three), $x^3 + x^2$ (three), $x^3 + x + 1$ (none), $x^3 + x^2 + 1$ (none), $x^3 + x^2 + x$ (one), $x^3 + x^2 + x + 1$ (three).

SECTION 45

45.1. If $a + bi$ is a root of multiplicity two, with $b \neq 0$, then $a - bi$ must also be a root of multiplicity two. (Write $f(x)$ as a product of factors irreducible over $\mathbb{R}$; there must be two quadratic factors.)

45.3. $x^3 + x + 1$

45.5. (a) $-1, \frac{3}{2}, -\frac{1}{2}$

(b) $2, \frac{1}{3}, \frac{1}{2}$

(c) $\frac{1}{2}$ (The other roots are imaginary: $\pm i 2\sqrt{2}$.)

45.7. (a) $(x - 1)(x - \sqrt{5})(x + \sqrt{5})$

(b) $(3x - 2)(x^2 + 1)$

(c) $x(x^2 - 2x + 2)$

SECTION 46

46.3. In one direction, use that $\mathbb{Q}(\sqrt{2}, i)$ contains $\sqrt{2}$ and $\sqrt{2}i$, and therefore their sum. In the other direction, use that $\mathbb{Q}(\sqrt{2} + \sqrt{2}i)$ contains $(\sqrt{2} + \sqrt{2}i)^2 = 4i$, and therefore it contains i; hence it contains $(\sqrt{2} + \sqrt{2}i)i = -\sqrt{2} + \sqrt{2}i$, and so on.

46.5. Assume $H_1 \subseteq H_2$ and $x \in E_{H_2}$. Then $\sigma(x) = x$ for all $\sigma \in H_2$ so $\sigma(x) = x$ for all $\sigma \in H_1$. Thus $x \in E_{H_1}$.

SECTION 47

47.1. Write Equation 46.1 as $f(x) = 0$. Then $f(a) = 0$ implies $\sigma(f(a)) = 0$ implies $f(\sigma(a)) = 0$.

47.3. Use the series $S_3 \rhd \langle (1\ \ 2\ \ 3) \rangle \rhd \langle (1) \rangle$.

SECTION 48

48.1. Start with a segment OA the length of the cosine. Construct a circle with unit radius and center at O. Let B denote an intersection of this circle and the perpendicular to OA at A. Angle AOB is the required angle.

48.3. Given a segment PQ, construct the circle through Q with center at P, and the circle through P with center at Q. These circles intersect at points R and S. The line through R and S is the perpendicular bisector of RS.

SECTION 49

49.1. For part (a), if d denotes the right-hand side of the displayed equation, then the displayed equation is equivalent to $dx + (-1)y + (y_1 - dx_1) = 0$. Each of d, -1 and $y_1 - dx_1$ is in F.

SECTION 50

50.1. Subtract one of the polynomials from the other. The result is a polynomial of smaller degree than that of the given polynomial, but with as many roots as the given polynomial. Apply Theorem 44.1.

SECTION 51

51.1. μ_1, μ_3, μ_5 and μ_6 all induce the identity. The other four elements induce $(H \quad V)$.

51.3. (a) $\{D_1, D_2\}$ \qquad\qquad (b) $\{D_1, D_2\}$ \qquad\qquad (c) $\{D_1\}, \{D_2\}$

51.5. (a) $\mathrm{Orb}(1) = \mathrm{Orb}(2) = \mathrm{Orb}(3) = \{1, 2, 3\}$
$\mathrm{Orb}(4) = \mathrm{Orb}(5) = \{4, 5\}$
(b) $G_1 = G_2 = G_3 = \langle (4, 5) \rangle$, $G_4 = G_5 = \langle (1 \quad 2 \quad 3) \rangle$
(c) Straightforward. For example, $|\mathrm{Orb}(1)| = 3$ and $|G|/|G_1| = 6/2 = 3$.

SECTION 52

52.1. $(1/6)(5 + 0 + 2 + 3 + 2 + 0) = 2$; for $g = (1 \quad 2 \quad 3)(4 \quad 5)$, the numbers in the sum are the values $\psi(g^k)$ for $k = 0, \ldots, 5$, in order. The two orbits are $\{1, 2, 3\}$ and $\{4, 5\}$.

52.3. (a) 24 \qquad (b) 6 \qquad (c) 3

52.5. $\frac{1}{4}(2^4 + 2^1 + 2^2 + 2^1) = 6$

52.7. $(\frac{1}{10})[5^5 + 4(5) + 5(5^3)] = 377$

SECTION 53

53.1. $2^2, 3^2, 5$ \qquad\qquad\qquad **53.3.** $2 \nmid 495$

53.5. Sylow 2-subgroups: $\langle (1 \quad 2) \rangle$, $\langle (2 \quad 3) \rangle$, $\langle (2 \quad 3) \rangle$. Sylow 3-subgroup: $\langle (1 \quad 2 \quad 3) \rangle$.

SECTION 54

54.1. D_4 **54.3.** D_3 **54.5.** D_1 **54.7.** D_2

SECTION 55

55.1. Top row, left to right: F_{VII}, F_V. Second row, left to right: F_{VI}, F_{III}, F_{IV}, F_{II}. Bottom row: F_I.

SECTION 56

56.1. Top row, left to right: $p1$, $p2$, pmm, pgg. Third row, left to right: pm, pg, pmg, cmm. Second row, left to right: cm, $p4g$, $p4m$, $p4$, $p3$. Bottom row, left to right: $p6m$, $p31m$, $p3m1$, $p6$.

SECTION 57

57.1. $-13, \sqrt{5}, 5\sqrt{2}, 9$

57.3. (a) Yes.   (b) Yes.

SECTION 58

58.1. The minimum distance is 3.

000	111
100	011
010	101
001	110

58.3. 11101, 11000, 10100, 01001, 00011, 11011, 10111, 10010

SECTION 59

59.1. Any set of four vectors that form a subspace of $\mathbb{Z}_2^5$.

59.3. $\begin{bmatrix} 0 & 0 & 0 & 0 \\ 1 & 0 & 0 & 0 \\ 0 & 1 & 1 & 0 \\ 1 & 1 & 1 & 0 \end{bmatrix}$

59.5. $\begin{bmatrix} 0 & 0 & 0 \\ 1 & 0 & 2 \\ 2 & 0 & 1 \\ 2 & 1 & 0 \\ 0 & 1 & 2 \\ 1 & 1 & 1 \\ 1 & 2 & 0 \\ 2 & 2 & 2 \\ 0 & 2 & 1 \end{bmatrix}$

SECTION 60

60.1. Here is one possibility.

0000	1100	0011	1111
1000	0100	1011	0111
0010	1110	0001	1101
1010	0110	1001	0101

60.3. Here is one possibility.

000	120	210	012	102	222	021	111	210
100	220	010	112	202	022	121	211	001
110	200	020	122	212	002	101	221	011

SECTION 61

61.1. (a) $(1/4)^3(3/4)^2 = 9/1024$
(b) $10(1/4)^2(3/4)p^3 = 135/512$

61.3. Let $a = 2p$, $b = 2q$. If $p < \frac{1}{2}$ and $w_1 < w_2$, then $a < 1$, $b > 1$, and $n - w_1 > n - w_2$, so $a^{w_1} > a^{w_2}$, $b^{n-w_1} > b^{n-w_2}$, and $p^{w_1} q^{n-w_1} = 2^{-n} a^{w_1} b^{n-w_1} > 2^{-n} a^{w_2} b^{n-w_2} = p^{w_2} q^{n-w_2}$.

61.5. 0001111

SECTION 62

62.1. Antisymmetry fails: $a \neq 0$, $a | -a$, and $-a | a$, but $a \neq -a$.

62.3. Like Figure 62.1; there are 16 subsets.

62.5.

SECTION 63

63.1. $a \wedge (b \vee c)) = a \wedge 1 = a$, but $(a \wedge b) \vee (a \wedge c) = 0 \vee 0 = 0$

63.3. Straightforward.

63.5. Both distributive and complemented.

63.7. Here are the number of subgroups of each order.

order	24	12	8	6	4	3	2	1
number	1	1	3	4	7	4	9	1

SECTION 64

64.1. (a) Use the suggestion.

(b) $a \vee 1 = (a \vee 1) \wedge (a \vee a') = a \vee (1 \wedge a') = a \vee a' = 1.$

64.3. Straightforward.

64.5. If $b \in B$, then $(0 \wedge b') \vee ((0' \wedge b) = 0 \vee b = b.$ In the other direction, take $b = 0$ to get $0 = (a \wedge 1) \vee (a' \wedge 0) = a \vee 0 = a.$

SECTION 65

65.1. Use $a \leq b$ iff $a \vee b = b$ iff $a \wedge b = a.$

65.3. Use Problem 65.2, $0 = \theta(0) = \theta(a \wedge a') = \theta(a) \wedge \theta(a')$, and $1 = \theta(1) = \theta(a \vee a') = \theta(a) \vee \theta(a').$

SECTION 66

66.1. The first distributive law is represented in Figure 66.3. For the commutative law for $\wedge$, use Figure 66.1 and then the same figure with a and b interchanged. For $\vee$, use the same idea with Figure 66.2.

PHOTO CREDIT LIST

FIGURE 1
From *Historic Ornament, A Pictorial Archive* by C. B. Griesbach, Dover Publications, Inc.

FIGURE 2
From *Historic Ornament, A Pictorial Archive* by C. B. Griesbach, Dover Publications, Inc.

FIGURE 3
From *Manual of Mineralogy* by C. S. Hurlbut and C. Klein, 19th ed., Wiley, ©1977.

FIGURE 54.4
R. B. Hoit/Photo Researchers.

FIGURE 55.8
Courtesy Bernard Quaritch, Ltd., London. Reprinted with permission.

FIGURE 56.6
From *Handbook of Regular Patterns*, by Peter S. Stevens, M.I.T Pres, 1981.
Designs and Historic Ornaments, by W. and G. Audsley, Dover Publications, Inc.
Chinese Lattice Designs, by D. S. Dye, Dover Publications, Inc.
Design Motifs of Ancient Mexico, by J. Enciso, Dover Publications, Inc.
Allover Patterns, By C. P. Hornung, Dover Publications, Inc.

INDEX